PRAISE FOR *PYTHON CRASH COURSE*

"It has been interesting to see No Starch Press producing future classics that should be alongside the more traditional programming books. *Python Crash Course* is one of those books."

—Greg Laden, ScienceBlogs

"Deals with some rather complex projects and lays them out in a consistent, logical, and pleasant manner that draws the reader into the subject."

—*Full Circle Magazine*

"Well presented with good explanations of the code snippets. The book works with you, one small step at a time, building more complex code, explaining what's going on all the way."

—FlickThrough Reviews

"Learning Python with *Python Crash Course* was an extremely positive experience! A great choice if you're new to Python."

—Mikke Goes Coding

"Does what it says on the tin, and does it really well. . . . Presents a large number of useful exercises as well as three challenging and entertaining projects."

—RealPython.com

"A fast-paced but comprehensive introduction to programming with Python, *Python Crash Course* is another superb book to add to your library and help you finally master Python."

—TutorialEdge.net

"A brilliant option for complete beginners without any coding experience. If you're looking for a solid, uncomplicated intro to this very deep language, I have to recommend this book."

—WhatPixel.com

"Contains literally everything you need to know about Python and even more."

—FireBearStudio.com

"While *Python Crash Course* uses Python to teach you to code, it also teaches clean programming skills that apply to most other languages."

—Great Lakes Geek

Printed in the United States of America

Fourth printing

27 26 25 24 23   4 5 6 7 8

ISBN-13: 978-1-7185-0270-3 (print)
ISBN-13: 978-1-7185-0271-0 (ebook)

Published by No Starch Press®, Inc.
245 8th Street, San Francisco, CA 94103
phone: +1.415.863.9900
www.nostarch.com; info@nostarch.com

Publisher: William Pollock
Managing Editor: Jill Franklin
Production Editor: Jennifer Kepler
Developmental Editor: Eva Morrow
Cover Illustrator: Josh Ellingson
Interior Design: Octopod Studios
Technical Reviewer: Kenneth Love
Copyeditor: Doug McNair
Compositor: Jeff Lytle, Happenstance Type-O-Rama
Proofreader: Scout Festa

The Library of Congress has catalogued the first edition as follows:

```
Matthes, Eric, 1972-
  Python crash course : a hands-on, project-based introduction to programming / by Eric Matthes.
       pages cm
  Includes index.
  Summary: "A project-based introduction to programming in Python, with exercises. Covers general
programming concepts, Python fundamentals, and problem solving. Includes three projects - how to
create a simple video game, use data visualization techniques to make graphs and charts, and build
an interactive web application"-- Provided by publisher.
  ISBN 978-1-59327-603-4 -- ISBN 1-59327-603-6
 1.  Python (Computer program language)  I. Title.
  QA76.73.P98M38 2015
  005.13'3--dc23
                                                                   2015018135
```

For customer service inquiries, please contact info@nostarch.com. For information on distribution, bulk sales, corporate sales, or translations: sales@nostarch.com. For permission to translate this work: rights@nostarch.com. To report counterfeit copies or piracy: counterfeit@nostarch.com.

[S]

For my father, who always made time to
answer my questions about programming,
and for Ever, who is just beginning to ask
me his questions

About the Author

Eric Matthes was a high school math and science teacher for 25 years, and he taught introductory Python classes whenever he could find a way to fit them into the curriculum. Eric is a full-time writer and programmer now, and he is involved in a number of open source projects. His projects have a diverse range of goals, from helping predict landslide activity in mountainous regions to simplifying the process of deploying Django projects. When he's not writing or programming, he enjoys climbing mountains and spending time with his family.

About the Technical Reviewer

Kenneth Love lives in the Pacific Northwest with their family and cats. Kenneth is a longtime Python programmer, open source contributor, teacher, and conference speaker.

BRIEF CONTENTS

Preface to the Third Edition . . . xxvii

Acknowledgments . . . xxxi

Introduction . . . xxxiii

PART I: BASICS . . . 1

Chapter 1: Getting Started . . . 3

Chapter 2: Variables and Simple Data Types . . . 15

Chapter 3: Introducing Lists . . . 33

Chapter 4: Working with Lists . . . 49

Chapter 5: if Statements . . . 71

Chapter 6: Dictionaries . . . 91

Chapter 7: User Input and while Loops . . . 113

Chapter 8: Functions . . . 129

Chapter 9: Classes . . . 157

Chapter 10: Files and Exceptions . . . 183

Chapter 11: Testing Your Code . . . 209

PART II: PROJECTS . . . 225

Project 1: Alien Invasion

Chapter 12: A Ship That Fires Bullets . . . 227

Chapter 13: Aliens! . . . 255

Chapter 14: Scoring . . . 277

Project 2: Data Visualization

Chapter 15: Generating Data. 301
Chapter 16: Downloading Data 329
Chapter 17: Working with APIs 355

Project 3: Web Applications

Chapter 18: Getting Started with Django. 373
Chapter 19: User Accounts 403
Chapter 20: Styling and Deploying an App 433

Appendix A: Installation and Troubleshooting 463
Appendix B: Text Editors and IDEs 469
Appendix C: Getting Help 477
Appendix D: Using Git for Version Control 483
Appendix E: Troubleshooting Deployments 493

Index 503

CONTENTS IN DETAIL

PREFACE TO THE THIRD EDITION xxvii

ACKNOWLEDGMENTS xxxi

INTRODUCTION xxxiii

Who Is This Book For? xxxiv
What Can You Expect to Learn? xxxiv
Online Resources xxxv
Why Python? xxxvi

PART I: BASICS 1

1
GETTING STARTED 3

Setting Up Your Programming Environment 3
  Python Versions 4
  Running Snippets of Python Code 4
  About the VS Code Editor 4
Python on Different Operating Systems 5
  Python on Windows 5
  Python on macOS 7
  Python on Linux 8
Running a Hello World Program 9
  Installing the Python Extension for VS Code 9
  Running hello_world.py 10
Troubleshooting 10
Running Python Programs from a Terminal 11
  On Windows 12
  On macOS and Linux 12
  Exercise 1-1: python.org 13
  Exercise 1-2: Hello World Typos 13
  Exercise 1-3: Infinite Skills 13
Summary 13

2
VARIABLES AND SIMPLE DATA TYPES 15

What Really Happens When You Run hello_world.py 15
Variables 16
  Naming and Using Variables 17
  Avoiding Name Errors When Using Variables 17
  Variables Are Labels 18
  Exercise 2-1: Simple Message 19
  Exercise 2-2: Simple Messages 19

Strings . . . 19
Changing Case in a String with Methods . . . 20
Using Variables in Strings . . . 20
Adding Whitespace to Strings with Tabs or Newlines . . . 21
Stripping Whitespace . . . 22
Removing Prefixes . . . 23
Avoiding Syntax Errors with Strings . . . 24
Exercise 2-3: Personal Message . . . 25
Exercise 2-4: Name Cases . . . 25
Exercise 2-5: Famous Quote . . . 25
Exercise 2-6: Famous Quote 2 . . . 25
Exercise 2-7: Stripping Names . . . 25
Exercise 2-8: File Extensions . . . 25
Numbers . . . 26
Integers . . . 26
Floats . . . 26
Integers and Floats . . . 27
Underscores in Numbers . . . 28
Multiple Assignment . . . 28
Constants . . . 28
Exercise 2-9: Number Eight . . . 29
Exercise 2-10: Favorite Number . . . 29
Comments . . . 29
How Do You Write Comments? . . . 29
What Kinds of Comments Should You Write? . . . 29
Exercise 2-11: Adding Comments . . . 30
The Zen of Python . . . 30
Exercise 2-12: Zen of Python . . . 31
Summary . . . 32

3
INTRODUCING LISTS 33

What Is a List? . . . 33
Accessing Elements in a List . . . 34
Index Positions Start at 0, Not 1 . . . 34
Using Individual Values from a List . . . 35
Exercise 3-1: Names . . . 36
Exercise 3-2: Greetings . . . 36
Exercise 3-3: Your Own List . . . 36
Modifying, Adding, and Removing Elements . . . 36
Modifying Elements in a List . . . 36
Adding Elements to a List . . . 37
Removing Elements from a List . . . 38
Exercise 3-4: Guest List . . . 41
Exercise 3-5: Changing Guest List . . . 42
Exercise 3-6: More Guests . . . 42
Exercise 3-7: Shrinking Guest List . . . 42
Organizing a List . . . 42
Sorting a List Permanently with the sort() Method . . . 43
Sorting a List Temporarily with the sorted() Function . . . 43
Printing a List in Reverse Order . . . 44

Finding the Length of a List . . . 44
Exercise 3-8: Seeing the World . . . 45
Exercise 3-9: Dinner Guests . . . 45
Exercise 3-10: Every Function . . . 45
Avoiding Index Errors When Working with Lists . . . 46
Exercise 3-11: Intentional Error . . . 47
Summary . . . 47

4 WORKING WITH LISTS 49

Looping Through an Entire List . . . 49
A Closer Look at Looping . . . 50
Doing More Work Within a for Loop . . . 51
Doing Something After a for Loop . . . 52
Avoiding Indentation Errors . . . 53
Forgetting to Indent . . . 53
Forgetting to Indent Additional Lines . . . 54
Indenting Unnecessarily . . . 54
Indenting Unnecessarily After the Loop . . . 55
Forgetting the Colon . . . 55
Exercise 4-1: Pizzas . . . 56
Exercise 4-2: Animals . . . 56
Making Numerical Lists . . . 56
Using the range() Function . . . 57
Using range() to Make a List of Numbers . . . 58
Simple Statistics with a List of Numbers . . . 59
List Comprehensions . . . 59
Exercise 4-3: Counting to Twenty . . . 60
Exercise 4-4: One Million . . . 60
Exercise 4-5: Summing a Million . . . 60
Exercise 4-6: Odd Numbers . . . 60
Exercise 4-7: Threes . . . 60
Exercise 4-8: Cubes . . . 60
Exercise 4-9: Cube Comprehension . . . 60
Working with Part of a List . . . 61
Slicing a List . . . 61
Looping Through a Slice . . . 62
Copying a List . . . 63
Exercise 4-10: Slices . . . 65
Exercise 4-11: My Pizzas, Your Pizzas . . . 65
Exercise 4-12: More Loops . . . 65
Tuples . . . 65
Defining a Tuple . . . 65
Looping Through All Values in a Tuple . . . 66
Writing Over a Tuple . . . 67
Exercise 4-13: Buffet . . . 67
Styling Your Code . . . 68
The Style Guide . . . 68
Indentation . . . 68
Line Length . . . 69
Blank Lines . . . 69

Other Style Guidelines . . . 69
Exercise 4-14: PEP 8 . . . 70
Exercise 4-15: Code Review . . . 70
Summary . . . 70

5
IF STATEMENTS 71

A Simple Example . . . 72
Conditional Tests . . . 72
Checking for Equality . . . 72
Ignoring Case When Checking for Equality . . . 73
Checking for Inequality . . . 74
Numerical Comparisons . . . 74
Checking Multiple Conditions . . . 75
Checking Whether a Value Is in a List . . . 76
Checking Whether a Value Is Not in a List . . . 76
Boolean Expressions . . . 77
Exercise 5-1: Conditional Tests . . . 77
Exercise 5-2: More Conditional Tests . . . 78
if Statements . . . 78
Simple if Statements . . . 78
if-else Statements . . . 79
The if-elif-else Chain . . . 80
Using Multiple elif Blocks . . . 81
Omitting the else Block . . . 82
Testing Multiple Conditions . . . 82
Exercise 5-3: Alien Colors #1 . . . 84
Exercise 5-4: Alien Colors #2 . . . 84
Exercise 5-5: Alien Colors #3 . . . 84
Exercise 5-6: Stages of Life . . . 84
Exercise 5-7: Favorite Fruit . . . 85
Using if Statements with Lists . . . 85
Checking for Special Items . . . 85
Checking That a List Is Not Empty . . . 86
Using Multiple Lists . . . 87
Exercise 5-8: Hello Admin . . . 88
Exercise 5-9: No Users . . . 88
Exercise 5-10: Checking Usernames . . . 88
Exercise 5-11: Ordinal Numbers . . . 88
Styling Your if Statements . . . 89
Exercise 5-12: Styling if Statements . . . 89
Exercise 5-13: Your Ideas . . . 89
Summary . . . 89

6
DICTIONARIES 91

A Simple Dictionary . . . 92
Working with Dictionaries . . . 92
Accessing Values in a Dictionary . . . 92
Adding New Key-Value Pairs . . . 93
Starting with an Empty Dictionary . . . 94

Modifying Values in a Dictionary 94
Removing Key-Value Pairs 96
A Dictionary of Similar Objects 96
Using get() to Access Values 97
Exercise 6-1: Person *98*
Exercise 6-2: Favorite Numbers *98*
Exercise 6-3: Glossary *99*
Looping Through a Dictionary 99
Looping Through All Key-Value Pairs 99
Looping Through All the Keys in a Dictionary 101
Looping Through a Dictionary's Keys in a Particular Order 102
Looping Through All Values in a Dictionary 103
Exercise 6-4: Glossary 2 . . . . . *104*
Exercise 6-5: Rivers *105*
Exercise 6-6: Polling *105*
Nesting 105
A List of Dictionaries 105
A List in a Dictionary 108
A Dictionary in a Dictionary 110
Exercise 6-7: People *111*
Exercise 6-8: Pets *111*
Exercise 6-9: Favorite Places *111*
Exercise 6-10: Favorite Numbers *111*
Exercise 6-11: Cities *111*
Exercise 6-12: Extensions *111*
Summary 111

7
USER INPUT AND WHILE LOOPS **113**

How the input() Function Works 114
Writing Clear Prompts 114
Using int() to Accept Numerical Input 115
The Modulo Operator 116
Exercise 7-1: Rental Car *117*
Exercise 7-2: Restaurant Seating *117*
Exercise 7-3: Multiples of Ten *117*
Introducing while Loops 117
The while Loop in Action 117
Letting the User Choose When to Quit 118
Using a Flag 120
Using break to Exit a Loop 121
Using continue in a Loop 122
Avoiding Infinite Loops 122
Exercise 7-4: Pizza Toppings *123*
Exercise 7-5: Movie Tickets *123*
Exercise 7-6: Three Exits *123*
Exercise 7-7: Infinity *123*
Using a while Loop with Lists and Dictionaries 124
Moving Items from One List to Another 124
Removing All Instances of Specific Values from a List 125
Filling a Dictionary with User Input 125
Exercise 7-8: Deli *127*

Exercise 7-9: No Pastrami 127
Exercise 7-10: Dream Vacation 127
Summary 127

8 FUNCTIONS 129

Defining a Function 130
Passing Information to a Function 130
Arguments and Parameters 131
Exercise 8-1: Message 131
Exercise 8-2: Favorite Book 131
Passing Arguments 131
Positional Arguments 132
Keyword Arguments 133
Default Values 134
Equivalent Function Calls 135
Avoiding Argument Errors 136
Exercise 8-3: T-Shirt 136
Exercise 8-4: Large Shirts 137
Exercise 8-5: Cities 137
Return Values 137
Returning a Simple Value 137
Making an Argument Optional 138
Returning a Dictionary 139
Using a Function with a while Loop 140
Exercise 8-6: City Names 141
Exercise 8-7: Album 142
Exercise 8-8: User Albums 142
Passing a List 142
Modifying a List in a Function 143
Preventing a Function from Modifying a List 145
Exercise 8-9: Messages 146
Exercise 8-10: Sending Messages 146
Exercise 8-11: Archived Messages 146
Passing an Arbitrary Number of Arguments 146
Mixing Positional and Arbitrary Arguments 147
Using Arbitrary Keyword Arguments 148
Exercise 8-12: Sandwiches 149
Exercise 8-13: User Profile 149
Exercise 8-14: Cars 149
Storing Your Functions in Modules 149
Importing an Entire Module 150
Importing Specific Functions 151
Using as to Give a Function an Alias 151
Using as to Give a Module an Alias 152
Importing All Functions in a Module 152
Styling Functions 153
Exercise 8-15: Printing Models 154
Exercise 8-16: Imports 154
Exercise 8-17: Styling Functions 154
Summary 154

9
CLASSES 157

Creating and Using a Class. 158
Creating the Dog Class 158
The __init__() Method 159
Making an Instance from a Class 159
Exercise 9-1: Restaurant. 162
Exercise 9-2: Three Restaurants. 162
Exercise 9-3: Users 162
Working with Classes and Instances. 162
The Car Class. 162
Setting a Default Value for an Attribute 163
Modifying Attribute Values 164
Exercise 9-4: Number Served. 166
Exercise 9-5: Login Attempts. 167
Inheritance 167
The __init__() Method for a Child Class 167
Defining Attributes and Methods for the Child Class 169
Overriding Methods from the Parent Class 170
Instances as Attributes 170
Modeling Real-World Objects. 172
Exercise 9-6: Ice Cream Stand 173
Exercise 9-7: Admin 173
Exercise 9-8: Privileges 173
Exercise 9-9: Battery Upgrade 173
Importing Classes. 173
Importing a Single Class 174
Storing Multiple Classes in a Module. 175
Importing Multiple Classes from a Module 176
Importing an Entire Module 176
Importing All Classes from a Module. 177
Importing a Module into a Module 177
Using Aliases 178
Finding Your Own Workflow 179
Exercise 9-10: Imported Restaurant 179
Exercise 9-11: Imported Admin. 179
Exercise 9-12: Multiple Modules. 179
The Python Standard Library 179
Exercise 9-13: Dice. 180
Exercise 9-14: Lottery 180
Exercise 9-15: Lottery Analysis 180
Exercise 9-16: Python Module of the Week 180
Styling Classes. 181
Summary 181

10
FILES AND EXCEPTIONS 183

Reading from a File 184
Reading the Contents of a File 184
Relative and Absolute File Paths 186
Accessing a File's Lines 186

Working with a File's Contents . . . 187
Large Files: One Million Digits . . . 188
Is Your Birthday Contained in Pi? . . . 189
Exercise 10-1: Learning Python . . . *189*
Exercise 10-2: Learning C . . . *190*
Exercise 10-3: Simpler Code . . . *190*
Writing to a File . . . 190
Writing a Single Line . . . 190
Writing Multiple Lines . . . 191
Exercise 10-4: Guest . . . *192*
Exercise 10-5: Guest Book . . . *192*
Exceptions . . . 192
Handling the ZeroDivisionError Exception . . . 192
Using try-except Blocks . . . 193
Using Exceptions to Prevent Crashes . . . 193
The else Block . . . 194
Handling the FileNotFoundError Exception . . . 195
Analyzing Text . . . 196
Working with Multiple Files . . . 197
Failing Silently . . . 198
Deciding Which Errors to Report . . . 199
Exercise 10-6: Addition . . . *200*
Exercise 10-7: Addition Calculator . . . *200*
Exercise 10-8: Cats and Dogs . . . *200*
Exercise 10-9: Silent Cats and Dogs . . . *200*
Exercise 10-10: Common Words . . . *200*
Storing Data . . . 201
Using json.dumps() and json.loads() . . . 201
Saving and Reading User-Generated Data . . . 202
Refactoring . . . 204
Exercise 10-11: Favorite Number . . . *206*
Exercise 10-12: Favorite Number Remembered . . . *206*
Exercise 10-13: User Dictionary . . . *206*
Exercise 10-14: Verify User . . . *206*
Summary . . . 207

11
TESTING YOUR CODE **209**

Installing pytest with pip . . . 210
Updating pip . . . 210
Installing pytest . . . 211
Testing a Function . . . 211
Unit Tests and Test Cases . . . 212
A Passing Test . . . 212
Running a Test . . . 213
A Failing Test . . . 214
Responding to a Failed Test . . . 215
Adding New Tests . . . 216
Exercise 11-1: City, Country . . . *217*
Exercise 11-2: Population . . . *217*

Testing a Class . . . 217
A Variety of Assertions . . . 217
A Class to Test . . . 218
Testing the AnonymousSurvey Class . . . 220
Using Fixtures . . . 221
Exercise 11-3: Employee . . . *223*
Summary . . . 223

PART II: PROJECTS 225

PROJECT 1: ALIEN INVASION

12 A SHIP THAT FIRES BULLETS 227

Planning Your Project . . . 228
Installing Pygame . . . 228
Starting the Game Project . . . 229
Creating a Pygame Window and Responding to User Input . . . 229
Controlling the Frame Rate . . . 230
Setting the Background Color . . . 231
Creating a Settings Class . . . 232
Adding the Ship Image . . . 233
Creating the Ship Class . . . 234
Drawing the Ship to the Screen . . . 235
Refactoring: The _check_events() and _update_screen() Methods . . . 237
The _check_events() Method . . . 237
The _update_screen() Method . . . 237
Exercise 12-1: Blue Sky . . . *238*
Exercise 12-2: Game Character . . . *238*
Piloting the Ship . . . 238
Responding to a Keypress . . . 238
Allowing Continuous Movement . . . 239
Moving Both Left and Right . . . 241
Adjusting the Ship's Speed . . . 242
Limiting the Ship's Range . . . 243
Refactoring _check_events() . . . 244
Pressing Q to Quit . . . 244
Running the Game in Fullscreen Mode . . . 245
A Quick Recap . . . 245
alien_invasion.py . . . 246
settings.py . . . 246
ship.py . . . 246
Exercise 12-3: Pygame Documentation . . . *246*
Exercise 12-4: Rocket . . . *246*
Exercise 12-5: Keys . . . *246*
Shooting Bullets . . . 247
Adding the Bullet Settings . . . 247

Creating the Bullet Class 247
Storing Bullets in a Group 248
Firing Bullets 249
Deleting Old Bullets 250
Limiting the Number of Bullets 251
Creating the _update_bullets() Method 252
Exercise 12-6: Sideways Shooter *253*
Summary 253

13
ALIENS! 255

Reviewing the Project 256
Creating the First Alien 256
Creating the Alien Class 257
Creating an Instance of the Alien 257
Building the Alien Fleet 259
Creating a Row of Aliens 259
Refactoring _create_fleet() 260
Adding Rows 261
Exercise 13-1: Stars *263*
Exercise 13-2: Better Stars *263*
Making the Fleet Move 263
Moving the Aliens Right 263
Creating Settings for Fleet Direction 264
Checking Whether an Alien Has Hit the Edge 265
Dropping the Fleet and Changing Direction 265
Exercise 13-3: Raindrops *266*
Exercise 13-4: Steady Rain *266*
Shooting Aliens 266
Detecting Bullet Collisions 267
Making Larger Bullets for Testing 268
Repopulating the Fleet 268
Speeding Up the Bullets 269
Refactoring _update_bullets() 269
Exercise 13-5: Sideways Shooter Part 2 . . . . . *270*
Ending the Game 270
Detecting Alien-Ship Collisions 270
Responding to Alien-Ship Collisions 271
Aliens That Reach the Bottom of the Screen 273
Game Over! 274
Identifying When Parts of the Game Should Run 275
Exercise 13-6: Game Over *275*
Summary 275

14
SCORING 277

Adding the Play Button 278
Creating a Button Class 278
Drawing the Button to the Screen 279
Starting the Game 281
Resetting the Game 281

Deactivating the Play Button . . . 282
Hiding the Mouse Cursor . . . 282
Exercise 14-1: Press P to Play . . . 283
Exercise 14-2: Target Practice . . . 283
Leveling Up . . . 283
Modifying the Speed Settings . . . 283
Resetting the Speed . . . 285
Exercise 14-3: Challenging Target Practice . . . 286
Exercise 14-4: Difficulty Levels . . . 286
Scoring . . . 286
Displaying the Score . . . 286
Making a Scoreboard . . . 287
Updating the Score as Aliens Are Shot Down . . . 289
Resetting the Score . . . 289
Making Sure to Score All Hits . . . 290
Increasing Point Values . . . 290
Rounding the Score . . . 291
High Scores . . . 292
Displaying the Level . . . 294
Displaying the Number of Ships . . . 296
Exercise 14-5: All-Time High Score . . . 299
Exercise 14-6: Refactoring . . . 299
Exercise 14-7: Expanding the Game . . . 299
Exercise 14-8: Sideways Shooter, Final Version . . . 299
Summary . . . 299

PROJECT 2: DATA VISUALIZATION

15
GENERATING DATA 301

Installing Matplotlib . . . 302
Plotting a Simple Line Graph . . . 302
Changing the Label Type and Line Thickness . . . 303
Correcting the Plot . . . 305
Using Built-in Styles . . . 306
Plotting and Styling Individual Points with scatter() . . . 306
Plotting a Series of Points with scatter() . . . 308
Calculating Data Automatically . . . 308
Customizing Tick Labels . . . 309
Defining Custom Colors . . . 310
Using a Colormap . . . 310
Saving Your Plots Automatically . . . 311
Exercise 15-1: Cubes . . . 311
Exercise 15-2: Colored Cubes . . . 311
Random Walks . . . 312
Creating the RandomWalk Class . . . 312
Choosing Directions . . . 312
Plotting the Random Walk . . . 313
Generating Multiple Random Walks . . . 314
Styling the Walk . . . 315

Exercise 15-3: Molecular Motion . . . 319
Exercise 15-4: Modified Random Walks . . . 319
Exercise 15-5: Refactoring . . . 319
Rolling Dice with Plotly . . . 319
Installing Plotly . . . 320
Creating the Die Class . . . 320
Rolling the Die . . . 320
Analyzing the Results . . . 321
Making a Histogram . . . 322
Customizing the Plot . . . 323
Rolling Two Dice . . . 324
Further Customizations . . . 325
Rolling Dice of Different Sizes . . . 326
Saving Figures . . . 327
Exercise 15-6: Two D8s . . . 328
Exercise 15-7: Three Dice . . . 328
Exercise 15-8: Multiplication . . . 328
Exercise 15-9: Die Comprehensions . . . 328
Exercise 15-10: Practicing with Both Libraries . . . 328
Summary . . . 328

16
DOWNLOADING DATA **329**

The CSV File Format . . . 330
Parsing the CSV File Headers . . . 330
Printing the Headers and Their Positions . . . 331
Extracting and Reading Data . . . 332
Plotting Data in a Temperature Chart . . . 332
The datetime Module . . . 333
Plotting Dates . . . 334
Plotting a Longer Timeframe . . . 336
Plotting a Second Data Series . . . 336
Shading an Area in the Chart . . . 337
Error Checking . . . 338
Downloading Your Own Data . . . 341
Exercise 16-1: Sitka Rainfall . . . 342
Exercise 16-2: Sitka–Death Valley Comparison . . . 342
Exercise 16-3: San Francisco . . . 342
Exercise 16-4: Automatic Indexes . . . 342
Exercise 16-5: Explore . . . 342
Mapping Global Datasets: GeoJSON Format . . . 342
Downloading Earthquake Data . . . 343
Examining GeoJSON Data . . . 343
Making a List of All Earthquakes . . . 345
Extracting Magnitudes . . . 346
Extracting Location Data . . . 346
Building a World Map . . . 347
Representing Magnitudes . . . 348
Customizing Marker Colors . . . 349
Other Color Scales . . . 350
Adding Hover Text . . . 350
Exercise 16-6: Refactoring . . . 352

Exercise 16-7: Automated Title *352*
Exercise 16-8: Recent Earthquakes *352*
Exercise 16-9: World Fires *352*
Summary 352

17
WORKING WITH APIS 355

Using an API 355
Git and GitHub 356
Requesting Data Using an API Call 356
Installing Requests 357
Processing an API Response 357
Working with the Response Dictionary 358
Summarizing the Top Repositories 361
Monitoring API Rate Limits 362
Visualizing Repositories Using Plotly 362
Styling the Chart 364
Adding Custom Tooltips 365
Adding Clickable Links 366
Customizing Marker Colors 367
More About Plotly and the GitHub API 368
The Hacker News API 368
Exercise 17-1: Other Languages *371*
Exercise 17-2: Active Discussions *371*
Exercise 17-3: Testing python_repos.py *372*
Exercise 17-4: Further Exploration *372*
Summary 372

PROJECT 3: WEB APPLICATIONS

18
GETTING STARTED WITH DJANGO 373

Setting Up a Project 374
Writing a Spec 374
Creating a Virtual Environment 374
Activating the Virtual Environment 375
Installing Django 375
Creating a Project in Django 376
Creating the Database 376
Viewing the Project 377
Exercise 18-1: New Projects *378*
Starting an App 379
Defining Models 379
Activating Models 380
The Django Admin Site 381
Defining the Entry Model 384
Migrating the Entry Model 385
Registering Entry with the Admin Site 385
The Django Shell 386

Exercise 18-2: Short Entries . . . 387
Exercise 18-3: The Django API . . . 388
Exercise 18-4: Pizzeria . . . 388
Making Pages: The Learning Log Home Page . . . 388
Mapping a URL . . . 388
Writing a View . . . 390
Writing a Template . . . 390
Exercise 18-5: Meal Planner . . . 392
Exercise 18-6: Pizzeria Home Page . . . 392
Building Additional Pages . . . 392
Template Inheritance . . . 392
The Topics Page . . . 394
Individual Topic Pages . . . 397
Exercise 18-7: Template Documentation . . . 400
Exercise 18-8: Pizzeria Pages . . . 400
Summary . . . 400

19
USER ACCOUNTS **403**

Allowing Users to Enter Data . . . 404
Adding New Topics . . . 404
Adding New Entries . . . 408
Editing Entries . . . 412
Exercise 19-1: Blog . . . 415
Setting Up User Accounts . . . 415
The accounts App . . . 415
The Login Page . . . 416
Logging Out . . . 419
The Registration Page . . . 420
Exercise 19-2: Blog Accounts . . . 423
Allowing Users to Own Their Data . . . 423
Restricting Access with @login_required . . . 423
Connecting Data to Certain Users . . . 425
Restricting Topics Access to Appropriate Users . . . 427
Protecting a User's Topics . . . 428
Protecting the edit_entry Page . . . 429
Associating New Topics with the Current User . . . 429
Exercise 19-3: Refactoring . . . 430
Exercise 19-4: Protecting new_entry . . . 430
Exercise 19-5: Protected Blog . . . 430
Summary . . . 430

20
STYLING AND DEPLOYING AN APP **433**

Styling Learning Log . . . 434
The django-bootstrap5 App . . . 434
Using Bootstrap to Style Learning Log . . . 434
Modifying base.html . . . 435
Styling the Home Page Using a Jumbotron . . . 440
Styling the Login Page . . . 441
Styling the Topics Page . . . 442

Styling the Entries on the Topic Page . . . 443
Exercise 20-1: Other Forms . . . 445
Exercise 20-2: Stylish Blog . . . 445
Deploying Learning Log . . . 445
Making a Platform.sh Account . . . 445
Installing the Platform.sh CLI . . . 446
Installing platformshconfig . . . 446
Creating a requirements.txt File . . . 446
Additional Deployment Requirements . . . 447
Adding Configuration Files . . . 447
Modifying settings.py for Platform.sh . . . 451
Using Git to Track the Project's Files . . . 451
Creating a Project on Platform.sh . . . 453
Pushing to Platform.sh . . . 455
Viewing the Live Project . . . 456
Refining the Platform.sh Deployment . . . 456
Creating Custom Error Pages . . . 459
Ongoing Development . . . 460
Deleting a Project on Platform.sh . . . 461
Exercise 20-3: Live Blog . . . 461
Exercise 20-4: Extended Learning Log . . . 462
Summary . . . 462

A
INSTALLATION AND TROUBLESHOOTING **463**

Python on Windows . . . 463
Using py Instead of python . . . 463
Rerunning the Installer . . . 464
Python on macOS . . . 464
Accidentally Installing Apple's Version of Python . . . 464
Python 2 on Older Versions of macOS . . . 465
Python on Linux . . . 465
Using the Default Python Installation . . . 465
Installing the Latest Version of Python . . . 465
Checking Which Version of Python You're Using . . . 466
Python Keywords and Built-in Functions . . . 466
Python Keywords . . . 466
Python Built-in Functions . . . 467

B
TEXT EDITORS AND IDES **469**

Working Efficiently with VS Code . . . 470
Configuring VS Code . . . 470
VS Code Shortcuts . . . 473
Other Text Editors and IDEs . . . 474
IDLE . . . 474
Geany . . . 474
Sublime Text . . . 474
Emacs and Vim . . . 475
PyCharm . . . 475
Jupyter Notebooks . . . 475

C
GETTING HELP 477

First Steps 477
Try It Again 478
Take a Break 478
Refer to This Book's Resources. 478
Searching Online. 479
Stack Overflow. 479
The Official Python Documentation 479
Official Library Documentation 480
r/learnpython. 480
Blog Posts 480
Discord. 480
Slack 481

D
USING GIT FOR VERSION CONTROL 483

Installing Git 484
Configuring Git. 484
Making a Project 484
Ignoring Files. 484
Initializing a Repository 485
Checking the Status 485
Adding Files to the Repository 486
Making a Commit 486
Checking the Log 487
The Second Commit 487
Abandoning Changes 488
Checking Out Previous Commits 489
Deleting the Repository. 491

E
TROUBLESHOOTING DEPLOYMENTS 493

Understanding Deployments 494
Basic Troubleshooting. 494
Follow Onscreen Suggestions. 495
Read the Log Output 496
OS-Specific Troubleshooting 497
Deploying from Windows 497
Deploying from macOS 499
Deploying from Linux. 499
Other Deployment Approaches 500

INDEX 503

PREFACE TO THE THIRD EDITION

The response to the first and second editions of *Python Crash Course* has been overwhelmingly positive. More than one million copies are in print, including translations in over 10 languages. I've received letters and emails from readers as young as 10, as well as from retirees who want to learn to program in their free time. *Python Crash Course* is being used in middle schools and high schools, and also in college classes. Students who are assigned more advanced textbooks are using *Python Crash Course* as a companion text for their classes and finding it a worthwhile supplement. People are using it to enhance their skills on the job, change careers, and start working on their own side projects. In short, people are using the book for the full range of purposes I had hoped they would, and much more.

The opportunity to write a third edition of *Python Crash Course* has been thoroughly enjoyable. Although Python is a mature language, it continues to evolve as every language does. My main goal in revising the book is to keep it a well-curated introductory Python course. By reading this book, you'll learn everything you need to start working on your own projects, and you'll build a solid foundation for all of your future learning as well. I've updated some sections to reflect newer, simpler ways of doing things in Python. I've also clarified some sections where certain details of the language were not presented as accurately as they could have been. All the projects have been completely updated using popular, well-maintained libraries that you can confidently use to build your own projects.

The following is a summary of specific changes that have been made in the third edition:

- Chapter 1 now features the text editor VS Code, which is popular among beginner and professional programmers and works well on all operating systems.
- Chapter 2 includes the new methods `removeprefix()` and `removesuffix()`, which are helpful when working with files and URLs. This chapter also features Python's newly improved error messages, which provide much more specific information to help you troubleshoot your code when something goes wrong.
- Chapter 10 uses the `pathlib` module for working with files. This is a much simpler approach to reading from and writing to files.
- Chapter 11 uses `pytest` to write automated tests for the code you write. The `pytest` library has become the industry standard tool for writing tests in Python. It's friendly enough to use for your first tests, and if you pursue a career as a Python programmer, you'll use it in professional settings as well.
- The *Alien Invasion* project in Chapters 12–14 includes a setting to control the frame rate, which makes the game run more consistently across different operating systems. A simpler approach is used to build the fleet of aliens, and the overall organization of the project has been cleaned up as well.
- The data visualization projects in Chapters 15–17 use the most recent features of Matplotlib and Plotly. The Matplotlib visualizations feature updated style settings. The random walk project has a small improvement that increases the accuracy of the plots, which means you'll see a wider variety of patterns emerge each time you generate a new walk. All the projects featuring Plotly now use the Plotly Express module, which lets you generate your initial visualizations with just a few lines of code. You can easily explore a variety of visualizations before committing to one kind of plot, and then focus on refining individual elements of that plot.
- The Learning Log project in Chapters 18–20 is built using the latest version of Django and styled using the latest version of Bootstrap. Some parts of the project have been renamed to make it easier to follow the overall organization of the project. The project is now deployed to Platform.sh, a modern hosting service for Django projects. The deployment process is controlled by YAML configuration files, which give you a great deal of control over how your project is deployed. This approach is consistent with how professional programmers deploy modern Django projects.
- Appendix A has been fully updated to recommend current best practices for installing Python on all major operating systems. Appendix B includes detailed instructions for setting up VS Code, and brief descriptions of most of the major text editors and IDEs in current use. Appendix C directs readers to several of the most popular online resources for getting

help. Appendix D continues to offer a mini crash course in using Git for version control. Appendix E is brand new for the third edition. Even with a good set of instructions for deploying the apps you create, there are many things that can go wrong. This appendix offers a detailed troubleshooting guide that you can use when the deployment process doesn't work on the first try.

- The index has been thoroughly updated to allow you to use *Python Crash Course* as a reference for all of your future Python projects.

Thank you for reading *Python Crash Course*! If you have any feedback or questions, please feel free to get in touch; I am *@ehmatthes* on Twitter.

ACKNOWLEDGMENTS

This book would not have been possible without the wonderful and extremely professional staff at No Starch Press. Bill Pollock invited me to write an introductory book, and I deeply appreciate that original offer. Liz Chadwick has worked on all three editions, and the book is better because of her ongoing involvement. Eva Morrow brought fresh eyes to this new edition, and her insights have improved the book as well. I appreciate Doug McNair's guidance in using proper grammar, without becoming overly formal. Jennifer Kepler has supervised the production work, which turns my many files into a polished final product.

There are many people at No Starch Press who have helped make this book a success but whom I haven't had the chance to work with directly. No Starch has a fantastic marketing team, who go beyond just selling books; they make sure readers find the books that are likely to work well for them and help them reach their goals. No Starch also has a strong foreign-rights department. *Python Crash Course* has reached readers around the world, in many languages, due to the diligence of this team. To all of these people whom I haven't worked with individually, thank you for helping *Python Crash Course* find its audience.

I'd like to thank Kenneth Love, the technical reviewer for all three editions of *Python Crash Course*. I met Kenneth at PyCon one year, and their enthusiasm for the language and the Python community has been a constant source of professional inspiration ever since. Kenneth, as always, went beyond simple fact-checking and reviewed the book with the goal of helping newer programmers develop a solid understanding of the Python language and programming in general. They also kept an eye out for areas that worked well enough in previous editions but could be improved upon, given the opportunity for a full rewrite. That said, any inaccuracies that remain are completely my own.

I'd also like to express my appreciation to all the readers who have shared their experience of working through *Python Crash Course*. Learning the basics of programming can change your perspective on the world, and sometimes this has a profound impact on people. It's deeply humbling to hear these stories, and I appreciate everyone who has shared their experiences so openly.

I'd like to thank my father for introducing me to programming at a young age and for not being afraid that I'd break his equipment. I'd like to thank my wife, Erin, for supporting and encouraging me through the writing of this book, and through all the work that goes into maintaining it through multiple editions. I'd also like to thank my son, Ever, whose curiosity continues to inspire me.

INTRODUCTION

Every programmer has a story about how they learned to write their first program. I started programming as a child, when my father was working for Digital Equipment Corporation, one of the pioneering companies of the modern computing era. I wrote my first program on a kit computer that my dad had assembled in our basement. The computer consisted of nothing more than a bare motherboard connected to a keyboard without a case, and its monitor was a bare cathode ray tube. My initial program was a simple number guessing game, which looked something like this:

```
I'm thinking of a number! Try to guess the number I'm thinking of: 25
Too low! Guess again: 50
Too high! Guess again: 42
That's it! Would you like to play again? (yes/no) no
Thanks for playing!
```

I'll always remember how satisfied I felt, watching my family play a game that I created and that worked as I intended it to.

That early experience had a lasting impact. There's real satisfaction in building something with a purpose, that solves a problem. The software I write now meets more significant needs than my childhood efforts did, but the sense of satisfaction I get from creating a program that works is still largely the same.

Who Is This Book For?

The goal of this book is to bring you up to speed with Python as quickly as possible so you can build programs that work—games, data visualizations, and web applications—while developing a foundation in programming that will serve you well for the rest of your life. *Python Crash Course* is written for people of any age who have never programmed in Python or have never programmed at all. This book is for those who want to learn the basics of programming quickly so they can focus on interesting projects, and those who like to test their understanding of new concepts by solving meaningful problems. *Python Crash Course* is also perfect for teachers at all levels who want to offer their students a project-based introduction to programming. If you're taking a college class and want a friendlier introduction to Python than the text you've been assigned, this book can make your class easier as well. If you're looking to change careers, *Python Crash Course* can help you make the transition into a more satisfying career track. It has worked well for a wide variety of readers, with a broad range of goals.

What Can You Expect to Learn?

The purpose of this book is to make you a good programmer in general and a good Python programmer in particular. You'll learn efficiently and adopt good habits as you gain a solid foundation in general programming concepts. After working your way through *Python Crash Course*, you should be ready to move on to more advanced Python techniques, and your next programming language will be even easier to grasp.

In Part I of this book, you'll learn basic programming concepts you need to know to write Python programs. These concepts are the same as those you'd learn when starting out in almost any programming language. You'll learn about different kinds of data and the ways you can store data in your programs. You'll build collections of data, such as lists and dictionaries, and you'll work through those collections in efficient ways. You'll learn to use `while` loops and `if` statements to test for certain conditions, so you can run specific sections of code while those conditions are true and run other sections when they're not—a technique that helps you automate many processes.

You'll learn to accept input from users to make your programs interactive, and to keep your programs running as long as the user wants. You'll explore how to write functions that make parts of your program reusable, so you only have to write blocks of code that perform certain actions once, while using that code as many times as you need. You'll then extend this concept to more complicated behavior with classes, making fairly simple programs respond to a variety of situations. You'll learn to write programs that handle common errors gracefully. After working through each of these basic concepts, you'll write a number of increasingly complex programs using what you've learned. Finally, you'll take your first step toward intermediate programming by learning how to write tests for your code, so you can develop your programs further without worrying about introducing bugs. All the information in Part I will prepare you for taking on larger, more complex projects.

In Part II, you'll apply what you learned in Part I to three projects. You can do any or all of these projects, in whichever order works best for you. In the first project, in Chapters 12–14, you'll create a *Space Invaders*–style shooting game called *Alien Invasion*, which includes several increasingly difficult levels of game play. After you've completed this project, you should be well on your way to being able to develop your own 2D games. Even if you don't aspire to become a game programmer, working through this project is an enjoyable way to tie together much of what you'll learn in Part I.

The second project, in Chapters 15–17, introduces you to data visualization. Data scientists use a variety of visualization techniques to help make sense of the vast amount of information available to them. You'll work with datasets that you generate through code, datasets that you download from online sources, and datasets your programs download automatically. After you've completed this project, you'll be able to write programs that sift through large datasets and make visual representations of many different kinds of information.

In the third project, in Chapters 18–20, you'll build a small web application called Learning Log. This project allows you to keep an organized journal of information you've learned about a specific topic. You'll be able to keep separate logs for different topics and allow others to create an account and start their own journals. You'll also learn how to deploy your project so anyone can access it online, from anywhere in the world.

Online Resources

No Starch Press has more information about this book available online at *https://nostarch.com/python-crash-course-3rd-edition*.

I also maintain an extensive set of supplementary resources at *https://ehmatthes.github.io/pcc_3e*. These resources include the following:

Setup instructions The setup instructions online are identical to what's in the book, but they include active links you can click for all the different steps. If you're having any setup issues, refer to this resource.

Updates Python, like all languages, is constantly evolving. I maintain a thorough set of updates, so if anything isn't working, check here to see whether instructions have changed.

Solutions to exercises You should spend significant time on your own attempting the exercises in the "Try It Yourself" sections. However, if you're stuck and can't make any progress, solutions to most of the exercises are online.

Cheat sheets A full set of downloadable cheat sheets for quick reference to major concepts is also online.

Why Python?

Every year, I consider whether to continue using Python or move on to a different language, perhaps one that's newer to the programming world. But I continue to focus on Python for many reasons. Python is an incredibly efficient language: your programs will do more in fewer lines of code than many other languages would require. Python's syntax will also help you write "clean" code. Your code will be easier to read, easier to debug, and easier to extend and build upon, compared to other languages.

People use Python for many purposes: to make games, build web applications, solve business problems, and develop internal tools at all kinds of interesting companies. Python is also used heavily in scientific fields, for academic research and applied work.

One of the most important reasons I continue to use Python is because of the Python community, which includes an incredibly diverse and welcoming group of people. Community is essential to programmers because programming isn't a solitary pursuit. Most of us, even the most experienced programmers, need to ask advice from others who have already solved similar problems. Having a well-connected and supportive community is critical to helping you solve problems, and the Python community is fully supportive of people who are learning Python as their first programming language or coming to Python with a background in other languages.

Python is a great language to learn, so let's get started!

PART I

BASICS

Part I of this book teaches you the basic concepts you'll need to write Python programs. Many of these concepts are common to all programming languages, so they'll be useful throughout your life as a programmer.

In **Chapter 1** you'll install Python on your computer and run your first program, which prints the message *Hello world!* to the screen.

In **Chapter 2** you'll learn to assign information to variables and work with text and numerical values.

Chapters 3 and **4** introduce lists. Lists can store as much information as you want in one place, allowing you to work with that data efficiently. You'll be able to work with hundreds, thousands, and even millions of values in just a few lines of code.

In **Chapter 5** you'll use `if` statements to write code that responds one way if certain conditions are true, and responds in a different way if those conditions are not true.

Chapter 6 shows you how to use Python's dictionaries, which let you make connections between different pieces of information. Like lists, dictionaries can contain as much information as you need to store.

In **Chapter 7** you'll learn how to accept input from users to make your programs interactive. You'll also learn about `while` loops, which run blocks of code repeatedly as long as certain conditions remain true.

In **Chapter 8** you'll write functions, which are named blocks of code that perform a specific task and can be run whenever you need them.

Chapter 9 introduces classes, which allow you to model real-world objects. You'll write code that represents dogs, cats, people, cars, rockets, and more.

Chapter 10 shows you how to work with files and handle errors so your programs won't crash unexpectedly. You'll store data before your program closes and read the data back in when the program runs again. You'll learn about Python's exceptions, which allow you to anticipate errors and make your programs handle those errors gracefully.

In **Chapter 11** you'll learn to write tests for your code, to check that your programs work the way you intend them to. As a result, you'll be able to expand your programs without worrying about introducing new bugs. Testing your code is one of the first skills that will help you transition from beginner to intermediate programmer.

1

GETTING STARTED

In this chapter, you'll run your first Python program, *hello_world.py*. First, you'll need to check whether a recent version of Python is installed on your computer; if it isn't, you'll install it. You'll also install a text editor to work with your Python programs. Text editors recognize Python code and highlight sections as you write, making it easy to understand your code's structure.

Setting Up Your Programming Environment

Python differs slightly on different operating systems, so you'll need to keep a few considerations in mind. In the following sections, we'll make sure Python is set up correctly on your system.

Python Versions

Every programming language evolves as new ideas and technologies emerge, and the developers of Python have continually made the language more versatile and powerful. As of this writing, the latest version is Python 3.11, but everything in this book should run on Python 3.9 or later. In this section, we'll find out if Python is already installed on your system and whether you need to install a newer version. Appendix A contains additional details about installing the latest version of Python on each major operating system as well.

Running Snippets of Python Code

You can run Python's interpreter in a terminal window, allowing you to try bits of Python code without having to save and run an entire program.

Throughout this book, you'll see code snippets that look like this:

```
>>> print("Hello Python interpreter!")
Hello Python interpreter!
```

The three angle brackets (`>>>`) prompt, which we'll refer to as a *Python prompt*, indicates that you should be using the terminal window. The bold text is the code you should type in and then execute by pressing ENTER. Most of the examples in this book are small, self-contained programs that you'll run from your text editor rather than the terminal, because you'll write most of your code in the text editor. But sometimes, basic concepts will be shown in a series of snippets run through a Python terminal session to demonstrate particular concepts more efficiently. When you see three angle brackets in a code listing, you're looking at code and output from a terminal session. We'll try coding in the interpreter on your system in a moment.

We'll also use a text editor to create a simple program called *Hello World!* that has become a staple of learning to program. There's a long-held tradition in the programming world that printing the message `Hello world!` to the screen as your first program in a new language will bring you good luck. Such a simple program serves a very real purpose. If it runs correctly on your system, then any Python program you write should work as well.

About the VS Code Editor

VS Code is a powerful, professional-quality text editor that's free and beginner-friendly. VS Code is great for both simple and complex projects, so if you become comfortable using it while learning Python, you can continue using it as you progress to larger and more complicated projects. VS Code can be installed on all modern operating systems, and it supports most programming languages, including Python.

Appendix B provides information on other text editors. If you're curious about the other options, you might want to skim that appendix at this point. If you want to begin programming quickly, you can use VS Code to start. Then you can consider other editors, once you've gained some experience as a programmer. In this chapter, I'll walk you through installing VS Code on your operating system.

NOTE *If you already have a text editor installed and you know how to configure it to run Python programs, you are welcome to use that editor instead.*

Python on Different Operating Systems

Python is a cross-platform programming language, which means it runs on all the major operating systems. Any Python program you write should run on any modern computer that has Python installed. However, the methods for setting up Python on different operating systems vary slightly.

In this section, you'll learn how to set up Python on your system. You'll first check whether a recent version of Python is installed on your system, and install it if it's not. Then you'll install VS Code. These are the only two steps that are different for each operating system.

In the sections that follow, you'll run *hello_world.py* and troubleshoot anything that doesn't work. I'll walk you through this process for each operating system, so you'll have a Python programming environment that you can rely on.

Python on Windows

Windows doesn't usually come with Python, so you'll probably need to install it and then install VS Code.

Installing Python

First, check whether Python is installed on your system. Open a command window by entering **`command`** into the Start menu and clicking the **Command Prompt** app. In the terminal window, enter **`python`** in lowercase. If you get a Python prompt (>>>) in response, Python is installed on your system. If you see an error message telling you that `python` is not a recognized command, or if the Microsoft store opens, Python isn't installed. Close the Microsoft store if it opens; it's better to download an official installer than to use Microsoft's version.

If Python is not installed on your system, or if you see a version earlier than Python 3.9, you need to download a Python installer for Windows. Go to *https://python.org* and hover over the **Downloads** link. You should see a button for downloading the latest version of Python. Click the button, which should automatically start downloading the correct installer for your system. After you've downloaded the file, run the installer. Make sure you select the option **Add Python to PATH**, which will make it easier to configure your system correctly. Figure 1-1 shows this option selected.

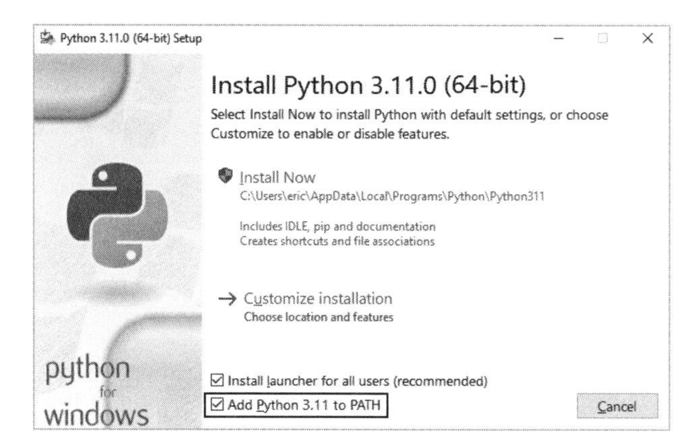

Figure 1-1: Make sure you select the checkbox labeled Add Python to PATH.

Running Python in a Terminal Session

Open a new command window and enter **python** in lowercase. You should see a Python prompt (>>>), which means Windows has found the version of Python you just installed.

```
C:\> python
Python 3.x.x (main, Jun . . . , 13:29:14) [MSC v.1932 64 bit (AMD64)] on win32
Type "help", "copyright", "credits" or "license" for more information.
>>>
```

NOTE *If you don't see this output or something similar, see the more detailed setup instructions in Appendix A.*

Enter the following line in your Python session:

```
>>> print("Hello Python interpreter!")
Hello Python interpreter!
>>>
```

You should see the output `Hello Python interpreter!` Anytime you want to run a snippet of Python code, open a command window and start a Python terminal session. To close the terminal session, press CTRL-Z and then press ENTER, or enter the command **exit()**.

Installing VS Code

You can download an installer for VS Code at *https://code.visualstudio.com*. Click the **Download for Windows** button and run the installer. Skip the following sections about macOS and Linux, and follow the steps in "Running a Hello World Program" on page 9.

Python on macOS

Python is not installed by default on the latest versions of macOS, so you'll need to install it if you haven't already done so. In this section, you'll install the latest version of Python, and then install VS Code and make sure it's configured correctly.

NOTE *Python 2 was included on older versions of macOS, but it's an outdated version that you shouldn't use.*

Checking Whether Python 3 Is Installed

Open a terminal window by going to **Applications ▸ Utilities ▸ Terminal**. You can also press ⌘-spacebar, type `terminal`, and then press ENTER. To see if you have a recent enough version of Python installed, enter **`python3`**. You'll most likely see a message about installing the *command line developer tools*. It's better to install these tools after installing Python, so if this message appears, cancel the pop-up window.

If the output shows you have Python 3.9 or a later version installed, you can skip the next section and go to "Running Python in a Terminal Session." If you see any version earlier than Python 3.9, follow the instructions in the next section to install the latest version.

Note that on macOS, whenever you see the `python` command in this book, you need to use the `python3` command instead to make sure you're using Python 3. On most macOS systems, the `python` command either points to an outdated version of Python that should only be used by internal system tools, or it points to nothing and generates an error message.

Installing the Latest Version of Python

You can find a Python installer for your system at *https://python.org*. Hover over the **Download** link, and you should see a button for downloading the latest version of Python. Click the button, which should automatically start downloading the correct installer for your system. After the file downloads, run the installer.

After the installer runs, a Finder window should appear. Double-click the *Install Certificates.command* file. Running this file will allow you to more easily install additional libraries that you'll need for real-world projects, including the projects in the second half of this book.

Running Python in a Terminal Session

You can now try running snippets of Python code by opening a new terminal window and typing **`python3`**:

```
$ python3
Python 3.x.x (v3.11.0:eb0004c271, Jun . . . , 10:03:01)
[Clang 13.0.0 (clang-1300.0.29.30)] on darwin
Type "help", "copyright", "credits" or "license" for more information.
>>>
```

This command starts a Python terminal session. You should see a Python prompt (>>>), which means macOS has found the version of Python you just installed.

Enter the following line in the terminal session:

```
>>> print("Hello Python interpreter!")
Hello Python interpreter!
>>>
```

You should see the message `Hello Python interpreter!`, which should print directly in the current terminal window. You can close the Python interpreter by pressing CTRL-D or by entering the command **`exit()`**.

NOTE *On newer macOS systems, you'll see a percent sign (%) as a terminal prompt instead of a dollar sign ($).*

Installing VS Code

To install the VS Code editor, you need to download the installer at *https://code.visualstudio.com*. Click the **Download** button, and then open a **Finder** window and go to the **Downloads** folder. Drag the **Visual Studio Code** installer to your Applications folder, then double-click the installer to run it.

Skip over the following section about Python on Linux, and follow the steps in "Running a Hello World Program" on page 9.

Python on Linux

Linux systems are designed for programming, so Python is already installed on most Linux computers. The people who write and maintain Linux expect you to do your own programming at some point, and encourage you to do so. For this reason, there's very little to install and only a few settings to change to start programming.

Checking Your Version of Python

Open a terminal window by running the Terminal application on your system (in Ubuntu, you can press CTRL-ALT-T). To find out which version of Python is installed, enter **`python3`** with a lowercase *p*. When Python is installed, this command starts the Python interpreter. You should see output indicating which version of Python is installed. You should also see a Python prompt (>>>) where you can start entering Python commands:

```
$ python3
Python 3.10.4 (main, Apr  . . . , 09:04:19) [GCC 11.2.0] on linux
Type "help", "copyright", "credits" or "license" for more information.
>>>
```

This output indicates that Python 3.10.4 is currently the default version of Python installed on this computer. When you've seen this output, press CTRL-D or enter **`exit()`** to leave the Python prompt and return to a terminal prompt. Whenever you see the `python` command in this book, enter **`python3`** instead.

You'll need Python 3.9 or later to run the code in this book. If the Python version installed on your system is earlier than Python 3.9, or if you want to update to the latest version currently available, refer to the instructions in Appendix A.

Running Python in a Terminal Session

You can try running snippets of Python code by opening a terminal and entering **`python3`**, as you did when checking your version. Do this again, and when you have Python running, enter the following line in the terminal session:

```
>>> print("Hello Python interpreter!")
Hello Python interpreter!
>>>
```

The message should print directly in the current terminal window. Remember that you can close the Python interpreter by pressing CTRL-D or by entering the command **`exit()`**.

Installing VS Code

On Ubuntu Linux, you can install VS Code from the Ubuntu Software Center. Click the Ubuntu Software icon in your menu and search for *vscode*. Click the app called **Visual Studio Code** (sometimes called *code*), and then click **Install**. Once it's installed, search your system for *VS Code* and launch the app.

Running a Hello World Program

With a recent version of Python and VS Code installed, you're almost ready to run your first Python program written in a text editor. But before doing so, you need to install the Python extension for VS Code.

Installing the Python Extension for VS Code

VS Code works with many different programming languages; to get the most out of it as a Python programmer, you'll need to install the Python extension. This extension adds support for writing, editing, and running Python programs.

To install the Python extension, click the Manage icon, which looks like a gear in the lower-left corner of the VS Code app. In the menu that appears, click **Extensions**. Enter **`python`** in the search box and click the **Python** extension. (If you see more than one extension named *Python*, choose the one supplied by Microsoft.) Click **Install** and install any additional tools that your system needs to complete the installation. If you see a message that you need to install Python, and you've already done so, you can ignore this message.

NOTE *If you're using macOS and a pop-up asks you to install the* command line developer tools, *click* **Install**. *You may see a message that it will take an excessively long time to install, but it should only take about 10 or 20 minutes on a reasonable internet connection.*

Running hello_world.py

Before you write your first program, make a folder called *python_work* on your desktop for your projects. It's best to use lowercase letters and underscores for spaces in file and folder names, because Python uses these naming conventions. You can make this folder somewhere other than the desktop, but it will be easier to follow some later steps if you save the *python_work* folder directly on your desktop.

Open VS Code, and close the **Get Started** tab if it's still open. Make a new file by clicking **File ▸ New File** or pressing CTRL-N (⌘-N on macOS). Save the file as *hello_world.py* in your *python_work* folder. The extension *.py* tells VS Code that your file is written in Python, and tells it how to run the program and highlight the text in a helpful way.

After you've saved your file, enter the following line in the editor:

hello_world.py

```
print("Hello Python world!")
```

To run your program, select **Run ▸ Run Without Debugging** or press CTRL-F5. A terminal screen should appear at the bottom of the VS Code window, showing your program's output:

```
Hello Python world!
```

You'll likely see some additional output showing the Python interpreter that was used to run your program. If you want to simplify the information that's displayed so you only see your program's output, see Appendix B. You can also find helpful suggestions about how to use VS Code more efficiently in Appendix B.

If you don't see this output, something might have gone wrong in the program. Check every character on the line you entered. Did you accidentally capitalize `print`? Did you forget one or both of the quotation marks or parentheses? Programming languages expect very specific syntax, and if you don't provide that, you'll get errors. If you can't get the program to run, see the suggestions in the next section.

Troubleshooting

If you can't get *hello_world.py* to run, here are a few remedies you can try that are also good general solutions for any programming problem:

- When a program contains a significant error, Python displays a *traceback*, which is an error report. Python looks through the file and tries to identify the problem. Check the traceback; it might give you a clue as to what issue is preventing the program from running.

- Step away from your computer, take a short break, and then try again. Remember that syntax is very important in programming, so something as simple as mismatched quotation marks or mismatched parentheses can prevent a program from running properly. Reread the relevant parts of this chapter, look over your code, and try to find the mistake.
- Start over again. You probably don't need to uninstall any software, but it might make sense to delete your *hello_world.py* file and re-create it from scratch.
- Ask someone else to follow the steps in this chapter, on your computer or a different one, and watch what they do carefully. You might have missed one small step that someone else happens to catch.
- See the additional installation instructions in Appendix A; some of the details included in the Appendix may help you solve your issue.
- Find someone who knows Python and ask them to help you get set up. If you ask around, you might find that you unexpectedly know someone who uses Python.
- The setup instructions in this chapter are also available through this book's companion website at *https://ehmatthes.github.io/pcc_3e*. The online version of these instructions might work better because you can simply cut and paste code and click links to the resources you need.
- Ask for help online. Appendix C provides a number of resources, such as forums and live chat sites, where you can ask for solutions from people who've already worked through the issue you're currently facing.

Never worry that you're bothering experienced programmers. Every programmer has been stuck at some point, and most programmers are happy to help you set up your system correctly. As long as you can state clearly what you're trying to do, what you've already tried, and the results you're getting, there's a good chance someone will be able to help you. As mentioned in the introduction, the Python community is very friendly and welcoming to beginners.

Python should run well on any modern computer. Early setup issues can be frustrating, but they're well worth sorting out. Once you get *hello_world.py* running, you can start to learn Python, and your programming work will become more interesting and satisfying.

Running Python Programs from a Terminal

You'll run most of your programs directly in your text editor. However, sometimes it's useful to run programs from a terminal instead. For example, you might want to run an existing program without opening it for editing.

You can do this on any system with Python installed if you know how to access the directory where the program file is stored. To try this, make sure you've saved the *hello_world.py* file in the *python_work* folder on your desktop.

On Windows

You can use the terminal command cd, for *change directory*, to navigate through your filesystem in a command window. The command dir, for *directory*, shows you all the files that exist in the current directory.

Open a new terminal window and enter the following commands to run *hello_world.py*:

```
C:\> cd Desktop\python_work
C:\Desktop\python_work> dir
hello_world.py
C:\Desktop\python_work> python hello_world.py
Hello Python world!
```

First, use the **cd** command to navigate to the *python_work* folder, which is in the *Desktop* folder. Next, use the **dir** command to make sure *hello_world.py* is in this folder. Then run the file using the command **python hello_world.py**.

Most of your programs will run fine directly from your editor. However, as your work becomes more complex, you'll want to run some of your programs from a terminal.

On macOS and Linux

Running a Python program from a terminal session is the same on Linux and macOS. You can use the terminal command cd, for *change directory*, to navigate through your filesystem in a terminal session. The command ls, for *list*, shows you all the nonhidden files that exist in the current directory.

Open a new terminal window and enter the following commands to run *hello_world.py*:

```
~$ cd Desktop/python_work/
~/Desktop/python_work$ ls
hello_world.py
~/Desktop/python_work$ python3 hello_world.py
Hello Python world!
```

First, use the **cd** command to navigate to the *python_work* folder, which is in the *Desktop* folder. Next, use the **ls** command to make sure *hello_world.py* is in this folder. Then run the file using the command **python3 hello_world.py**.

Most of your programs will run fine directly from your editor. But as your work becomes more complex, you'll want to run some of your programs from a terminal.

TRY IT YOURSELF

The exercises in this chapter are exploratory in nature. Starting in Chapter 2, the challenges you'll solve will be based on what you've learned.

1-1. python.org: Explore the Python home page (*https://python.org*) to find topics that interest you. As you become familiar with Python, different parts of the site will be more useful to you.

1-2. Hello World Typos: Open the *hello_world.py* file you just created. Make a typo somewhere in the line and run the program again. Can you make a typo that generates an error? Can you make sense of the error message? Can you make a typo that doesn't generate an error? Why do you think it didn't make an error?

1-3. Infinite Skills: If you had infinite programming skills, what would you build? You're about to learn how to program. If you have an end goal in mind, you'll have an immediate use for your new skills; now is a great time to write brief descriptions of what you want to create. It's a good habit to keep an "ideas" notebook that you can refer to whenever you want to start a new project. Take a few minutes now to describe three programs you want to create.

Summary

In this chapter, you learned a bit about Python in general, and you installed Python on your system if it wasn't already there. You also installed a text editor to make it easier to write Python code. You ran snippets of Python code in a terminal session, and you ran your first program, *hello_world.py*. You probably learned a bit about troubleshooting as well.

In the next chapter, you'll learn about the different kinds of data you can work with in your Python programs, and you'll start to use variables as well.

2

VARIABLES AND SIMPLE DATA TYPES

In this chapter you'll learn about the different kinds of data you can work with in your Python programs. You'll also learn how to use variables to represent data in your programs.

What Really Happens When You Run hello_world.py

Let's take a closer look at what Python does when you run *hello_world.py*. As it turns out, Python does a fair amount of work, even when it runs a simple program:

hello_world.py

```
print("Hello Python world!")
```

When you run this code, you should see the following output:

```
Hello Python world!
```

When you run the file *hello_world.py*, the ending *.py* indicates that the file is a Python program. Your editor then runs the file through the *Python interpreter*, which reads through the program and determines what each word in the program means. For example, when the interpreter sees the word `print` followed by parentheses, it prints to the screen whatever is inside the parentheses.

As you write your programs, your editor highlights different parts of your program in different ways. For example, it recognizes that `print()` is the name of a function and displays that word in one color. It recognizes that `"Hello Python world!"` is not Python code, and displays that phrase in a different color. This feature is called *syntax highlighting* and is quite useful as you start to write your own programs.

Variables

Let's try using a variable in *hello_world.py*. Add a new line at the beginning of the file, and modify the second line:

hello_world.py

```
message = "Hello Python world!"
print(message)
```

Run this program to see what happens. You should see the same output you saw previously:

```
Hello Python world!
```

We've added a *variable* named `message`. Every variable is connected to a *value*, which is the information associated with that variable. In this case the value is the `"Hello Python world!"` text.

Adding a variable makes a little more work for the Python interpreter. When it processes the first line, it associates the variable `message` with the `"Hello Python world!"` text. When it reaches the second line, it prints the value associated with `message` to the screen.

Let's expand on this program by modifying *hello_world.py* to print a second message. Add a blank line to *hello_world.py*, and then add two new lines of code:

```
message = "Hello Python world!"
print(message)

message = "Hello Python Crash Course world!"
print(message)
```

Now when you run *hello_world.py*, you should see two lines of output:

```
Hello Python world!
Hello Python Crash Course world!
```

You can change the value of a variable in your program at any time, and Python will always keep track of its current value.

Naming and Using Variables

When you're using variables in Python, you need to adhere to a few rules and guidelines. Breaking some of these rules will cause errors; other guidelines just help you write code that's easier to read and understand. Be sure to keep the following rules in mind when working with variables:

- Variable names can contain only letters, numbers, and underscores. They can start with a letter or an underscore, but not with a number. For instance, you can call a variable `message_1` but not `1_message`.
- Spaces are not allowed in variable names, but underscores can be used to separate words in variable names. For example, `greeting_message` works but `greeting message` will cause errors.
- Avoid using Python keywords and function names as variable names. For example, do not use the word `print` as a variable name; Python has reserved it for a particular programmatic purpose. (See "Python Keywords and Built-in Functions" on page 466.)
- Variable names should be short but descriptive. For example, `name` is better than `n`, `student_name` is better than `s_n`, and `name_length` is better than `length_of_persons_name`.
- Be careful when using the lowercase letter *l* and the uppercase letter *O* because they could be confused with the numbers *1* and *0*.

It can take some practice to learn how to create good variable names, especially as your programs become more interesting and complicated. As you write more programs and start to read through other people's code, you'll get better at coming up with meaningful names.

NOTE *The Python variables you're using at this point should be lowercase. You won't get errors if you use uppercase letters, but uppercase letters in variable names have special meanings that we'll discuss in later chapters.*

Avoiding Name Errors When Using Variables

Every programmer makes mistakes, and most make mistakes every day. Although good programmers might create errors, they also know how to respond to those errors efficiently. Let's look at an error you're likely to make early on and learn how to fix it.

We'll write some code that generates an error on purpose. Enter the following code, including the misspelled word `mesage`, which is shown in bold:

```
message = "Hello Python Crash Course reader!"
print(mesage)
```

When an error occurs in your program, the Python interpreter does its best to help you figure out where the problem is. The interpreter provides a traceback when a program cannot run successfully. A *traceback* is a record of where the interpreter ran into trouble when trying to execute your code.

Here's an example of the traceback that Python provides after you've accidentally misspelled a variable's name:

```
Traceback (most recent call last):
❶  File "hello_world.py", line 2, in <module>
❷    print(mesage)
           ^^^^^^
❸ NameError: name 'mesage' is not defined. Did you mean: 'message'?
```

The output reports that an error occurs in line 2 of the file *hello_world.py* ❶. The interpreter shows this line ❷ to help us spot the error quickly and tells us what kind of error it found ❸. In this case it found a *name error* and reports that the variable being printed, `mesage`, has not been defined. Python can't identify the variable name provided. A name error usually means we either forgot to set a variable's value before using it, or we made a spelling mistake when entering the variable's name. If Python finds a variable name that's similar to the one it doesn't recognize, it will ask if that's the name you meant to use.

In this example we omitted the letter *s* in the variable name `message` in the second line. The Python interpreter doesn't spellcheck your code, but it does ensure that variable names are spelled consistently. For example, watch what happens when we spell *message* incorrectly in the line that defines the variable:

```
mesage = "Hello Python Crash Course reader!"
print(mesage)
```

In this case, the program runs successfully!

```
Hello Python Crash Course reader!
```

The variable names match, so Python sees no issue. Programming languages are strict, but they disregard good and bad spelling. As a result, you don't need to consider English spelling and grammar rules when you're trying to create variable names and writing code.

Many programming errors are simple, single-character typos in one line of a program. If you find yourself spending a long time searching for one of these errors, know that you're in good company. Many experienced and talented programmers spend hours hunting down these kinds of tiny errors. Try to laugh about it and move on, knowing it will happen frequently throughout your programming life.

Variables Are Labels

Variables are often described as boxes you can store values in. This idea can be helpful the first few times you use a variable, but it isn't an accurate way to describe how variables are represented internally in Python. It's much better to think of variables as labels that you can assign to values. You can also say that a variable references a certain value.

This distinction probably won't matter much in your initial programs, but it's worth learning earlier rather than later. At some point, you'll see unexpected behavior from a variable, and an accurate understanding of how variables work will help you identify what's happening in your code.

NOTE *The best way to understand new programming concepts is to try using them in your programs. If you get stuck while working on an exercise in this book, try doing something else for a while. If you're still stuck, review the relevant part of that chapter. If you still need help, see the suggestions in Appendix C.*

TRY IT YOURSELF

Write a separate program to accomplish each of these exercises. Save each program with a filename that follows standard Python conventions, using lowercase letters and underscores, such as *simple_message.py* and *simple_messages.py*.

2-1. Simple Message: Assign a message to a variable, and then print that message.

2-2. Simple Messages: Assign a message to a variable, and print that message. Then change the value of the variable to a new message, and print the new message.

Strings

Because most programs define and gather some sort of data and then do something useful with it, it helps to classify different types of data. The first data type we'll look at is the string. Strings are quite simple at first glance, but you can use them in many different ways.

A *string* is a series of characters. Anything inside quotes is considered a string in Python, and you can use single or double quotes around your strings like this:

```
"This is a string."
'This is also a string.'
```

This flexibility allows you to use quotes and apostrophes within your strings:

```
'I told my friend, "Python is my favorite language!"'
"The language 'Python' is named after Monty Python, not the snake."
"One of Python's strengths is its diverse and supportive community."
```

Let's explore some of the ways you can use strings.

Changing Case in a String with Methods

One of the simplest tasks you can do with strings is change the case of the words in a string. Look at the following code, and try to determine what's happening:

name.py

```
name = "ada lovelace"
print(name.title())
```

Save this file as *name.py* and then run it. You should see this output:

```
Ada Lovelace
```

In this example, the variable `name` refers to the lowercase string `"ada lovelace"`. The method `title()` appears after the variable in the `print()` call. A *method* is an action that Python can perform on a piece of data. The dot (`.`) after `name` in `name.title()` tells Python to make the `title()` method act on the variable `name`. Every method is followed by a set of parentheses, because methods often need additional information to do their work. That information is provided inside the parentheses. The `title()` function doesn't need any additional information, so its parentheses are empty.

The `title()` method changes each word to title case, where each word begins with a capital letter. This is useful because you'll often want to think of a name as a piece of information. For example, you might want your program to recognize the input values `Ada`, `ADA`, and `ada` as the same name, and display all of them as `Ada`.

Several other useful methods are available for dealing with case as well. For example, you can change a string to all uppercase or all lowercase letters like this:

```
name = "Ada Lovelace"
print(name.upper())
print(name.lower())
```

This will display the following:

```
ADA LOVELACE
ada lovelace
```

The `lower()` method is particularly useful for storing data. You typically won't want to trust the capitalization that your users provide, so you'll convert strings to lowercase before storing them. Then when you want to display the information, you'll use the case that makes the most sense for each string.

Using Variables in Strings

In some situations, you'll want to use a variable's value inside a string. For example, you might want to use two variables to represent a first name and a last name, respectively, and then combine those values to display someone's full name:

full_name.py

```
first_name = "ada"
last_name = "lovelace"
❶ full_name = f"{first_name} {last_name}"
print(full_name)
```

To insert a variable's value into a string, place the letter `f` immediately before the opening quotation mark ❶. Put braces around the name or names of any variable you want to use inside the string. Python will replace each variable with its value when the string is displayed.

These strings are called *f-strings*. The *f* is for *format*, because Python formats the string by replacing the name of any variable in braces with its value. The output from the previous code is:

```
ada lovelace
```

You can do a lot with f-strings. For example, you can use f-strings to compose complete messages using the information associated with a variable, as shown here:

```
first_name = "ada"
last_name = "lovelace"
full_name = f"{first_name} {last_name}"
❶ print(f"Hello, {full_name.title()}!")
```

The full name is used in a sentence that greets the user ❶, and the `title()` method changes the name to title case. This code returns a simple but nicely formatted greeting:

```
Hello, Ada Lovelace!
```

You can also use f-strings to compose a message, and then assign the entire message to a variable:

```
first_name = "ada"
last_name = "lovelace"
full_name = f"{first_name} {last_name}"
❶ message = f"Hello, {full_name.title()}!"
❷ print(message)
```

This code displays the message `Hello, Ada Lovelace!` as well, but by assigning the message to a variable ❶ we make the final `print()` call much simpler ❷.

Adding Whitespace to Strings with Tabs or Newlines

In programming, *whitespace* refers to any nonprinting characters, such as spaces, tabs, and end-of-line symbols. You can use whitespace to organize your output so it's easier for users to read.

To add a tab to your text, use the character combination \t:

```
>>> print("Python")
Python
>>> print("\tPython")
    Python
```

To add a newline in a string, use the character combination \n:

```
>>> print("Languages:\nPython\nC\nJavaScript")
Languages:
Python
C
JavaScript
```

You can also combine tabs and newlines in a single string. The string "\n\t" tells Python to move to a new line, and start the next line with a tab. The following example shows how you can use a one-line string to generate four lines of output:

```
>>> print("Languages:\n\tPython\n\tC\n\tJavaScript")
Languages:
    Python
    C
    JavaScript
```

Newlines and tabs will be very useful in the next two chapters, when you start to produce many lines of output from just a few lines of code.

Stripping Whitespace

Extra whitespace can be confusing in your programs. To programmers, 'python' and 'python ' look pretty much the same. But to a program, they are two different strings. Python detects the extra space in 'python ' and considers it significant unless you tell it otherwise.

It's important to think about whitespace, because often you'll want to compare two strings to determine whether they are the same. For example, one important instance might involve checking people's usernames when they log in to a website. Extra whitespace can be confusing in much simpler situations as well. Fortunately, Python makes it easy to eliminate extra whitespace from data that people enter.

Python can look for extra whitespace on the right and left sides of a string. To ensure that no whitespace exists at the right side of a string, use the rstrip() method:

```
❶ >>> favorite_language = 'python '
❷ >>> favorite_language
  'python '
❸ >>> favorite_language.rstrip()
  'python'
❹ >>> favorite_language
  'python '
```

The value associated with favorite_language ❶ contains extra whitespace at the end of the string. When you ask Python for this value in a terminal session, you can see the space at the end of the value ❷. When the rstrip() method acts on the variable favorite_language ❸, this extra space is removed. However, it is only removed temporarily. If you ask for the value of favorite_language again, the string looks the same as when it was entered, including the extra whitespace ❹.

To remove the whitespace from the string permanently, you have to associate the stripped value with the variable name:

```
>>> favorite_language = 'python '
❶ >>> favorite_language = favorite_language.rstrip()
>>> favorite_language
'python'
```

To remove the whitespace from the string, you strip the whitespace from the right side of the string and then associate this new value with the original variable ❶. Changing a variable's value is done often in programming. This is how a variable's value can be updated as a program is executed or in response to user input.

You can also strip whitespace from the left side of a string using the lstrip() method, or from both sides at once using strip():

```
❶ >>> favorite_language = ' python '
❷ >>> favorite_language.rstrip()
' python'
❸ >>> favorite_language.lstrip()
'python '
❹ >>> favorite_language.strip()
'python'
```

In this example, we start with a value that has whitespace at the beginning and the end ❶. We then remove the extra space from the right side ❷, from the left side ❸, and from both sides ❹. Experimenting with these stripping functions can help you become familiar with manipulating strings. In the real world, these stripping functions are used most often to clean up user input before it's stored in a program.

Removing Prefixes

When working with strings, another common task is to remove a prefix. Consider a URL with the common prefix *https://*. We want to remove this prefix, so we can focus on just the part of the URL that users need to enter into an address bar. Here's how to do that:

```
>>> nostarch_url = 'https://nostarch.com'
>>> nostarch_url.removeprefix('https://')
'nostarch.com'
```

Enter the name of the variable followed by a dot, and then the method `removeprefix()`. Inside the parentheses, enter the prefix you want to remove from the original string.

Like the methods for removing whitespace, `removeprefix()` leaves the original string unchanged. If you want to keep the new value with the prefix removed, either reassign it to the original variable or assign it to a new variable:

```
>>> simple_url = nostarch_url.removeprefix('https://')
```

When you see a URL in an address bar and the *https://* part isn't shown, the browser is probably using a method like `removeprefix()` behind the scenes.

Avoiding Syntax Errors with Strings

One kind of error that you might see with some regularity is a syntax error. A *syntax error* occurs when Python doesn't recognize a section of your program as valid Python code. For example, if you use an apostrophe within single quotes, you'll produce an error. This happens because Python interprets everything between the first single quote and the apostrophe as a string. It then tries to interpret the rest of the text as Python code, which causes errors.

Here's how to use single and double quotes correctly. Save this program as *apostrophe.py* and then run it:

apostrophe.py

```
message = "One of Python's strengths is its diverse community."
print(message)
```

The apostrophe appears inside a set of double quotes, so the Python interpreter has no trouble reading the string correctly:

```
One of Python's strengths is its diverse community.
```

However, if you use single quotes, Python can't identify where the string should end:

```
message = 'One of Python's strengths is its diverse community.'
print(message)
```

You'll see the following output:

```
  File "apostrophe.py", line 1
    message = 'One of Python's strengths is its diverse community.'
                                                                  ❶ ^
SyntaxError: unterminated string literal (detected at line 1)
```

In the output you can see that the error occurs right after the final single quote ❶. This syntax error indicates that the interpreter doesn't recognize something in the code as valid Python code, and it thinks the problem might be a string that's not quoted correctly. Errors can come from a variety of sources, and I'll point out some common ones as they arise. You might see syntax errors often as you learn to write proper Python code. Syntax errors are also the least specific kind of error, so they can be difficult and frustrating to identify and correct. If you get stuck on a particularly stubborn error, see the suggestions in Appendix C.

NOTE *Your editor's syntax highlighting feature should help you spot some syntax errors quickly as you write your programs. If you see Python code highlighted as if it's English or English highlighted as if it's Python code, you probably have a mismatched quotation mark somewhere in your file.*

TRY IT YOURSELF

Save each of the following exercises as a separate file, with a name like *name_cases.py*. If you get stuck, take a break or see the suggestions in Appendix C.

2-3. Personal Message: Use a variable to represent a person's name, and print a message to that person. Your message should be simple, such as, "Hello Eric, would you like to learn some Python today?"

2-4. Name Cases: Use a variable to represent a person's name, and then print that person's name in lowercase, uppercase, and title case.

2-5. Famous Quote: Find a quote from a famous person you admire. Print the quote and the name of its author. Your output should look something like the following, including the quotation marks:

> *Albert Einstein once said, "A person who never made a mistake never tried anything new."*

2-6. Famous Quote 2: Repeat Exercise 2-5, but this time, represent the famous person's name using a variable called `famous_person`. Then compose your message and represent it with a new variable called `message`. Print your message.

2-7. Stripping Names: Use a variable to represent a person's name, and include some whitespace characters at the beginning and end of the name. Make sure you use each character combination, `"\t"` and `"\n"`, at least once.

Print the name once, so the whitespace around the name is displayed. Then print the name using each of the three stripping functions, `lstrip()`, `rstrip()`, and `strip()`.

2-8. File Extensions: Python has a `removesuffix()` method that works exactly like `removeprefix()`. Assign the value `'python_notes.txt'` to a variable called `filename`. Then use the `removesuffix()` method to display the filename without the file extension, like some file browsers do.

Numbers

Numbers are used quite often in programming to keep score in games, represent data in visualizations, store information in web applications, and so on. Python treats numbers in several different ways, depending on how they're being used. Let's first look at how Python manages integers, because they're the simplest to work with.

Integers

You can add (+), subtract (-), multiply (*), and divide (/) integers in Python.

```
>>> 2 + 3
5
>>> 3 - 2
1
>>> 2 * 3
6
>>> 3 / 2
1.5
```

In a terminal session, Python simply returns the result of the operation. Python uses two multiplication symbols to represent exponents:

```
>>> 3 ** 2
9
>>> 3 ** 3
27
>>> 10 ** 6
1000000
```

Python supports the order of operations too, so you can use multiple operations in one expression. You can also use parentheses to modify the order of operations so Python can evaluate your expression in the order you specify. For example:

```
>>> 2 + 3*4
14
>>> (2 + 3) * 4
20
```

The spacing in these examples has no effect on how Python evaluates the expressions; it simply helps you more quickly spot the operations that have priority when you're reading through the code.

Floats

Python calls any number with a decimal point a *float*. This term is used in most programming languages, and it refers to the fact that a decimal point can appear at any position in a number. Every programming language must be carefully designed to properly manage decimal numbers so numbers behave appropriately, no matter where the decimal point appears.

For the most part, you can use floats without worrying about how they behave. Simply enter the numbers you want to use, and Python will most likely do what you expect:

```
>>> 0.1 + 0.1
0.2
>>> 0.2 + 0.2
0.4
>>> 2 * 0.1
0.2
>>> 2 * 0.2
0.4
```

However, be aware that you can sometimes get an arbitrary number of decimal places in your answer:

```
>>> 0.2 + 0.1
0.30000000000000004
>>> 3 * 0.1
0.30000000000000004
```

This happens in all languages and is of little concern. Python tries to find a way to represent the result as precisely as possible, which is sometimes difficult given how computers have to represent numbers internally. Just ignore the extra decimal places for now; you'll learn ways to deal with the extra places when you need to in the projects in Part II.

Integers and Floats

When you divide any two numbers, even if they are integers that result in a whole number, you'll always get a float:

```
>>> 4/2
2.0
```

If you mix an integer and a float in any other operation, you'll get a float as well:

```
>>> 1 + 2.0
3.0
>>> 2 * 3.0
6.0
>>> 3.0 ** 2
9.0
```

Python defaults to a float in any operation that uses a float, even if the output is a whole number.

Underscores in Numbers

When you're writing long numbers, you can group digits using underscores to make large numbers more readable:

```
>>> universe_age = 14_000_000_000
```

When you print a number that was defined using underscores, Python prints only the digits:

```
>>> print(universe_age)
14000000000
```

Python ignores the underscores when storing these kinds of values. Even if you don't group the digits in threes, the value will still be unaffected. To Python, `1000` is the same as `1_000`, which is the same as `10_00`. This feature works for both integers and floats.

Multiple Assignment

You can assign values to more than one variable using just a single line of code. This can help shorten your programs and make them easier to read; you'll use this technique most often when initializing a set of numbers.

For example, here's how you can initialize the variables `x`, `y`, and `z` to zero:

```
>>> x, y, z = 0, 0, 0
```

You need to separate the variable names with commas, and do the same with the values, and Python will assign each value to its respective variable. As long as the number of values matches the number of variables, Python will match them up correctly.

Constants

A *constant* is a variable whose value stays the same throughout the life of a program. Python doesn't have built-in constant types, but Python programmers use all capital letters to indicate a variable should be treated as a constant and never be changed:

```
MAX_CONNECTIONS = 5000
```

When you want to treat a variable as a constant in your code, write the name of the variable in all capital letters.

TRY IT YOURSELF

2-9. Number Eight: Write addition, subtraction, multiplication, and division operations that each result in the number 8. Be sure to enclose your operations in print() calls to see the results. You should create four lines that look like this:

```
print(5+3)
```

Your output should be four lines, with the number 8 appearing once on each line.

2-10. Favorite Number: Use a variable to represent your favorite number. Then, using that variable, create a message that reveals your favorite number. Print that message.

Comments

Comments are an extremely useful feature in most programming languages. Everything you've written in your programs so far is Python code. As your programs become longer and more complicated, you should add notes within your programs that describe your overall approach to the problem you're solving. A *comment* allows you to write notes in your spoken language, within your programs.

How Do You Write Comments?

In Python, the hash mark (#) indicates a comment. Anything following a hash mark in your code is ignored by the Python interpreter. For example:

comment.py

```
# Say hello to everyone.
print("Hello Python people!")
```

Python ignores the first line and executes the second line.

```
Hello Python people!
```

What Kinds of Comments Should You Write?

The main reason to write comments is to explain what your code is supposed to do and how you are making it work. When you're in the middle of working on a project, you understand how all of the pieces fit together. But when you return to a project after some time away, you'll likely have forgotten some of the details. You can always study your code for a while and figure out how segments were supposed to work, but writing good comments can save you time by summarizing your overall approach clearly.

If you want to become a professional programmer or collaborate with other programmers, you should write meaningful comments. Today, most software is written collaboratively, whether by a group of employees at one company or a group of people working together on an open source project. Skilled programmers expect to see comments in code, so it's best to start adding descriptive comments to your programs now. Writing clear, concise comments in your code is one of the most beneficial habits you can form as a new programmer.

When you're deciding whether to write a comment, ask yourself if you had to consider several approaches before coming up with a reasonable way to make something work; if so, write a comment about your solution. It's much easier to delete extra comments later than to go back and write comments for a sparsely commented program. From now on, I'll use comments in examples throughout this book to help explain sections of code.

TRY IT YOURSELF

2-11. Adding Comments: Choose two of the programs you've written, and add at least one comment to each. If you don't have anything specific to write because your programs are too simple at this point, just add your name and the current date at the top of each program file. Then write one sentence describing what the program does.

The Zen of Python

Experienced Python programmers will encourage you to avoid complexity and aim for simplicity whenever possible. The Python community's philosophy is contained in "The Zen of Python" by Tim Peters. You can access this brief set of principles for writing good Python code by entering `import this` into your interpreter. I won't reproduce the entire "Zen of Python" here, but I'll share a few lines to help you understand why they should be important to you as a beginning Python programmer.

```
>>> import this
The Zen of Python, by Tim Peters
Beautiful is better than ugly.
```

Python programmers embrace the notion that code can be beautiful and elegant. In programming, people solve problems. Programmers have always respected well-designed, efficient, and even beautiful solutions to problems. As you learn more about Python and use it to write more code, someone might look over your shoulder one day and say, "Wow, that's some beautiful code!"

```
Simple is better than complex.
```

If you have a choice between a simple and a complex solution, and both work, use the simple solution. Your code will be easier to maintain, and it will be easier for you and others to build on that code later on.

```
Complex is better than complicated.
```

Real life is messy, and sometimes a simple solution to a problem is unattainable. In that case, use the simplest solution that works.

```
Readability counts.
```

Even when your code is complex, aim to make it readable. When you're working on a project that involves complex coding, focus on writing informative comments for that code.

```
There should be one-- and preferably only one --obvious way to do it.
```

If two Python programmers are asked to solve the same problem, they should come up with fairly compatible solutions. This is not to say there's no room for creativity in programming. On the contrary, there is plenty of room for creativity! However, much of programming consists of using small, common approaches to simple situations within a larger, more creative project. The nuts and bolts of your programs should make sense to other Python programmers.

```
Now is better than never.
```

You could spend the rest of your life learning all the intricacies of Python and of programming in general, but then you'd never complete any projects. Don't try to write perfect code; write code that works, and then decide whether to improve your code for that project or move on to something new.

As you continue to the next chapter and start digging into more involved topics, try to keep this philosophy of simplicity and clarity in mind. Experienced programmers will respect your code more and will be happy to give you feedback and collaborate with you on interesting projects.

TRY IT YOURSELF

2-12. Zen of Python: Enter `import this` into a Python terminal session and skim through the additional principles.

Summary

In this chapter you learned how to work with variables. You learned to use descriptive variable names and resolve name errors and syntax errors when they arise. You learned what strings are and how to display them using lowercase, uppercase, and title case. You started using whitespace to organize output neatly, and you learned how to remove unneeded elements from a string. You started working with integers and floats, and you learned some of the ways you can work with numerical data. You also learned to write explanatory comments to make your code easier for you and others to read. Finally, you read about the philosophy of keeping your code as simple as possible, whenever possible.

In Chapter 3, you'll learn how to store collections of information in data structures called *lists*. You'll also learn how to work through a list, manipulating any information in that list.

3

INTRODUCING LISTS

In this chapter and the next you'll learn what lists are and how to start working with the elements in a list. Lists allow you to store sets of information in one place, whether you have just a few items or millions of items. Lists are one of Python's most powerful features readily accessible to new programmers, and they tie together many important concepts in programming.

What Is a List?

A *list* is a collection of items in a particular order. You can make a list that includes the letters of the alphabet, the digits from 0 to 9, or the names of all the people in your family. You can put anything you want into a list, and the items in your list don't have to be related in any particular way. Because a list usually contains more than one element, it's a good idea to make the name of your list plural, such as `letters`, `digits`, or `names`.

In Python, square brackets (`[]`) indicate a list, and individual elements in the list are separated by commas. Here's a simple example of a list that contains a few kinds of bicycles:

bicycles.py

```
bicycles = ['trek', 'cannondale', 'redline', 'specialized']
print(bicycles)
```

If you ask Python to print a list, Python returns its representation of the list, including the square brackets:

```
['trek', 'cannondale', 'redline', 'specialized']
```

Because this isn't the output you want your users to see, let's learn how to access the individual items in a list.

Accessing Elements in a List

Lists are ordered collections, so you can access any element in a list by telling Python the position, or *index*, of the item desired. To access an element in a list, write the name of the list followed by the index of the item enclosed in square brackets.

For example, let's pull out the first bicycle in the list `bicycles`:

```
bicycles = ['trek', 'cannondale', 'redline', 'specialized']
print(bicycles[0])
```

When we ask for a single item from a list, Python returns just that element without square brackets:

```
trek
```

This is the result you want your users to see: clean, neatly formatted output.

You can also use the string methods from Chapter 2 on any element in this list. For example, you can format the element `'trek'` to look more presentable by using the `title()` method:

```
bicycles = ['trek', 'cannondale', 'redline', 'specialized']
print(bicycles[0].title())
```

This example produces the same output as the preceding example, except `'Trek'` is capitalized.

Index Positions Start at 0, Not 1

Python considers the first item in a list to be at position 0, not position 1. This is true of most programming languages, and the reason has to do with how the list operations are implemented at a lower level. If you're receiving unexpected results, ask yourself if you're making a simple but common off-by-one error.

The second item in a list has an index of 1. Using this counting system, you can get any element you want from a list by subtracting one from its position in the list. For instance, to access the fourth item in a list, you request the item at index 3.

The following asks for the bicycles at index `1` and index `3`:

```
bicycles = ['trek', 'cannondale', 'redline', 'specialized']
print(bicycles[1])
print(bicycles[3])
```

This code returns the second and fourth bicycles in the list:

```
cannondale
specialized
```

Python has a special syntax for accessing the last element in a list. If you ask for the item at index `-1`, Python always returns the last item in the list:

```
bicycles = ['trek', 'cannondale', 'redline', 'specialized']
print(bicycles[-1])
```

This code returns the value `'specialized'`. This syntax is quite useful, because you'll often want to access the last items in a list without knowing exactly how long the list is. This convention extends to other negative index values as well. The index `-2` returns the second item from the end of the list, the index `-3` returns the third item from the end, and so forth.

Using Individual Values from a List

You can use individual values from a list just as you would any other variable. For example, you can use f-strings to create a message based on a value from a list.

Let's try pulling the first bicycle from the list and composing a message using that value:

```
bicycles = ['trek', 'cannondale', 'redline', 'specialized']
message = f"My first bicycle was a {bicycles[0].title()}."

print(message)
```

We build a sentence using the value at `bicycles[0]` and assign it to the variable `message`. The output is a simple sentence about the first bicycle in the list:

```
My first bicycle was a Trek.
```

TRY IT YOURSELF

Try these short programs to get some firsthand experience with Python's lists. You might want to create a new folder for each chapter's exercises, to keep them organized.

3-1. Names: Store the names of a few of your friends in a list called `names`. Print each person's name by accessing each element in the list, one at a time.

3-2. Greetings: Start with the list you used in Exercise 3-1, but instead of just printing each person's name, print a message to them. The text of each message should be the same, but each message should be personalized with the person's name.

3-3. Your Own List: Think of your favorite mode of transportation, such as a motorcycle or a car, and make a list that stores several examples. Use your list to print a series of statements about these items, such as "I would like to own a Honda motorcycle."

Modifying, Adding, and Removing Elements

Most lists you create will be *dynamic*, meaning you'll build a list and then add and remove elements from it as your program runs its course. For example, you might create a game in which a player has to shoot aliens out of the sky. You could store the initial set of aliens in a list and then remove an alien from the list each time one is shot down. Each time a new alien appears on the screen, you add it to the list. Your list of aliens will increase and decrease in length throughout the course of the game.

Modifying Elements in a List

The syntax for modifying an element is similar to the syntax for accessing an element in a list. To change an element, use the name of the list followed by the index of the element you want to change, and then provide the new value you want that item to have.

For example, say we have a list of motorcycles and the first item in the list is `'honda'`. We can change the value of this first item after the list has been created:

motorcycles.py

```
motorcycles = ['honda', 'yamaha', 'suzuki']
print(motorcycles)

motorcycles[0] = 'ducati'
print(motorcycles)
```

Here we define the list motorcycles, with 'honda' as the first element. Then we change the value of the first item to 'ducati'. The output shows that the first item has been changed, while the rest of the list stays the same:

```
['honda', 'yamaha', 'suzuki']
['ducati', 'yamaha', 'suzuki']
```

You can change the value of any item in a list, not just the first item.

Adding Elements to a List

You might want to add a new element to a list for many reasons. For example, you might want to make new aliens appear in a game, add new data to a visualization, or add new registered users to a website you've built. Python provides several ways to add new data to existing lists.

Appending Elements to the End of a List

The simplest way to add a new element to a list is to *append* the item to the list. When you append an item to a list, the new element is added to the end of the list. Using the same list we had in the previous example, we'll add the new element 'ducati' to the end of the list:

```
motorcycles = ['honda', 'yamaha', 'suzuki']
print(motorcycles)

motorcycles.append('ducati')
print(motorcycles)
```

Here the append() method adds 'ducati' to the end of the list, without affecting any of the other elements in the list:

```
['honda', 'yamaha', 'suzuki']
['honda', 'yamaha', 'suzuki', 'ducati']
```

The append() method makes it easy to build lists dynamically. For example, you can start with an empty list and then add items to the list using a series of append() calls. Using an empty list, let's add the elements 'honda', 'yamaha', and 'suzuki' to the list:

```
motorcycles = []

motorcycles.append('honda')
motorcycles.append('yamaha')
motorcycles.append('suzuki')

print(motorcycles)
```

The resulting list looks exactly the same as the lists in the previous examples:

```
['honda', 'yamaha', 'suzuki']
```

Building lists this way is very common, because you often won't know the data your users want to store in a program until after the program is running. To put your users in control, start by defining an empty list that will hold the users' values. Then append each new value provided to the list you just created.

Inserting Elements into a List

You can add a new element at any position in your list by using the `insert()` method. You do this by specifying the index of the new element and the value of the new item:

```
motorcycles = ['honda', 'yamaha', 'suzuki']

motorcycles.insert(0, 'ducati')
print(motorcycles)
```

In this example, we insert the value `'ducati'` at the beginning of the list. The `insert()` method opens a space at position `0` and stores the value `'ducati'` at that location:

```
['ducati', 'honda', 'yamaha', 'suzuki']
```

This operation shifts every other value in the list one position to the right.

Removing Elements from a List

Often, you'll want to remove an item or a set of items from a list. For example, when a player shoots down an alien from the sky, you'll most likely want to remove it from the list of active aliens. Or when a user decides to cancel their account on a web application you created, you'll want to remove that user from the list of active users. You can remove an item according to its position in the list or according to its value.

Removing an Item Using the del Statement

If you know the position of the item you want to remove from a list, you can use the `del` statement:

```
motorcycles = ['honda', 'yamaha', 'suzuki']
print(motorcycles)

del motorcycles[0]
print(motorcycles)
```

Here we use the `del` statement to remove the first item, `'honda'`, from the list of motorcycles:

```
['honda', 'yamaha', 'suzuki']
['yamaha', 'suzuki']
```

You can remove an item from any position in a list using the del statement if you know its index. For example, here's how to remove the second item, 'yamaha', from the list:

```
motorcycles = ['honda', 'yamaha', 'suzuki']
print(motorcycles)

del motorcycles[1]
print(motorcycles)
```

The second motorcycle is deleted from the list:

```
['honda', 'yamaha', 'suzuki']
['honda', 'suzuki']
```

In both examples, you can no longer access the value that was removed from the list after the del statement is used.

Removing an Item Using the pop() Method

Sometimes you'll want to use the value of an item after you remove it from a list. For example, you might want to get the *x* and *y* position of an alien that was just shot down, so you can draw an explosion at that position. In a web application, you might want to remove a user from a list of active members and then add that user to a list of inactive members.

The pop() method removes the last item in a list, but it lets you work with that item after removing it. The term *pop* comes from thinking of a list as a stack of items and popping one item off the top of the stack. In this analogy, the top of a stack corresponds to the end of a list.

Let's pop a motorcycle from the list of motorcycles:

```
❶ motorcycles = ['honda', 'yamaha', 'suzuki']
  print(motorcycles)

❷ popped_motorcycle = motorcycles.pop()
❸ print(motorcycles)
❹ print(popped_motorcycle)
```

We start by defining and printing the list motorcycles ❶. Then we pop a value from the list, and assign that value to the variable popped_motorcycle ❷. We print the list ❸ to show that a value has been removed from the list. Then we print the popped value ❹ to prove that we still have access to the value that was removed.

The output shows that the value 'suzuki' was removed from the end of the list and is now assigned to the variable popped_motorcycle:

```
['honda', 'yamaha', 'suzuki']
['honda', 'yamaha']
suzuki
```

How might this pop() method be useful? Imagine that the motorcycles in the list are stored in chronological order, according to when we owned them. If this is the case, we can use the pop() method to print a statement about the last motorcycle we bought:

```
motorcycles = ['honda', 'yamaha', 'suzuki']

last_owned = motorcycles.pop()
print(f"The last motorcycle I owned was a {last_owned.title()}.")
```

The output is a simple sentence about the most recent motorcycle we owned:

```
The last motorcycle I owned was a Suzuki.
```

Popping Items from Any Position in a List

You can use pop() to remove an item from any position in a list by including the index of the item you want to remove in parentheses:

```
motorcycles = ['honda', 'yamaha', 'suzuki']

first_owned = motorcycles.pop(0)
print(f"The first motorcycle I owned was a {first_owned.title()}.")
```

We start by popping the first motorcycle in the list, and then we print a message about that motorcycle. The output is a simple sentence describing the first motorcycle I ever owned:

```
The first motorcycle I owned was a Honda.
```

Remember that each time you use pop(), the item you work with is no longer stored in the list.

If you're unsure whether to use the del statement or the pop() method, here's a simple way to decide: when you want to delete an item from a list and not use that item in any way, use the del statement; if you want to use an item as you remove it, use the pop() method.

Removing an Item by Value

Sometimes you won't know the position of the value you want to remove from a list. If you only know the value of the item you want to remove, you can use the remove() method.

For example, say we want to remove the value 'ducati' from the list of motorcycles:

```
motorcycles = ['honda', 'yamaha', 'suzuki', 'ducati']
print(motorcycles)

motorcycles.remove('ducati')
print(motorcycles)
```

Here the remove() method tells Python to figure out where 'ducati' appears in the list and remove that element:

```
['honda', 'yamaha', 'suzuki', 'ducati']
['honda', 'yamaha', 'suzuki']
```

You can also use the remove() method to work with a value that's being removed from a list. Let's remove the value 'ducati' and print a reason for removing it from the list:

```
❶ motorcycles = ['honda', 'yamaha', 'suzuki', 'ducati']
  print(motorcycles)

❷ too_expensive = 'ducati'
❸ motorcycles.remove(too_expensive)
  print(motorcycles)
❹ print(f"\nA {too_expensive.title()} is too expensive for me.")
```

After defining the list ❶, we assign the value 'ducati' to a variable called too_expensive ❷. We then use this variable to tell Python which value to remove from the list ❸. The value 'ducati' has been removed from the list ❹ but is still accessible through the variable too_expensive, allowing us to print a statement about why we removed 'ducati' from the list of motorcycles:

```
['honda', 'yamaha', 'suzuki', 'ducati']
['honda', 'yamaha', 'suzuki']

A Ducati is too expensive for me.
```

NOTE *The remove() method deletes only the first occurrence of the value you specify. If there's a possibility the value appears more than once in the list, you'll need to use a loop to make sure all occurrences of the value are removed. You'll learn how to do this in Chapter 7.*

TRY IT YOURSELF

The following exercises are a bit more complex than those in Chapter 2, but they give you an opportunity to use lists in all of the ways described.

3-4. Guest List: If you could invite anyone, living or deceased, to dinner, who would you invite? Make a list that includes at least three people you'd like to invite to dinner. Then use your list to print a message to each person, inviting them to dinner.

(continued)

3-5. Changing Guest List: You just heard that one of your guests can't make the dinner, so you need to send out a new set of invitations. You'll have to think of someone else to invite.

- Start with your program from Exercise 3-4. Add a `print()` call at the end of your program, stating the name of the guest who can't make it.
- Modify your list, replacing the name of the guest who can't make it with the name of the new person you are inviting.
- Print a second set of invitation messages, one for each person who is still in your list.

3-6. More Guests: You just found a bigger dinner table, so now more space is available. Think of three more guests to invite to dinner.

- Start with your program from Exercise 3-4 or 3-5. Add a `print()` call to the end of your program, informing people that you found a bigger table.
- Use `insert()` to add one new guest to the beginning of your list.
- Use `insert()` to add one new guest to the middle of your list.
- Use `append()` to add one new guest to the end of your list.
- Print a new set of invitation messages, one for each person in your list.

3-7. Shrinking Guest List: You just found out that your new dinner table won't arrive in time for the dinner, and now you have space for only two guests.

- Start with your program from Exercise 3-6. Add a new line that prints a message saying that you can invite only two people for dinner.
- Use `pop()` to remove guests from your list one at a time until only two names remain in your list. Each time you pop a name from your list, print a message to that person letting them know you're sorry you can't invite them to dinner.
- Print a message to each of the two people still on your list, letting them know they're still invited.
- Use `del` to remove the last two names from your list, so you have an empty list. Print your list to make sure you actually have an empty list at the end of your program.

Organizing a List

Often, your lists will be created in an unpredictable order, because you can't always control the order in which your users provide their data. Although this is unavoidable in most circumstances, you'll frequently want to present your information in a particular order. Sometimes you'll want to preserve the original order of your list, and other times you'll want to change the original order. Python provides a number of different ways to organize your lists, depending on the situation.

Sorting a List Permanently with the sort() Method

Python's sort() method makes it relatively easy to sort a list. Imagine we have a list of cars and want to change the order of the list to store them alphabetically. To keep the task simple, let's assume that all the values in the list are lowercase:

cars.py

```
cars = ['bmw', 'audi', 'toyota', 'subaru']
cars.sort()
print(cars)
```

The sort() method changes the order of the list permanently. The cars are now in alphabetical order, and we can never revert to the original order:

```
['audi', 'bmw', 'subaru', 'toyota']
```

You can also sort this list in reverse-alphabetical order by passing the argument reverse=True to the sort() method. The following example sorts the list of cars in reverse-alphabetical order:

```
cars = ['bmw', 'audi', 'toyota', 'subaru']
cars.sort(reverse=True)
print(cars)
```

Again, the order of the list is permanently changed:

```
['toyota', 'subaru', 'bmw', 'audi']
```

Sorting a List Temporarily with the sorted() Function

To maintain the original order of a list but present it in a sorted order, you can use the sorted() function. The sorted() function lets you display your list in a particular order, but doesn't affect the actual order of the list.

Let's try this function on the list of cars.

```
cars = ['bmw', 'audi', 'toyota', 'subaru']

❶ print("Here is the original list:")
print(cars)

❷ print("\nHere is the sorted list:")
print(sorted(cars))

❸ print("\nHere is the original list again:")
print(cars)
```

We first print the list in its original order ❶ and then in alphabetical order ❷. After the list is displayed in the new order, we show that the list is still stored in its original order ❸:

```
Here is the original list:
['bmw', 'audi', 'toyota', 'subaru']

Here is the sorted list:
['audi', 'bmw', 'subaru', 'toyota']

❶ Here is the original list again:
['bmw', 'audi', 'toyota', 'subaru']
```

Notice that the list still exists in its original order ❶ after the `sorted()` function has been used. The `sorted()` function can also accept a `reverse=True` argument if you want to display a list in reverse-alphabetical order.

NOTE *Sorting a list alphabetically is a bit more complicated when all the values are not in lowercase. There are several ways to interpret capital letters when determining a sort order, and specifying the exact order can be more complex than we want to deal with at this time. However, most approaches to sorting will build directly on what you learned in this section.*

Printing a List in Reverse Order

To reverse the original order of a list, you can use the `reverse()` method. If we originally stored the list of cars in chronological order according to when we owned them, we could easily rearrange the list into reverse-chronological order:

```
cars = ['bmw', 'audi', 'toyota', 'subaru']
print(cars)

cars.reverse()
print(cars)
```

Notice that `reverse()` doesn't sort backward alphabetically; it simply reverses the order of the list:

```
['bmw', 'audi', 'toyota', 'subaru']
['subaru', 'toyota', 'audi', 'bmw']
```

The `reverse()` method changes the order of a list permanently, but you can revert to the original order anytime by applying `reverse()` to the same list a second time.

Finding the Length of a List

You can quickly find the length of a list by using the `len()` function. The list in this example has four items, so its length is `4`:

```
>>> cars = ['bmw', 'audi', 'toyota', 'subaru']
>>> len(cars)
4
```

You'll find `len()` useful when you need to identify the number of aliens that still need to be shot down in a game, determine the amount of data you have to manage in a visualization, or figure out the number of registered users on a website, among other tasks.

NOTE *Python counts the items in a list starting with one, so you shouldn't run into any off-by-one errors when determining the length of a list.*

TRY IT YOURSELF

3-8. Seeing the World: Think of at least five places in the world you'd like to visit.

- Store the locations in a list. Make sure the list is not in alphabetical order.
- Print your list in its original order. Don't worry about printing the list neatly; just print it as a raw Python list.
- Use `sorted()` to print your list in alphabetical order without modifying the actual list.
- Show that your list is still in its original order by printing it.
- Use `sorted()` to print your list in reverse-alphabetical order without changing the order of the original list.
- Show that your list is still in its original order by printing it again.
- Use `reverse()` to change the order of your list. Print the list to show that its order has changed.
- Use `reverse()` to change the order of your list again. Print the list to show it's back to its original order.
- Use `sort()` to change your list so it's stored in alphabetical order. Print the list to show that its order has been changed.
- Use `sort()` to change your list so it's stored in reverse-alphabetical order. Print the list to show that its order has changed.

3-9. Dinner Guests: Working with one of the programs from Exercises 3-4 through 3-7 (pages 41–42), use `len()` to print a message indicating the number of people you're inviting to dinner.

3-10. Every Function: Think of things you could store in a list. For example, you could make a list of mountains, rivers, countries, cities, languages, or anything else you'd like. Write a program that creates a list containing these items and then uses each function introduced in this chapter at least once.

Avoiding Index Errors When Working with Lists

There's one type of error that's common to see when you're working with lists for the first time. Let's say you have a list with three items, and you ask for the fourth item:

motorcycles.py

```
motorcycles = ['honda', 'yamaha', 'suzuki']
print(motorcycles[3])
```

This example results in an *index error*:

```
Traceback (most recent call last):
  File "motorcycles.py", line 2, in <module>
    print(motorcycles[3])
          ~~~~~~~~~~~^^^
IndexError: list index out of range
```

Python attempts to give you the item at index 3. But when it searches the list, no item in `motorcycles` has an index of 3. Because of the off-by-one nature of indexing in lists, this error is typical. People think the third item is item number 3, because they start counting at 1. But in Python the third item is number 2, because it starts indexing at 0.

An index error means Python can't find an item at the index you requested. If an index error occurs in your program, try adjusting the index you're asking for by one. Then run the program again to see if the results are correct.

Keep in mind that whenever you want to access the last item in a list, you should use the index `-1`. This will always work, even if your list has changed size since the last time you accessed it:

```
motorcycles = ['honda', 'yamaha', 'suzuki']
print(motorcycles[-1])
```

The index `-1` always returns the last item in a list, in this case the value `'suzuki'`:

```
suzuki
```

The only time this approach will cause an error is when you request the last item from an empty list:

```
motorcycles = []
print(motorcycles[-1])
```

No items are in `motorcycles`, so Python returns another index error:

```
Traceback (most recent call last):
  File "motorcycles.py", line 3, in <module>
    print(motorcycles[-1])
          ~~~~~~~~~~~^^^^
IndexError: list index out of range
```

If an index error occurs and you can't figure out how to resolve it, try printing your list or just printing the length of your list. Your list might look much different than you thought it did, especially if it has been managed dynamically by your program. Seeing the actual list, or the exact number of items in your list, can help you sort out such logical errors.

TRY IT YOURSELF

3-11. Intentional Error: If you haven't received an index error in one of your programs yet, try to make one happen. Change an index in one of your programs to produce an index error. Make sure you correct the error before closing the program.

Summary

In this chapter, you learned what lists are and how to work with the individual items in a list. You learned how to define a list and how to add and remove elements. You learned how to sort lists permanently and temporarily for display purposes. You also learned how to find the length of a list and how to avoid index errors when you're working with lists.

In Chapter 4 you'll learn how to work with items in a list more efficiently. By looping through each item in a list using just a few lines of code you'll be able to work efficiently, even when your list contains thousands or millions of items.

4

WORKING WITH LISTS

In Chapter 3 you learned how to make a simple list, and you learned to work with the individual elements in a list. In this chapter you'll learn how to loop through an entire list using just a few lines of code, regardless of how long the list is. *Looping* allows you to take the same action, or set of actions, with every item in a list. As a result, you'll be able to work efficiently with lists of any length, including those with thousands or even millions of items.

Looping Through an Entire List

You'll often want to run through all entries in a list, performing the same task with each item. For example, in a game you might want to move every element on the screen by the same amount. In a list of numbers, you might want to perform the same statistical operation on every element.

Or perhaps you'll want to display each headline from a list of articles on a website. When you want to do the same action with every item in a list, you can use Python's for loop.

Say we have a list of magicians' names, and we want to print out each name in the list. We could do this by retrieving each name from the list individually, but this approach could cause several problems. For one, it would be repetitive to do this with a long list of names. Also, we'd have to change our code each time the list's length changed. Using a for loop avoids both of these issues by letting Python manage these issues internally.

Let's use a for loop to print out each name in a list of magicians:

magicians.py

```
magicians = ['alice', 'david', 'carolina']
for magician in magicians:
    print(magician)
```

We begin by defining a list, just as we did in Chapter 3. Then we define a for loop. This line tells Python to pull a name from the list magicians, and associate it with the variable magician. Next, we tell Python to print the name that's just been assigned to magician. Python then repeats these last two lines, once for each name in the list. It might help to read this code as "For every magician in the list of magicians, print the magician's name." The output is a simple printout of each name in the list:

```
alice
david
carolina
```

A Closer Look at Looping

Looping is important because it's one of the most common ways a computer automates repetitive tasks. For example, in a simple loop like we used in *magicians.py*, Python initially reads the first line of the loop:

```
for magician in magicians:
```

This line tells Python to retrieve the first value from the list magicians and associate it with the variable magician. This first value is 'alice'. Python then reads the next line:

```
    print(magician)
```

Python prints the current value of magician, which is still 'alice'. Because the list contains more values, Python returns to the first line of the loop:

```
for magician in magicians:
```

Python retrieves the next name in the list, 'david', and associates that value with the variable magician. Python then executes the line:

```
    print(magician)
```

Python prints the current value of `magician` again, which is now `'david'`. Python repeats the entire loop once more with the last value in the list, `'carolina'`. Because no more values are in the list, Python moves on to the next line in the program. In this case nothing comes after the `for` loop, so the program ends.

When you're using loops for the first time, keep in mind that the set of steps is repeated once for each item in the list, no matter how many items are in the list. If you have a million items in your list, Python repeats these steps a million times—and usually very quickly.

Also keep in mind when writing your own `for` loops that you can choose any name you want for the temporary variable that will be associated with each value in the list. However, it's helpful to choose a meaningful name that represents a single item from the list. For example, here's a good way to start a `for` loop for a list of cats, a list of dogs, and a general list of items:

```
for cat in cats:
for dog in dogs:
for item in list_of_items:
```

These naming conventions can help you follow the action being done on each item within a `for` loop. Using singular and plural names can help you identify whether a section of code is working with a single element from the list or the entire list.

Doing More Work Within a for Loop

You can do just about anything with each item in a `for` loop. Let's build on the previous example by printing a message to each magician, telling them that they performed a great trick:

magicians.py

```
magicians = ['alice', 'david', 'carolina']
for magician in magicians:
    print(f"{magician.title()}, that was a great trick!")
```

The only difference in this code is where we compose a message to each magician, starting with that magician's name. The first time through the loop the value of `magician` is `'alice'`, so Python starts the first message with the name `'Alice'`. The second time through, the message will begin with `'David'`, and the third time through, the message will begin with `'Carolina'`.

The output shows a personalized message for each magician in the list:

```
Alice, that was a great trick!
David, that was a great trick!
Carolina, that was a great trick!
```

You can also write as many lines of code as you like in the `for` loop. Every indented line following the line `for magician in magicians` is considered *inside the loop*, and each indented line is executed once for each value in the list. Therefore, you can do as much work as you like with each value in the list.

Let's add a second line to our message, telling each magician that we're looking forward to their next trick:

```
magicians = ['alice', 'david', 'carolina']
for magician in magicians:
    print(f"{magician.title()}, that was a great trick!")
    print(f"I can't wait to see your next trick, {magician.title()}.\n")
```

Because we have indented both calls to `print()`, each line will be executed once for every magician in the list. The newline (`"\n"`) in the second `print()` call inserts a blank line after each pass through the loop. This creates a set of messages that are neatly grouped for each person in the list:

```
Alice, that was a great trick!
I can't wait to see your next trick, Alice.

David, that was a great trick!
I can't wait to see your next trick, David.

Carolina, that was a great trick!
I can't wait to see your next trick, Carolina.
```

You can use as many lines as you like in your `for` loops. In practice, you'll often find it useful to do a number of different operations with each item in a list when you use a `for` loop.

Doing Something After a for Loop

What happens once a `for` loop has finished executing? Usually, you'll want to summarize a block of output or move on to other work that your program must accomplish.

Any lines of code after the `for` loop that are not indented are executed once without repetition. Let's write a thank you to the group of magicians as a whole, thanking them for putting on an excellent show. To display this group message after all of the individual messages have been printed, we place the thank you message after the `for` loop, without indentation:

```
magicians = ['alice', 'david', 'carolina']
for magician in magicians:
    print(f"{magician.title()}, that was a great trick!")
    print(f"I can't wait to see your next trick, {magician.title()}.\n")

print("Thank you, everyone. That was a great magic show!")
```

The first two calls to `print()` are repeated once for each magician in the list, as you saw earlier. However, because the last line is not indented, it's printed only once:

```
Alice, that was a great trick!
I can't wait to see your next trick, Alice.

David, that was a great trick!
```

```
I can't wait to see your next trick, David.

Carolina, that was a great trick!
I can't wait to see your next trick, Carolina.

Thank you, everyone. That was a great magic show!
```

When you're processing data using a `for` loop, you'll find that this is a good way to summarize an operation that was performed on an entire dataset. For example, you might use a `for` loop to initialize a game by running through a list of characters and displaying each character on the screen. You might then write some additional code after this loop that displays a *Play Now* button after all the characters have been drawn to the screen.

Avoiding Indentation Errors

Python uses indentation to determine how a line, or group of lines, is related to the rest of the program. In the previous examples, the lines that printed messages to individual magicians were part of the `for` loop because they were indented. Python's use of indentation makes code very easy to read. Basically, it uses whitespace to force you to write neatly formatted code with a clear visual structure. In longer Python programs, you'll notice blocks of code indented at a few different levels. These indentation levels help you gain a general sense of the overall program's organization.

As you begin to write code that relies on proper indentation, you'll need to watch for a few common *indentation errors*. For example, people sometimes indent lines of code that don't need to be indented or forget to indent lines that need to be indented. Seeing examples of these errors now will help you avoid them in the future and correct them when they do appear in your own programs.

Let's examine some of the more common indentation errors.

Forgetting to Indent

Always indent the line after the `for` statement in a loop. If you forget, Python will remind you:

magicians.py

```
magicians = ['alice', 'david', 'carolina']
for magician in magicians:
❶ print(magician)
```

The call to `print()` ❶ should be indented, but it's not. When Python expects an indented block and doesn't find one, it lets you know which line it had a problem with:

```
  File "magicians.py", line 3
    print(magician)
    ^
IndentationError: expected an indented block after 'for' statement on line 2
```

You can usually resolve this kind of indentation error by indenting the line or lines immediately after the `for` statement.

Forgetting to Indent Additional Lines

Sometimes your loop will run without any errors but won't produce the expected result. This can happen when you're trying to do several tasks in a loop and you forget to indent some of its lines.

For example, this is what happens when we forget to indent the second line in the loop that tells each magician we're looking forward to their next trick:

```
magicians = ['alice', 'david', 'carolina']
for magician in magicians:
    print(f"{magician.title()}, that was a great trick!")
❶ print(f"I can't wait to see your next trick, {magician.title()}.\n")
```

The second call to `print()` ❶ is supposed to be indented, but because Python finds at least one indented line after the `for` statement, it doesn't report an error. As a result, the first `print()` call is executed once for each name in the list because it is indented. The second `print()` call is not indented, so it is executed only once after the loop has finished running. Because the final value associated with `magician` is `'carolina'`, she is the only one who receives the "looking forward to the next trick" message:

```
Alice, that was a great trick!
David, that was a great trick!
Carolina, that was a great trick!
I can't wait to see your next trick, Carolina.
```

This is a *logical error*. The syntax is valid Python code, but the code does not produce the desired result because a problem occurs in its logic. If you expect to see a certain action repeated once for each item in a list and it's executed only once, determine whether you need to simply indent a line or a group of lines.

Indenting Unnecessarily

If you accidentally indent a line that doesn't need to be indented, Python informs you about the unexpected indent:

hello_world.py

```
message = "Hello Python world!"
    print(message)
```

We don't need to indent the `print()` call, because it isn't part of a loop; hence, Python reports that error:

```
  File "hello_world.py", line 2
    print(message)
    ^
IndentationError: unexpected indent
```

You can avoid unexpected indentation errors by indenting only when you have a specific reason to do so. In the programs you're writing at this point, the only lines you should indent are the actions you want to repeat for each item in a for loop.

Indenting Unnecessarily After the Loop

If you accidentally indent code that should run after a loop has finished, that code will be repeated once for each item in the list. Sometimes this prompts Python to report an error, but often this will result in a logical error.

For example, let's see what happens when we accidentally indent the line that thanked the magicians as a group for putting on a good show:

magicians.py

```
magicians = ['alice', 'david', 'carolina']
for magician in magicians:
    print(f"{magician.title()}, that was a great trick!")
    print(f"I can't wait to see your next trick, {magician.title()}.\n")

❶   print("Thank you everyone, that was a great magic show!")
```

Because the last line ❶ is indented, it's printed once for each person in the list:

```
Alice, that was a great trick!
I can't wait to see your next trick, Alice.

Thank you everyone, that was a great magic show!
David, that was a great trick!
I can't wait to see your next trick, David.

Thank you everyone, that was a great magic show!
Carolina, that was a great trick!
I can't wait to see your next trick, Carolina.

Thank you everyone, that was a great magic show!
```

This is another logical error, similar to the one in "Forgetting to Indent Additional Lines" on page 54. Because Python doesn't know what you're trying to accomplish with your code, it will run all code that is written in valid syntax. If an action is repeated many times when it should be executed only once, you probably need to unindent the code for that action.

Forgetting the Colon

The colon at the end of a for statement tells Python to interpret the next line as the start of a loop.

```
magicians = ['alice', 'david', 'carolina']
❶ for magician in magicians
    print(magician)
```

If you accidentally forget the colon ❶, you'll get a syntax error because Python doesn't know exactly what you're trying to do:

```
  File "magicians.py", line 2
    for magician in magicians
                             ^
SyntaxError: expected ':'
```

Python doesn't know if you simply forgot the colon, or if you meant to write additional code to set up a more complex loop. If the interpreter can identify a possible fix it will suggest one, like adding a colon at the end of a line, as it does here with the response `expected ':'`. Some errors have easy, obvious fixes, thanks to the suggestions in Python's tracebacks. Some errors are much harder to resolve, even when the eventual fix only involves a single character. Don't feel bad when a small fix takes a long time to find; you are absolutely not alone in this experience.

TRY IT YOURSELF

4-1. Pizzas: Think of at least three kinds of your favorite pizza. Store these pizza names in a list, and then use a `for` loop to print the name of each pizza.

- Modify your `for` loop to print a sentence using the name of the pizza, instead of printing just the name of the pizza. For each pizza, you should have one line of output containing a simple statement like *I like pepperoni pizza*.
- Add a line at the end of your program, outside the `for` loop, that states how much you like pizza. The output should consist of three or more lines about the kinds of pizza you like and then an additional sentence, such as *I really love pizza!*

4-2. Animals: Think of at least three different animals that have a common characteristic. Store the names of these animals in a list, and then use a `for` loop to print out the name of each animal.

- Modify your program to print a statement about each animal, such as *A dog would make a great pet.*
- Add a line at the end of your program, stating what these animals have in common. You could print a sentence, such as *Any of these animals would make a great pet!*

Making Numerical Lists

Many reasons exist to store a set of numbers. For example, you'll need to keep track of the positions of each character in a game, and you might want to keep track of a player's high scores as well. In data visualizations, you'll almost always work with sets of numbers, such as temperatures, distances, population sizes, or latitude and longitude values, among other types of numerical sets.

Lists are ideal for storing sets of numbers, and Python provides a variety of tools to help you work efficiently with lists of numbers. Once you understand how to use these tools effectively, your code will work well even when your lists contain millions of items.

Using the range() Function

Python's `range()` function makes it easy to generate a series of numbers. For example, you can use the `range()` function to print a series of numbers like this:

first_numbers.py

```
for value in range(1, 5):
    print(value)
```

Although this code looks like it should print the numbers from 1 to 5, it doesn't print the number 5:

```
1
2
3
4
```

In this example, `range()` prints only the numbers 1 through 4. This is another result of the off-by-one behavior you'll see often in programming languages. The `range()` function causes Python to start counting at the first value you give it, and it stops when it reaches the second value you provide. Because it stops at that second value, the output never contains the end value, which would have been 5 in this case.

To print the numbers from 1 to 5, you would use `range(1, 6)`:

```
for value in range(1, 6):
    print(value)
```

This time the output starts at 1 and ends at 5:

```
1
2
3
4
5
```

If your output is different from what you expect when you're using `range()`, try adjusting your end value by 1.

You can also pass `range()` only one argument, and it will start the sequence of numbers at 0. For example, `range(6)` would return the numbers from 0 through 5.

Using range() to Make a List of Numbers

If you want to make a list of numbers, you can convert the results of range() directly into a list using the list() function. When you wrap list() around a call to the range() function, the output will be a list of numbers.

In the example in the previous section, we simply printed out a series of numbers. We can use list() to convert that same set of numbers into a list:

```
numbers = list(range(1, 6))
print(numbers)
```

This is the result:

```
[1, 2, 3, 4, 5]
```

We can also use the range() function to tell Python to skip numbers in a given range. If you pass a third argument to range(), Python uses that value as a step size when generating numbers.

For example, here's how to list the even numbers between 1 and 10:

even_numbers.py

```
even_numbers = list(range(2, 11, 2))
print(even_numbers)
```

In this example, the range() function starts with the value 2 and then adds 2 to that value. It adds 2 repeatedly until it reaches or passes the end value, 11, and produces this result:

```
[2, 4, 6, 8, 10]
```

You can create almost any set of numbers you want to using the range() function. For example, consider how you might make a list of the first 10 square numbers (that is, the square of each integer from 1 through 10). In Python, two asterisks (**) represent exponents. Here's how you might put the first 10 square numbers into a list:

square_numbers.py

```
squares = []
for value in range(1, 11):
❶    square = value ** 2
❷    squares.append(square)

print(squares)
```

We start with an empty list called squares. Then, we tell Python to loop through each value from 1 to 10 using the range() function. Inside the loop, the current value is raised to the second power and assigned to the variable square ❶. Each new value of square is then appended to the list squares ❷. Finally, when the loop has finished running, the list of squares is printed:

```
[1, 4, 9, 16, 25, 36, 49, 64, 81, 100]
```

To write this code more concisely, omit the temporary variable square and append each new value directly to the list:

```
squares = []
for value in range(1,11):
    squares.append(value**2)

print(squares)
```

This line does the same work as the lines inside the for loop in the previous listing. Each value in the loop is raised to the second power and then immediately appended to the list of squares.

You can use either of these approaches when you're making more complex lists. Sometimes using a temporary variable makes your code easier to read; other times it makes the code unnecessarily long. Focus first on writing code that you understand clearly, and does what you want it to do. Then look for more efficient approaches as you review your code.

Simple Statistics with a List of Numbers

A few Python functions are helpful when working with lists of numbers. For example, you can easily find the minimum, maximum, and sum of a list of numbers:

```
>>> digits = [1, 2, 3, 4, 5, 6, 7, 8, 9, 0]
>>> min(digits)
0
>>> max(digits)
9
>>> sum(digits)
45
```

NOTE *The examples in this section use short lists of numbers that fit easily on the page. They would work just as well if your list contained a million or more numbers.*

List Comprehensions

The approach described earlier for generating the list squares consisted of using three or four lines of code. A *list comprehension* allows you to generate this same list in just one line of code. A list comprehension combines the for loop and the creation of new elements into one line, and automatically appends each new element. List comprehensions are not always presented to beginners, but I've included them here because you'll most likely see them as soon as you start looking at other people's code.

The following example builds the same list of square numbers you saw earlier but uses a list comprehension:

squares.py

```
squares = [value**2 for value in range(1, 11)]
print(squares)
```

To use this syntax, begin with a descriptive name for the list, such as `squares`. Next, open a set of square brackets and define the expression for the values you want to store in the new list. In this example the expression is `value**2`, which raises the value to the second power. Then, write a `for` loop to generate the numbers you want to feed into the expression, and close the square brackets. The `for` loop in this example is `for value in range(1, 11)`, which feeds the values 1 through 10 into the expression `value**2`. Note that no colon is used at the end of the `for` statement.

The result is the same list of square numbers you saw earlier:

```
[1, 4, 9, 16, 25, 36, 49, 64, 81, 100]
```

It takes practice to write your own list comprehensions, but you'll find them worthwhile once you become comfortable creating ordinary lists. When you're writing three or four lines of code to generate lists and it begins to feel repetitive, consider writing your own list comprehensions.

TRY IT YOURSELF

4-3. Counting to Twenty: Use a `for` loop to print the numbers from 1 to 20, inclusive.

4-4. One Million: Make a list of the numbers from one to one million, and then use a `for` loop to print the numbers. (If the output is taking too long, stop it by pressing CTRL-C or by closing the output window.)

4-5. Summing a Million: Make a list of the numbers from one to one million, and then use `min()` and `max()` to make sure your list actually starts at one and ends at one million. Also, use the `sum()` function to see how quickly Python can add a million numbers.

4-6. Odd Numbers: Use the third argument of the `range()` function to make a list of the odd numbers from 1 to 20. Use a `for` loop to print each number.

4-7. Threes: Make a list of the multiples of 3, from 3 to 30. Use a `for` loop to print the numbers in your list.

4-8. Cubes: A number raised to the third power is called a *cube*. For example, the cube of 2 is written as `2**3` in Python. Make a list of the first 10 cubes (that is, the cube of each integer from 1 through 10), and use a `for` loop to print out the value of each cube.

4-9. Cube Comprehension: Use a list comprehension to generate a list of the first 10 cubes.

Working with Part of a List

In Chapter 3 you learned how to access single elements in a list, and in this chapter you've been learning how to work through all the elements in a list. You can also work with a specific group of items in a list, called a *slice* in Python.

Slicing a List

To make a slice, you specify the index of the first and last elements you want to work with. As with the `range()` function, Python stops one item before the second index you specify. To output the first three elements in a list, you would request indices `0` through `3`, which would return elements `0`, `1`, and `2`.

The following example involves a list of players on a team:

players.py

```
players = ['charles', 'martina', 'michael', 'florence', 'eli']
print(players[0:3])
```

This code prints a slice of the list. The output retains the structure of the list, and includes the first three players in the list:

```
['charles', 'martina', 'michael']
```

You can generate any subset of a list. For example, if you want the second, third, and fourth items in a list, you would start the slice at index `1` and end it at index `4`:

```
players = ['charles', 'martina', 'michael', 'florence', 'eli']
print(players[1:4])
```

This time the slice starts with `'martina'` and ends with `'florence'`:

```
['martina', 'michael', 'florence']
```

If you omit the first index in a slice, Python automatically starts your slice at the beginning of the list:

```
players = ['charles', 'martina', 'michael', 'florence', 'eli']
print(players[:4])
```

Without a starting index, Python starts at the beginning of the list:

```
['charles', 'martina', 'michael', 'florence']
```

A similar syntax works if you want a slice that includes the end of a list. For example, if you want all items from the third item through the last item, you can start with index `2` and omit the second index:

```
players = ['charles', 'martina', 'michael', 'florence', 'eli']
print(players[2:])
```

Python returns all items from the third item through the end of the list:

```
['michael', 'florence', 'eli']
```

This syntax allows you to output all of the elements from any point in your list to the end, regardless of the length of the list. Recall that a negative index returns an element a certain distance from the end of a list; therefore, you can output any slice from the end of a list. For example, if we want to output the last three players on the roster, we can use the slice `players[-3:]`:

```
players = ['charles', 'martina', 'michael', 'florence', 'eli']
print(players[-3:])
```

This prints the names of the last three players and will continue to work as the list of players changes in size.

NOTE *You can include a third value in the brackets indicating a slice. If a third value is included, this tells Python how many items to skip between items in the specified range.*

Looping Through a Slice

You can use a slice in a `for` loop if you want to loop through a subset of the elements in a list. In the next example, we loop through the first three players and print their names as part of a simple roster:

```
players = ['charles', 'martina', 'michael', 'florence', 'eli']

print("Here are the first three players on my team:")
❶ for player in players[:3]:
    print(player.title())
```

Instead of looping through the entire list of players, Python loops through only the first three names ❶:

```
Here are the first three players on my team:
Charles
Martina
Michael
```

Slices are very useful in a number of situations. For instance, when you're creating a game, you could add a player's final score to a list every time that player finishes playing. You could then get a player's top three scores by sorting the list in decreasing order and taking a slice that includes just the first three scores. When you're working with data, you can use slices to process your data in chunks of a specific size. Or, when you're building a web application, you could use slices to display information in a series of pages with an appropriate amount of information on each page.

Copying a List

Often, you'll want to start with an existing list and make an entirely new list based on the first one. Let's explore how copying a list works and examine one situation in which copying a list is useful.

To copy a list, you can make a slice that includes the entire original list by omitting the first index and the second index ([:]). This tells Python to make a slice that starts at the first item and ends with the last item, producing a copy of the entire list.

For example, imagine we have a list of our favorite foods and want to make a separate list of foods that a friend likes. This friend likes everything in our list so far, so we can create their list by copying ours:

foods.py

```
  my_foods = ['pizza', 'falafel', 'carrot cake']
❶ friend_foods = my_foods[:]

  print("My favorite foods are:")
  print(my_foods)

  print("\nMy friend's favorite foods are:")
  print(friend_foods)
```

First, we make a list of the foods we like called `my_foods`. Then we make a new list called `friend_foods`. We make a copy of `my_foods` by asking for a slice of `my_foods` without specifying any indices ❶, and assign the copy to `friend_foods`. When we print each list, we see that they both contain the same foods:

```
My favorite foods are:
['pizza', 'falafel', 'carrot cake']

My friend's favorite foods are:
['pizza', 'falafel', 'carrot cake']
```

To prove that we actually have two separate lists, we'll add a new food to each list and show that each list keeps track of the appropriate person's favorite foods:

```
  my_foods = ['pizza', 'falafel', 'carrot cake']
❶ friend_foods = my_foods[:]

❷ my_foods.append('cannoli')
❸ friend_foods.append('ice cream')

  print("My favorite foods are:")
  print(my_foods)

  print("\nMy friend's favorite foods are:")
  print(friend_foods)
```

We copy the original items in my_foods to the new list friend_foods, as we did in the previous example ❶. Next, we add a new food to each list: we add 'cannoli' to my_foods ❷, and we add 'ice cream' to friend_foods ❸. We then print the two lists to see whether each of these foods is in the appropriate list:

```
My favorite foods are:
['pizza', 'falafel', 'carrot cake', 'cannoli']

My friend's favorite foods are:
['pizza', 'falafel', 'carrot cake', 'ice cream']
```

The output shows that 'cannoli' now appears in our list of favorite foods but 'ice cream' does not. We can see that 'ice cream' now appears in our friend's list but 'cannoli' does not. If we had simply set friend_foods equal to my_foods, we would not produce two separate lists. For example, here's what happens when you try to copy a list without using a slice:

```
my_foods = ['pizza', 'falafel', 'carrot cake']

# This doesn't work:
friend_foods = my_foods

my_foods.append('cannoli')
friend_foods.append('ice cream')

print("My favorite foods are:")
print(my_foods)

print("\nMy friend's favorite foods are:")
print(friend_foods)
```

Instead of assigning a copy of my_foods to friend_foods, we set friend_foods equal to my_foods. This syntax actually tells Python to associate the new variable friend_foods with the list that is already associated with my_foods, so now both variables point to the same list. As a result, when we add 'cannoli' to my_foods, it will also appear in friend_foods. Likewise 'ice cream' will appear in both lists, even though it appears to be added only to friend_foods.

The output shows that both lists are the same now, which is not what we wanted:

```
My favorite foods are:
['pizza', 'falafel', 'carrot cake', 'cannoli', 'ice cream']

My friend's favorite foods are:
['pizza', 'falafel', 'carrot cake', 'cannoli', 'ice cream']
```

NOTE *Don't worry about the details in this example for now. If you're trying to work with a copy of a list and you see unexpected behavior, make sure you are copying the list using a slice, as we did in the first example.*

TRY IT YOURSELF

4-10. Slices: Using one of the programs you wrote in this chapter, add several lines to the end of the program that do the following:

- Print the message *The first three items in the list are:*. Then use a slice to print the first three items from that program's list.
- Print the message *Three items from the middle of the list are:*. Then use a slice to print three items from the middle of the list.
- Print the message *The last three items in the list are:*. Then use a slice to print the last three items in the list.

4-11. My Pizzas, Your Pizzas: Start with your program from Exercise 4-1 (page 56). Make a copy of the list of pizzas, and call it `friend_pizzas`. Then, do the following:

- Add a new pizza to the original list.
- Add a different pizza to the list `friend_pizzas`.
- Prove that you have two separate lists. Print the message *My favorite pizzas are:*, and then use a `for` loop to print the first list. Print the message *My friend's favorite pizzas are:*, and then use a `for` loop to print the second list. Make sure each new pizza is stored in the appropriate list.

4-12. More Loops: All versions of *foods.py* in this section have avoided using `for` loops when printing, to save space. Choose a version of *foods.py*, and write two `for` loops to print each list of foods.

Tuples

Lists work well for storing collections of items that can change throughout the life of a program. The ability to modify lists is particularly important when you're working with a list of users on a website or a list of characters in a game. However, sometimes you'll want to create a list of items that cannot change. Tuples allow you to do just that. Python refers to values that cannot change as *immutable*, and an immutable list is called a *tuple*.

Defining a Tuple

A tuple looks just like a list, except you use parentheses instead of square brackets. Once you define a tuple, you can access individual elements by using each item's index, just as you would for a list.

For example, if we have a rectangle that should always be a certain size, we can ensure that its size doesn't change by putting the dimensions into a tuple:

dimensions.py

```
dimensions = (200, 50)
print(dimensions[0])
print(dimensions[1])
```

We define the tuple `dimensions`, using parentheses instead of square brackets. Then we print each element in the tuple individually, using the same syntax we've been using to access elements in a list:

```
200
50
```

Let's see what happens if we try to change one of the items in the tuple `dimensions`:

```
dimensions = (200, 50)
dimensions[0] = 250
```

This code tries to change the value of the first dimension, but Python returns a type error. Because we're trying to alter a tuple, which can't be done to that type of object, Python tells us we can't assign a new value to an item in a tuple:

```
Traceback (most recent call last):
  File "dimensions.py", line 2, in <module>
    dimensions[0] = 250
TypeError: 'tuple' object does not support item assignment
```

This is beneficial because we want Python to raise an error when a line of code tries to change the dimensions of the rectangle.

NOTE *Tuples are technically defined by the presence of a comma; the parentheses make them look neater and more readable. If you want to define a tuple with one element, you need to include a trailing comma:*

```
my_t = (3,)
```

It doesn't often make sense to build a tuple with one element, but this can happen when tuples are generated automatically.

Looping Through All Values in a Tuple

You can loop over all the values in a tuple using a `for` loop, just as you did with a list:

```
dimensions = (200, 50)
for dimension in dimensions:
    print(dimension)
```

Python returns all the elements in the tuple, just as it would for a list:

```
200
50
```

Writing Over a Tuple

Although you can't modify a tuple, you can assign a new value to a variable that represents a tuple. For example, if we wanted to change the dimensions of this rectangle, we could redefine the entire tuple:

```
dimensions = (200, 50)
print("Original dimensions:")
for dimension in dimensions:
    print(dimension)

dimensions = (400, 100)
print("\nModified dimensions:")
for dimension in dimensions:
    print(dimension)
```

The first four lines define the original tuple and print the initial dimensions. We then associate a new tuple with the variable `dimensions`, and print the new values. Python doesn't raise any errors this time, because reassigning a variable is valid:

```
Original dimensions:
200
50

Modified dimensions:
400
100
```

When compared with lists, tuples are simple data structures. Use them when you want to store a set of values that should not be changed throughout the life of a program.

TRY IT YOURSELF

4-13. Buffet: A buffet-style restaurant offers only five basic foods. Think of five simple foods, and store them in a tuple.

- Use a `for` loop to print each food the restaurant offers.
- Try to modify one of the items, and make sure that Python rejects the change.
- The restaurant changes its menu, replacing two of the items with different foods. Add a line that rewrites the tuple, and then use a `for` loop to print each of the items on the revised menu.

Styling Your Code

Now that you're writing longer programs, it's a good idea to learn how to style your code consistently. Take the time to make your code as easy as possible to read. Writing easy-to-read code helps you keep track of what your programs are doing and helps others understand your code as well.

Python programmers have agreed on a number of styling conventions to ensure that everyone's code is structured in roughly the same way. Once you've learned to write clean Python code, you should be able to understand the overall structure of anyone else's Python code, as long as they follow the same guidelines. If you're hoping to become a professional programmer at some point, you should begin following these guidelines as soon as possible to develop good habits.

The Style Guide

When someone wants to make a change to the Python language, they write a *Python Enhancement Proposal (PEP)*. One of the oldest PEPs is *PEP 8*, which instructs Python programmers on how to style their code. PEP 8 is fairly lengthy, but much of it relates to more complex coding structures than what you've seen so far.

The Python style guide was written with the understanding that code is read more often than it is written. You'll write your code once and then start reading it as you begin debugging. When you add features to a program, you'll spend more time reading your code. When you share your code with other programmers, they'll read your code as well.

Given the choice between writing code that's easier to write or code that's easier to read, Python programmers will almost always encourage you to write code that's easier to read. The following guidelines will help you write clear code from the start.

Indentation

PEP 8 recommends that you use four spaces per indentation level. Using four spaces improves readability while leaving room for multiple levels of indentation on each line.

In a word processing document, people often use tabs rather than spaces to indent. This works well for word processing documents, but the Python interpreter gets confused when tabs are mixed with spaces. Every text editor provides a setting that lets you use the TAB key but then converts each tab to a set number of spaces. You should definitely use your TAB key, but also make sure your editor is set to insert spaces rather than tabs into your document.

Mixing tabs and spaces in your file can cause problems that are very difficult to diagnose. If you think you have a mix of tabs and spaces, you can convert all tabs in a file to spaces in most editors.

Line Length

Many Python programmers recommend that each line should be less than 80 characters. Historically, this guideline developed because most computers could fit only 79 characters on a single line in a terminal window. Currently, people can fit much longer lines on their screens, but other reasons exist to adhere to the 79-character standard line length.

Professional programmers often have several files open on the same screen, and using the standard line length allows them to see entire lines in two or three files that are open side by side onscreen. PEP 8 also recommends that you limit all of your comments to 72 characters per line, because some of the tools that generate automatic documentation for larger projects add formatting characters at the beginning of each commented line.

The PEP 8 guidelines for line length are not set in stone, and some teams prefer a 99-character limit. Don't worry too much about line length in your code as you're learning, but be aware that people who are working collaboratively almost always follow the PEP 8 guidelines. Most editors allow you to set up a visual cue, usually a vertical line on your screen, that shows you where these limits are.

NOTE *Appendix B shows you how to configure your text editor so it always inserts four spaces each time you press the TAB key and shows a vertical guideline to help you follow the 79-character limit.*

Blank Lines

To group parts of your program visually, use blank lines. You should use blank lines to organize your files, but don't do so excessively. By following the examples provided in this book, you should strike the right balance. For example, if you have five lines of code that build a list and then another three lines that do something with that list, it's appropriate to place a blank line between the two sections. However, you should not place three or four blank lines between the two sections.

Blank lines won't affect how your code runs, but they will affect the readability of your code. The Python interpreter uses horizontal indentation to interpret the meaning of your code, but it disregards vertical spacing.

Other Style Guidelines

PEP 8 has many additional styling recommendations, but most of the guidelines refer to more complex programs than what you're writing at this point. As you learn more complex Python structures, I'll share the relevant parts of the PEP 8 guidelines.

TRY IT YOURSELF

4-14. PEP 8: Look through the original PEP 8 style guide at *https://python.org/dev/peps/pep-0008*. You won't use much of it now, but it might be interesting to skim through it.

4-15. Code Review: Choose three of the programs you've written in this chapter and modify each one to comply with PEP 8.

- Use four spaces for each indentation level. Set your text editor to insert four spaces every time you press the TAB key, if you haven't already done so (see Appendix B for instructions on how to do this).
- Use less than 80 characters on each line, and set your editor to show a vertical guideline at the 80th character position.
- Don't use blank lines excessively in your program files.

Summary

In this chapter, you learned how to work efficiently with the elements in a list. You learned how to work through a list using a `for` loop, how Python uses indentation to structure a program, and how to avoid some common indentation errors. You learned to make simple numerical lists, as well as a few operations you can perform on numerical lists. You learned how to slice a list to work with a subset of items and how to copy lists properly using a slice. You also learned about tuples, which provide a degree of protection to a set of values that shouldn't change, and how to style your increasingly complex code to make it easy to read.

In Chapter 5, you'll learn to respond appropriately to different conditions by using `if` statements. You'll learn to string together relatively complex sets of conditional tests to respond appropriately to exactly the kind of situation or information you're looking for. You'll also learn to use `if` statements while looping through a list to take specific actions with selected elements from a list.

5

IF STATEMENTS

Programming often involves examining a set of conditions and deciding which action to take based on those conditions. Python's `if` statement allows you to examine the current state of a program and respond appropriately to that state.

In this chapter, you'll learn to write conditional tests, which allow you to check any condition of interest. You'll learn to write simple `if` statements, and you'll learn how to create a more complex series of `if` statements to identify when the exact conditions you want are present. You'll then apply this concept to lists, so you'll be able to write a `for` loop that handles most items in a list one way but handles certain items with specific values in a different way.

A Simple Example

The following example shows how `if` tests let you respond to special situations correctly. Imagine you have a list of cars and you want to print out the name of each car. Car names are proper names, so the names of most cars should be printed in title case. However, the value `'bmw'` should be printed in all uppercase. The following code loops through a list of car names and looks for the value `'bmw'`. Whenever the value is `'bmw'`, it's printed in uppercase instead of title case:

cars.py

```
cars = ['audi', 'bmw', 'subaru', 'toyota']

for car in cars:
❶    if car == 'bmw':
        print(car.upper())
    else:
        print(car.title())
```

The loop in this example first checks if the current value of `car` is `'bmw'` ❶. If it is, the value is printed in uppercase. If the value of `car` is anything other than `'bmw'`, it's printed in title case:

```
Audi
BMW
Subaru
Toyota
```

This example combines a number of the concepts you'll learn about in this chapter. Let's begin by looking at the kinds of tests you can use to examine the conditions in your program.

Conditional Tests

At the heart of every `if` statement is an expression that can be evaluated as `True` or `False` and is called a *conditional test*. Python uses the values `True` and `False` to decide whether the code in an `if` statement should be executed. If a conditional test evaluates to `True`, Python executes the code following the `if` statement. If the test evaluates to `False`, Python ignores the code following the `if` statement.

Checking for Equality

Most conditional tests compare the current value of a variable to a specific value of interest. The simplest conditional test checks whether the value of a variable is equal to the value of interest:

```
>>> car = 'bmw'
>>> car == 'bmw'
True
```

The first line sets the value of car to 'bmw' using a single equal sign, as you've seen many times already. The next line checks whether the value of car is 'bmw' by using a double equal sign (==). This *equality operator* returns True if the values on the left and right side of the operator match, and False if they don't match. The values in this example match, so Python returns True.

When the value of car is anything other than 'bmw', this test returns False:

```
>>> car = 'audi'
>>> car == 'bmw'
False
```

A single equal sign is really a statement; you might read the first line of code here as "Set the value of car equal to 'audi'." On the other hand, a double equal sign asks a question: "Is the value of car equal to 'bmw'?" Most programming languages use equal signs in this way.

Ignoring Case When Checking for Equality

Testing for equality is case sensitive in Python. For example, two values with different capitalization are not considered equal:

```
>>> car = 'Audi'
>>> car == 'audi'
False
```

If case matters, this behavior is advantageous. But if case doesn't matter and instead you just want to test the value of a variable, you can convert the variable's value to lowercase before doing the comparison:

```
>>> car = 'Audi'
>>> car.lower() == 'audi'
True
```

This test will return True no matter how the value 'Audi' is formatted because the test is now case insensitive. The lower() method doesn't change the value that was originally stored in car, so you can do this kind of comparison without affecting the original variable:

```
>>> car = 'Audi'
>>> car.lower() == 'audi'
True
>>> car
'Audi'
```

We first assign the capitalized string 'Audi' to the variable car. Then, we convert the value of car to lowercase and compare the lowercase value to the string 'audi'. The two strings match, so Python returns True. We can see that the value stored in car has not been affected by the lower() method.

Websites enforce certain rules for the data that users enter in a manner similar to this. For example, a site might use a conditional test like this to ensure that every user has a truly unique username, not just a variation on the capitalization of another person's username. When someone submits a new username, that new username is converted to lowercase and compared to the lowercase versions of all existing usernames. During this check, a username like 'John' will be rejected if any variation of 'john' is already in use.

Checking for Inequality

When you want to determine whether two values are not equal, you can use the *inequality operator* (!=). Let's use another if statement to examine how to use the inequality operator. We'll store a requested pizza topping in a variable and then print a message if the person did not order anchovies:

toppings.py

```
requested_topping = 'mushrooms'

if requested_topping != 'anchovies':
    print("Hold the anchovies!")
```

This code compares the value of requested_topping to the value 'anchovies'. If these two values do not match, Python returns True and executes the code following the if statement. If the two values match, Python returns False and does not run the code following the if statement.

Because the value of requested_topping is not 'anchovies', the print() function is executed:

```
Hold the anchovies!
```

Most of the conditional expressions you write will test for equality, but sometimes you'll find it more efficient to test for inequality.

Numerical Comparisons

Testing numerical values is pretty straightforward. For example, the following code checks whether a person is 18 years old:

```
>>> age = 18
>>> age == 18
True
```

You can also test to see if two numbers are not equal. For example, the following code prints a message if the given answer is not correct:

magic_number.py

```
answer = 17
if answer != 42:
    print("That is not the correct answer. Please try again!")
```

The conditional test passes, because the value of answer (17) is not equal to 42. Because the test passes, the indented code block is executed:

```
That is not the correct answer. Please try again!
```

You can include various mathematical comparisons in your conditional statements as well, such as less than, less than or equal to, greater than, and greater than or equal to:

```
>>> age = 19
>>> age < 21
True
>>> age <= 21
True
>>> age > 21
False
>>> age >= 21
False
```

Each mathematical comparison can be used as part of an `if` statement, which can help you detect the exact conditions of interest.

Checking Multiple Conditions

You may want to check multiple conditions at the same time. For example, sometimes you might need two conditions to be `True` to take an action. Other times, you might be satisfied with just one condition being `True`. The keywords `and` and `or` can help you in these situations.

Using and to Check Multiple Conditions

To check whether two conditions are both `True` simultaneously, use the keyword `and` to combine the two conditional tests; if each test passes, the overall expression evaluates to `True`. If either test fails or if both tests fail, the expression evaluates to `False`.

For example, you can check whether two people are both over 21 by using the following test:

```
>>> age_0 = 22
>>> age_1 = 18
❶ >>> age_0 >= 21 and age_1 >= 21
False
❷ >>> age_1 = 22
>>> age_0 >= 21 and age_1 >= 21
True
```

First, we define two ages, `age_0` and `age_1`. Then we check whether both ages are 21 or older ❶. The test on the left passes, but the test on the right fails, so the overall conditional expression evaluates to `False`. We then change `age_1` to 22 ❷. The value of `age_1` is now greater than 21, so both individual tests pass, causing the overall conditional expression to evaluate as `True`.

To improve readability, you can use parentheses around the individual tests, but they are not required. If you use parentheses, your test would look like this:

```
(age_0 >= 21) and (age_1 >= 21)
```

Using or to Check Multiple Conditions

The keyword or allows you to check multiple conditions as well, but it passes when either or both of the individual tests pass. An or expression fails only when both individual tests fail.

Let's consider two ages again, but this time we'll look for only one person to be over 21:

```
>>> age_0 = 22
>>> age_1 = 18
❶ >>> age_0 >= 21 or age_1 >= 21
True
❷ >>> age_0 = 18
>>> age_0 >= 21 or age_1 >= 21
False
```

We start with two age variables again. Because the test for age_0 ❶ passes, the overall expression evaluates to True. We then lower age_0 to 18. In the final test ❷, both tests now fail and the overall expression evaluates to False.

Checking Whether a Value Is in a List

Sometimes it's important to check whether a list contains a certain value before taking an action. For example, you might want to check whether a new username already exists in a list of current usernames before completing someone's registration on a website. In a mapping project, you might want to check whether a submitted location already exists in a list of known locations.

To find out whether a particular value is already in a list, use the keyword in. Let's consider some code you might write for a pizzeria. We'll make a list of toppings a customer has requested for a pizza and then check whether certain toppings are in the list.

```
>>> requested_toppings = ['mushrooms', 'onions', 'pineapple']
>>> 'mushrooms' in requested_toppings
True
>>> 'pepperoni' in requested_toppings
False
```

The keyword in tells Python to check for the existence of 'mushrooms' and 'pepperoni' in the list requested_toppings. This technique is quite powerful because you can create a list of essential values, and then easily check whether the value you're testing matches one of the values in the list.

Checking Whether a Value Is Not in a List

Other times, it's important to know if a value does not appear in a list. You can use the keyword not in this situation. For example, consider a list of users who are banned from commenting in a forum. You can check whether a user has been banned before allowing that person to submit a comment:

banned_users.py

```
banned_users = ['andrew', 'carolina', 'david']
user = 'marie'
```

```
if user not in banned_users:
    print(f"{user.title()}, you can post a response if you wish.")
```

The `if` statement here reads quite clearly. If the value of `user` is not in the list `banned_users`, Python returns `True` and executes the indented line.

The user `'marie'` is not in the list `banned_users`, so she sees a message inviting her to post a response:

```
Marie, you can post a response if you wish.
```

Boolean Expressions

As you learn more about programming, you'll hear the term *Boolean expression* at some point. A Boolean expression is just another name for a conditional test. A *Boolean value* is either `True` or `False`, just like the value of a conditional expression after it has been evaluated.

Boolean values are often used to keep track of certain conditions, such as whether a game is running or whether a user can edit certain content on a website:

```
game_active = True
can_edit = False
```

Boolean values provide an efficient way to track the state of a program or a particular condition that is important in your program.

TRY IT YOURSELF

5-1. Conditional Tests: Write a series of conditional tests. Print a statement describing each test and your prediction for the results of each test. Your code should look something like this:

```
car = 'subaru'
print("Is car == 'subaru'? I predict True.")
print(car == 'subaru')

print("\nIs car == 'audi'? I predict False.")
print(car == 'audi')
```

- Look closely at your results, and make sure you understand why each line evaluates to `True` or `False`.
- Create at least 10 tests. Have at least 5 tests evaluate to `True` and another 5 tests evaluate to `False`.

(continued)

5-2. More Conditional Tests: You don't have to limit the number of tests you create to 10. If you want to try more comparisons, write more tests and add them to *conditional_tests.py*. Have at least one `True` and one `False` result for each of the following:

- Tests for equality and inequality with strings
- Tests using the `lower()` method
- Numerical tests involving equality and inequality, greater than and less than, greater than or equal to, and less than or equal to
- Tests using the `and` keyword and the `or` keyword
- Test whether an item is in a list
- Test whether an item is not in a list

if Statements

When you understand conditional tests, you can start writing `if` statements. Several different kinds of `if` statements exist, and your choice of which to use depends on the number of conditions you need to test. You saw several examples of `if` statements in the discussion about conditional tests, but now let's dig deeper into the topic.

Simple if Statements

The simplest kind of `if` statement has one test and one action:

```
if conditional_test:
    do something
```

You can put any conditional test in the first line and just about any action in the indented block following the test. If the conditional test evaluates to `True`, Python executes the code following the `if` statement. If the test evaluates to `False`, Python ignores the code following the `if` statement.

Let's say we have a variable representing a person's age, and we want to know if that person is old enough to vote. The following code tests whether the person can vote:

voting.py

```
age = 19
if age >= 18:
    print("You are old enough to vote!")
```

Python checks to see whether the value of `age` is greater than or equal to 18. It is, so Python executes the indented `print()` call:

```
You are old enough to vote!
```

Indentation plays the same role in `if` statements as it did in `for` loops. All indented lines after an `if` statement will be executed if the test passes, and the entire block of indented lines will be ignored if the test does not pass.

You can have as many lines of code as you want in the block following the `if` statement. Let's add another line of output if the person is old enough to vote, asking if the individual has registered to vote yet:

```
age = 19
if age >= 18:
    print("You are old enough to vote!")
    print("Have you registered to vote yet?")
```

The conditional test passes, and both `print()` calls are indented, so both lines are printed:

```
You are old enough to vote!
Have you registered to vote yet?
```

If the value of `age` is less than 18, this program would produce no output.

if-else Statements

Often, you'll want to take one action when a conditional test passes and a different action in all other cases. Python's `if-else` syntax makes this possible. An `if-else` block is similar to a simple `if` statement, but the `else` statement allows you to define an action or set of actions that are executed when the conditional test fails.

We'll display the same message we had previously if the person is old enough to vote, but this time we'll add a message for anyone who is not old enough to vote:

```
age = 17
❶ if age >= 18:
    print("You are old enough to vote!")
    print("Have you registered to vote yet?")
❷ else:
    print("Sorry, you are too young to vote.")
    print("Please register to vote as soon as you turn 18!")
```

If the conditional test ❶ passes, the first block of indented `print()` calls is executed. If the test evaluates to `False`, the `else` block ❷ is executed. Because `age` is less than 18 this time, the conditional test fails and the code in the `else` block is executed:

```
Sorry, you are too young to vote.
Please register to vote as soon as you turn 18!
```

This code works because it has only two possible situations to evaluate: a person is either old enough to vote or not old enough to vote. The `if-else` structure works well in situations in which you want Python to always execute one of two possible actions. In a simple `if-else` chain like this, one of the two actions will always be executed.

The if-elif-else Chain

Often, you'll need to test more than two possible situations, and to evaluate these you can use Python's `if-elif-else` syntax. Python executes only one block in an `if-elif-else` chain. It runs each conditional test in order, until one passes. When a test passes, the code following that test is executed and Python skips the rest of the tests.

Many real-world situations involve more than two possible conditions. For example, consider an amusement park that charges different rates for different age groups:

- Admission for anyone under age 4 is free.
- Admission for anyone between the ages of 4 and 18 is $25.
- Admission for anyone age 18 or older is $40.

How can we use an `if` statement to determine a person's admission rate? The following code tests for the age group of a person and then prints an admission price message:

amusement_park.py

```
  age = 12
❶ if age < 4:
      print("Your admission cost is $0.")
❷ elif age < 18:
      print("Your admission cost is $25.")
❸ else:
      print("Your admission cost is $40.")
```

The `if` test ❶ checks whether a person is under 4 years old. When the test passes, an appropriate message is printed and Python skips the rest of the tests. The `elif` line ❷ is really another `if` test, which runs only if the previous test failed. At this point in the chain, we know the person is at least 4 years old because the first test failed. If the person is under 18, an appropriate message is printed and Python skips the `else` block. If both the `if` and `elif` tests fail, Python runs the code in the `else` block ❸.

In this example the `if` test ❶ evaluates to `False`, so its code block is not executed. However, the `elif` test evaluates to `True` (12 is less than 18) so its code is executed. The output is one sentence, informing the user of the admission cost:

```
Your admission cost is $25.
```

Any age greater than 17 would cause the first two tests to fail. In these situations, the `else` block would be executed and the admission price would be $40.

Rather than printing the admission price within the `if-elif-else` block, it would be more concise to set just the price inside the `if-elif-else` chain and then have a single print() call that runs after the chain has been evaluated:

```
age = 12

if age < 4:
    price = 0
elif age < 18:
    price = 25
else:
    price = 40

print(f"Your admission cost is ${price}.")
```

The indented lines set the value of price according to the person's age, as in the previous example. After the price is set by the if-elif-else chain, a separate unindented print() call uses this value to display a message reporting the person's admission price.

This code produces the same output as the previous example, but the purpose of the if-elif-else chain is narrower. Instead of determining a price and displaying a message, it simply determines the admission price. In addition to being more efficient, this revised code is easier to modify than the original approach. To change the text of the output message, you would need to change only one print() call rather than three separate print() calls.

Using Multiple elif Blocks

You can use as many elif blocks in your code as you like. For example, if the amusement park were to implement a discount for seniors, you could add one more conditional test to the code to determine whether someone qualifies for the senior discount. Let's say that anyone 65 or older pays half the regular admission, or $20:

```
age = 12

if age < 4:
    price = 0
elif age < 18:
    price = 25
elif age < 65:
    price = 40
else:
    price = 20

print(f"Your admission cost is ${price}.")
```

Most of this code is unchanged. The second elif block now checks to make sure a person is less than age 65 before assigning them the full admission rate of $40. Notice that the value assigned in the else block needs to be changed to $20, because the only ages that make it to this block are for people 65 or older.

Omitting the else Block

Python does not require an else block at the end of an if-elif chain. Sometimes, an else block is useful. Other times, it's clearer to use an additional elif statement that catches the specific condition of interest:

```
age = 12

if age < 4:
    price = 0
elif age < 18:
    price = 25
elif age < 65:
    price = 40
elif age >= 65:
    price = 20

print(f"Your admission cost is ${price}.")
```

The final elif block assigns a price of $20 when the person is 65 or older, which is a little clearer than the general else block. With this change, every block of code must pass a specific test in order to be executed.

The else block is a catchall statement. It matches any condition that wasn't matched by a specific if or elif test, and that can sometimes include invalid or even malicious data. If you have a specific final condition you're testing for, consider using a final elif block and omit the else block. As a result, you'll be more confident that your code will run only under the correct conditions.

Testing Multiple Conditions

The if-elif-else chain is powerful, but it's only appropriate to use when you just need one test to pass. As soon as Python finds one test that passes, it skips the rest of the tests. This behavior is beneficial, because it's efficient and allows you to test for one specific condition.

However, sometimes it's important to check all conditions of interest. In this case, you should use a series of simple if statements with no elif or else blocks. This technique makes sense when more than one condition could be True, and you want to act on every condition that is True.

Let's reconsider the pizzeria example. If someone requests a two-topping pizza, you'll need to be sure to include both toppings on their pizza:

toppings.py

```
requested_toppings = ['mushrooms', 'extra cheese']

if 'mushrooms' in requested_toppings:
    print("Adding mushrooms.")
❶ if 'pepperoni' in requested_toppings:
    print("Adding pepperoni.")
```

```
if 'extra cheese' in requested_toppings:
    print("Adding extra cheese.")

print("\nFinished making your pizza!")
```

We start with a list containing the requested toppings. The first `if` statement checks to see whether the person requested mushrooms on their pizza. If so, a message is printed confirming that topping. The test for pepperoni ❶ is another simple `if` statement, not an `elif` or `else` statement, so this test is run regardless of whether the previous test passed or not. The last `if` statement checks whether extra cheese was requested, regardless of the results from the first two tests. These three independent tests are executed every time this program is run.

Because every condition in this example is evaluated, both mushrooms and extra cheese are added to the pizza:

```
Adding mushrooms.
Adding extra cheese.

Finished making your pizza!
```

This code would not work properly if we used an `if-elif-else` block, because the code would stop running after only one test passes. Here's what that would look like:

```
requested_toppings = ['mushrooms', 'extra cheese']

if 'mushrooms' in requested_toppings:
    print("Adding mushrooms.")
elif 'pepperoni' in requested_toppings:
    print("Adding pepperoni.")
elif 'extra cheese' in requested_toppings:
    print("Adding extra cheese.")

print("\nFinished making your pizza!")
```

The test for `'mushrooms'` is the first test to pass, so mushrooms are added to the pizza. However, the values `'extra cheese'` and `'pepperoni'` are never checked, because Python doesn't run any tests beyond the first test that passes in an `if-elif-else` chain. The customer's first topping will be added, but all of their other toppings will be missed:

```
Adding mushrooms.

Finished making your pizza!
```

In summary, if you want only one block of code to run, use an `if-elif-else` chain. If more than one block of code needs to run, use a series of independent `if` statements.

TRY IT YOURSELF

5-3. Alien Colors #1: Imagine an alien was just shot down in a game. Create a variable called `alien_color` and assign it a value of `'green'`, `'yellow'`, or `'red'`.

- Write an `if` statement to test whether the alien's color is green. If it is, print a message that the player just earned 5 points.
- Write one version of this program that passes the `if` test and another that fails. (The version that fails will have no output.)

5-4. Alien Colors #2: Choose a color for an alien as you did in Exercise 5-3, and write an `if-else` chain.

- If the alien's color is green, print a statement that the player just earned 5 points for shooting the alien.
- If the alien's color isn't green, print a statement that the player just earned 10 points.
- Write one version of this program that runs the `if` block and another that runs the `else` block.

5-5. Alien Colors #3: Turn your `if-else` chain from Exercise 5-4 into an `if-elif-else` chain.

- If the alien is green, print a message that the player earned 5 points.
- If the alien is yellow, print a message that the player earned 10 points.
- If the alien is red, print a message that the player earned 15 points.
- Write three versions of this program, making sure each message is printed for the appropriate color alien.

5-6. Stages of Life: Write an `if-elif-else` chain that determines a person's stage of life. Set a value for the variable age, and then:

- If the person is less than 2 years old, print a message that the person is a baby.
- If the person is at least 2 years old but less than 4, print a message that the person is a toddler.
- If the person is at least 4 years old but less than 13, print a message that the person is a kid.
- If the person is at least 13 years old but less than 20, print a message that the person is a teenager.
- If the person is at least 20 years old but less than 65, print a message that the person is an adult.
- If the person is age 65 or older, print a message that the person is an elder.

5-7. Favorite Fruit: Make a list of your favorite fruits, and then write a series of independent `if` statements that check for certain fruits in your list.

- Make a list of your three favorite fruits and call it `favorite_fruits`.
- Write five `if` statements. Each should check whether a certain kind of fruit is in your list. If the fruit is in your list, the `if` block should print a statement, such as *You really like bananas!*

Using if Statements with Lists

You can do some interesting work when you combine lists and `if` statements. You can watch for special values that need to be treated differently than other values in the list. You can efficiently manage changing conditions, such as the availability of certain items in a restaurant throughout a shift. You can also begin to prove that your code works as you expect it to in all possible situations.

Checking for Special Items

This chapter began with a simple example that showed how to handle a special value like `'bmw'`, which needed to be printed in a different format than other values in the list. Now that you have a basic understanding of conditional tests and `if` statements, let's take a closer look at how you can watch for special values in a list and handle those values appropriately.

Let's continue with the pizzeria example. The pizzeria displays a message whenever a topping is added to your pizza, as it's being made. The code for this action can be written very efficiently by making a list of toppings the customer has requested and using a loop to announce each topping as it's added to the pizza:

toppings.py

```
requested_toppings = ['mushrooms', 'green peppers', 'extra cheese']

for requested_topping in requested_toppings:
    print(f"Adding {requested_topping}.")

print("\nFinished making your pizza!")
```

The output is straightforward because this code is just a simple `for` loop:

```
Adding mushrooms.
Adding green peppers.
Adding extra cheese.

Finished making your pizza!
```

But what if the pizzeria runs out of green peppers? An `if` statement inside the `for` loop can handle this situation appropriately:

```
requested_toppings = ['mushrooms', 'green peppers', 'extra cheese']

for requested_topping in requested_toppings:
    if requested_topping == 'green peppers':
        print("Sorry, we are out of green peppers right now.")
    else:
        print(f"Adding {requested_topping}.")

print("\nFinished making your pizza!")
```

This time, we check each requested item before adding it to the pizza. The `if` statement checks to see if the person requested green peppers. If so, we display a message informing them why they can't have green peppers. The `else` block ensures that all other toppings will be added to the pizza.

The output shows that each requested topping is handled appropriately.

```
Adding mushrooms.
Sorry, we are out of green peppers right now.
Adding extra cheese.

Finished making your pizza!
```

Checking That a List Is Not Empty

We've made a simple assumption about every list we've worked with so far: we've assumed that each list has at least one item in it. Soon we'll let users provide the information that's stored in a list, so we won't be able to assume that a list has any items in it each time a loop is run. In this situation, it's useful to check whether a list is empty before running a `for` loop.

As an example, let's check whether the list of requested toppings is empty before building the pizza. If the list is empty, we'll prompt the user and make sure they want a plain pizza. If the list is not empty, we'll build the pizza just as we did in the previous examples:

```
requested_toppings = []

if requested_toppings:
    for requested_topping in requested_toppings:
        print(f"Adding {requested_topping}.")
    print("\nFinished making your pizza!")
else:
    print("Are you sure you want a plain pizza?")
```

This time we start out with an empty list of requested toppings. Instead of jumping right into a `for` loop, we do a quick check first. When the name of a list is used in an `if` statement, Python returns `True` if the list contains at least one item; an empty list evaluates to `False`. If `requested_toppings` passes the conditional test, we run the same `for` loop we used in the previous example. If the conditional test fails, we print a message asking the customer if they really want a plain pizza with no toppings.

The list is empty in this case, so the output asks if the user really wants a plain pizza:

```
Are you sure you want a plain pizza?
```

If the list is not empty, the output will show each requested topping being added to the pizza.

Using Multiple Lists

People will ask for just about anything, especially when it comes to pizza toppings. What if a customer actually wants french fries on their pizza? You can use lists and `if` statements to make sure your input makes sense before you act on it.

Let's watch out for unusual topping requests before we build a pizza. The following example defines two lists. The first is a list of available toppings at the pizzeria, and the second is the list of toppings that the user has requested. This time, each item in `requested_toppings` is checked against the list of available toppings before it's added to the pizza:

```
available_toppings = ['mushrooms', 'olives', 'green peppers',
                      'pepperoni', 'pineapple', 'extra cheese']

❶ requested_toppings = ['mushrooms', 'french fries', 'extra cheese']

for requested_topping in requested_toppings:
❷     if requested_topping in available_toppings:
          print(f"Adding {requested_topping}.")
❸     else:
          print(f"Sorry, we don't have {requested_topping}.")

print("\nFinished making your pizza!")
```

First, we define a list of available toppings at this pizzeria. Note that this could be a tuple if the pizzeria has a stable selection of toppings. Then, we make a list of toppings that a customer has requested. There's an unusual request for a topping in this example: `'french fries'` ❶. Next we loop through the list of requested toppings. Inside the loop, we check to see if each requested topping is actually in the list of available toppings ❷. If it is, we add that topping to the pizza. If the requested topping is not in the list of available toppings, the else block will run ❸. The else block prints a message telling the user which toppings are unavailable.

This code syntax produces clean, informative output:

```
Adding mushrooms.
Sorry, we don't have french fries.
Adding extra cheese.

Finished making your pizza!
```

In just a few lines of code, we've managed a real-world situation pretty effectively!

> **TRY IT YOURSELF**
>
> **5-8. Hello Admin:** Make a list of five or more usernames, including the name `'admin'`. Imagine you are writing code that will print a greeting to each user after they log in to a website. Loop through the list, and print a greeting to each user.
>
> - If the username is `'admin'`, print a special greeting, such as *Hello admin, would you like to see a status report?*
> - Otherwise, print a generic greeting, such as *Hello Jaden, thank you for logging in again.*
>
> **5-9. No Users:** Add an `if` test to *hello_admin.py* to make sure the list of users is not empty.
>
> - If the list is empty, print the message *We need to find some users!*
> - Remove all of the usernames from your list, and make sure the correct message is printed.
>
> **5-10. Checking Usernames:** Do the following to create a program that simulates how websites ensure that everyone has a unique username.
>
> - Make a list of five or more usernames called `current_users`.
> - Make another list of five usernames called `new_users`. Make sure one or two of the new usernames are also in the `current_users` list.
> - Loop through the `new_users` list to see if each new username has already been used. If it has, print a message that the person will need to enter a new username. If a username has not been used, print a message saying that the username is available.
> - Make sure your comparison is case insensitive. If `'John'` has been used, `'JOHN'` should not be accepted. (To do this, you'll need to make a copy of `current_users` containing the lowercase versions of all existing users.)
>
> **5-11. Ordinal Numbers:** Ordinal numbers indicate their position in a list, such as *1st* or *2nd*. Most ordinal numbers end in *th*, except 1, 2, and 3.
>
> - Store the numbers 1 through 9 in a list.
> - Loop through the list.
> - Use an `if-elif-else` chain inside the loop to print the proper ordinal ending for each number. Your output should read `"1st 2nd 3rd 4th 5th 6th 7th 8th 9th"`, and each result should be on a separate line.

Styling Your if Statements

In every example in this chapter, you've seen good styling habits. The only recommendation PEP 8 provides for styling conditional tests is to use a single space around comparison operators, such as `==`, `>=`, and `<=`. For example:

```
if age < 4:
```

is better than:

```
if age<4:
```

Such spacing does not affect the way Python interprets your code; it just makes your code easier for you and others to read.

> **TRY IT YOURSELF**
>
> **5-12. Styling `if` Statements:** Review the programs you wrote in this chapter, and make sure you styled your conditional tests appropriately.
>
> **5-13. Your Ideas:** At this point, you're a more capable programmer than you were when you started this book. Now that you have a better sense of how real-world situations are modeled in programs, you might be thinking of some problems you could solve with your own programs. Record any new ideas you have about problems you might want to solve as your programming skills continue to improve. Consider games you might want to write, datasets you might want to explore, and web applications you'd like to create.

Summary

In this chapter you learned how to write conditional tests, which always evaluate to `True` or `False`. You learned to write simple `if` statements, `if-else` chains, and `if-elif-else` chains. You began using these structures to identify particular conditions you need to test and to know when those conditions have been met in your programs. You learned to handle certain items in a list differently than all other items while continuing to utilize the efficiency of a `for` loop. You also revisited Python's style recommendations to ensure that your increasingly complex programs are still relatively easy to read and understand.

In Chapter 6 you'll learn about Python's dictionaries. A dictionary is similar to a list, but it allows you to connect pieces of information. You'll learn how to build dictionaries, loop through them, and use them in combination with lists and `if` statements. Learning about dictionaries will enable you to model an even wider variety of real-world situations.

6

DICTIONARIES

In this chapter you'll learn how to use Python's dictionaries, which allow you to connect pieces of related information. You'll learn how to access the information once it's in a dictionary and how to modify that information. Because dictionaries can store an almost limitless amount of information, I'll show you how to loop through the data in a dictionary. Additionally, you'll learn to nest dictionaries inside lists, lists inside dictionaries, and even dictionaries inside other dictionaries.

Understanding dictionaries allows you to model a variety of real-world objects more accurately. You'll be able to create a dictionary representing a person and then store as much information as you want about that person. You can store their name, age, location, profession, and any other aspect of a person you can describe. You'll be able to store any two kinds of information that can be matched up, such as a list of words and their meanings, a list of people's names and their favorite numbers, a list of mountains and their elevations, and so forth.

A Simple Dictionary

Consider a game featuring aliens that can have different colors and point values. This simple dictionary stores information about a particular alien:

alien.py

```
alien_0 = {'color': 'green', 'points': 5}

print(alien_0['color'])
print(alien_0['points'])
```

The dictionary `alien_0` stores the alien's color and point value. The last two lines access and display that information, as shown here:

```
green
5
```

As with most new programming concepts, using dictionaries takes practice. Once you've worked with dictionaries for a bit, you'll see how effectively they can model real-world situations.

Working with Dictionaries

A *dictionary* in Python is a collection of *key-value pairs*. Each *key* is connected to a value, and you can use a key to access the value associated with that key. A key's value can be a number, a string, a list, or even another dictionary. In fact, you can use any object that you can create in Python as a value in a dictionary.

In Python, a dictionary is wrapped in braces ({}) with a series of key-value pairs inside the braces, as shown in the earlier example:

```
alien_0 = {'color': 'green', 'points': 5}
```

A *key-value pair* is a set of values associated with each other. When you provide a key, Python returns the value associated with that key. Every key is connected to its value by a colon, and individual key-value pairs are separated by commas. You can store as many key-value pairs as you want in a dictionary.

The simplest dictionary has exactly one key-value pair, as shown in this modified version of the `alien_0` dictionary:

```
alien_0 = {'color': 'green'}
```

This dictionary stores one piece of information about `alien_0`: the alien's color. The string `'color'` is a key in this dictionary, and its associated value is `'green'`.

Accessing Values in a Dictionary

To get the value associated with a key, give the name of the dictionary and then place the key inside a set of square brackets, as shown here:

alien.py

```
alien_0 = {'color': 'green'}
print(alien_0['color'])
```

This returns the value associated with the key 'color' from the dictionary alien_0:

```
green
```

You can have an unlimited number of key-value pairs in a dictionary. For example, here's the original alien_0 dictionary with two key-value pairs:

```
alien_0 = {'color': 'green', 'points': 5}
```

Now you can access either the color or the point value of alien_0. If a player shoots down this alien, you can look up how many points they should earn using code like this:

```
alien_0 = {'color': 'green', 'points': 5}

new_points = alien_0['points']
print(f"You just earned {new_points} points!")
```

Once the dictionary has been defined, we pull the value associated with the key 'points' from the dictionary. This value is then assigned to the variable new_points. The last line prints a statement about how many points the player just earned:

```
You just earned 5 points!
```

If you run this code every time an alien is shot down, the alien's point value will be retrieved.

Adding New Key-Value Pairs

Dictionaries are dynamic structures, and you can add new key-value pairs to a dictionary at any time. To add a new key-value pair, you would give the name of the dictionary followed by the new key in square brackets, along with the new value.

Let's add two new pieces of information to the alien_0 dictionary: the alien's *x*- and *y*-coordinates, which will help us display the alien at a particular position on the screen. Let's place the alien on the left edge of the screen, 25 pixels down from the top. Because screen coordinates usually start at the upper-left corner of the screen, we'll place the alien on the left edge of the screen by setting the *x*-coordinate to 0 and 25 pixels from the top by setting its *y*-coordinate to positive 25, as shown here:

alien.py

```
alien_0 = {'color': 'green', 'points': 5}
print(alien_0)

alien_0['x_position'] = 0
alien_0['y_position'] = 25
print(alien_0)
```

We start by defining the same dictionary that we've been working with. We then print this dictionary, displaying a snapshot of its information. Next, we add a new key-value pair to the dictionary: the key `'x_position'` and the value `0`. We do the same for the key `'y_position'`. When we print the modified dictionary, we see the two additional key-value pairs:

```
{'color': 'green', 'points': 5}
{'color': 'green', 'points': 5, 'x_position': 0, 'y_position': 25}
```

The final version of the dictionary contains four key-value pairs. The original two specify color and point value, and two more specify the alien's position.

Dictionaries retain the order in which they were defined. When you print a dictionary or loop through its elements, you will see the elements in the same order they were added to the dictionary.

Starting with an Empty Dictionary

It's sometimes convenient, or even necessary, to start with an empty dictionary and then add each new item to it. To start filling an empty dictionary, define a dictionary with an empty set of braces and then add each key-value pair on its own line. For example, here's how to build the `alien_0` dictionary using this approach:

alien.py

```
alien_0 = {}

alien_0['color'] = 'green'
alien_0['points'] = 5

print(alien_0)
```

We first define an empty `alien_0` dictionary, and then add color and point values to it. The result is the dictionary we've been using in previous examples:

```
{'color': 'green', 'points': 5}
```

Typically, you'll use empty dictionaries when storing user-supplied data in a dictionary or when writing code that generates a large number of key-value pairs automatically.

Modifying Values in a Dictionary

To modify a value in a dictionary, give the name of the dictionary with the key in square brackets and then the new value you want associated with that key. For example, consider an alien that changes from green to yellow as a game progresses:

alien.py

```
alien_0 = {'color': 'green'}
print(f"The alien is {alien_0['color']}.")
```

```
alien_0['color'] = 'yellow'
print(f"The alien is now {alien_0['color']}.")
```

We first define a dictionary for alien_0 that contains only the alien's color; then we change the value associated with the key 'color' to 'yellow'. The output shows that the alien has indeed changed from green to yellow:

```
The alien is green.
The alien is now yellow.
```

For a more interesting example, let's track the position of an alien that can move at different speeds. We'll store a value representing the alien's current speed and then use it to determine how far to the right the alien should move:

```
alien_0 = {'x_position': 0, 'y_position': 25, 'speed': 'medium'}
print(f"Original position: {alien_0['x_position']}")

# Move the alien to the right.
# Determine how far to move the alien based on its current speed.
❶ if alien_0['speed'] == 'slow':
    x_increment = 1
elif alien_0['speed'] == 'medium':
    x_increment = 2
else:
    # This must be a fast alien.
    x_increment = 3

# The new position is the old position plus the increment.
❷ alien_0['x_position'] = alien_0['x_position'] + x_increment

print(f"New position: {alien_0['x_position']}")
```

We start by defining an alien with an initial *x* position and *y* position, and a speed of 'medium'. We've omitted the color and point values for the sake of simplicity, but this example would work the same way if you included those key-value pairs as well. We also print the original value of x_position to see how far the alien moves to the right.

An if-elif-else chain determines how far the alien should move to the right, and assigns this value to the variable x_increment ❶. If the alien's speed is 'slow', it moves one unit to the right; if the speed is 'medium', it moves two units to the right; and if it's 'fast', it moves three units to the right. Once the increment has been calculated, it's added to the value of x_position ❷, and the result is stored in the dictionary's x_position.

Because this is a medium-speed alien, its position shifts two units to the right:

```
Original position: 0
New position: 2
```

This technique is pretty cool: by changing one value in the alien's dictionary, you can change the overall behavior of the alien. For example, to turn this medium-speed alien into a fast alien, you would add this line:

```
alien_0['speed'] = 'fast'
```

The `if-elif-else` block would then assign a larger value to `x_increment` the next time the code runs.

Removing Key-Value Pairs

When you no longer need a piece of information that's stored in a dictionary, you can use the `del` statement to completely remove a key-value pair. All `del` needs is the name of the dictionary and the key that you want to remove.

For example, let's remove the key `'points'` from the `alien_0` dictionary, along with its value:

alien.py

```
alien_0 = {'color': 'green', 'points': 5}
print(alien_0)

❶ del alien_0['points']
print(alien_0)
```

The `del` statement ❶ tells Python to delete the key `'points'` from the dictionary `alien_0` and to remove the value associated with that key as well. The output shows that the key `'points'` and its value of `5` are deleted from the dictionary, but the rest of the dictionary is unaffected:

```
{'color': 'green', 'points': 5}
{'color': 'green'}
```

NOTE *Be aware that the deleted key-value pair is removed permanently.*

A Dictionary of Similar Objects

The previous example involved storing different kinds of information about one object, an alien in a game. You can also use a dictionary to store one kind of information about many objects. For example, say you want to poll a number of people and ask them what their favorite programming language is. A dictionary is useful for storing the results of a simple poll, like this:

favorite_languages.py

```
favorite_languages = {
    'jen': 'python',
    'sarah': 'c',
    'edward': 'rust',
    'phil': 'python',
    }
```

As you can see, we've broken a larger dictionary into several lines. Each key is the name of a person who responded to the poll, and each value is their language choice. When you know you'll need more than one line to define a dictionary, press ENTER after the opening brace. Then indent the next line one level (four spaces) and write the first key-value pair, followed by a comma. From this point forward when you press ENTER, your text editor should automatically indent all subsequent key-value pairs to match the first key-value pair.

Once you've finished defining the dictionary, add a closing brace on a new line after the last key-value pair, and indent it one level so it aligns with the keys in the dictionary. It's good practice to include a comma after the last key-value pair as well, so you're ready to add a new key-value pair on the next line.

NOTE *Most editors have some functionality that helps you format extended lists and dictionaries in a similar manner to this example. Other acceptable ways to format long dictionaries are available as well, so you may see slightly different formatting in your editor, or in other sources.*

To use this dictionary, given the name of a person who took the poll, you can easily look up their favorite language:

favorite_languages.py

```
favorite_languages = {
    'jen': 'python',
    'sarah': 'c',
    'edward': 'rust',
    'phil': 'python',
    }

❶ language = favorite_languages['sarah'].title()
print(f"Sarah's favorite language is {language}.")
```

To see which language Sarah chose, we ask for the value at:

```
favorite_languages['sarah']
```

We use this syntax to pull Sarah's favorite language from the dictionary ❶ and assign it to the variable language. Creating a new variable here makes for a much cleaner print() call. The output shows Sarah's favorite language:

```
Sarah's favorite language is C.
```

You could use this same syntax with any individual represented in the dictionary.

Using get() to Access Values

Using keys in square brackets to retrieve the value you're interested in from a dictionary might cause one potential problem: if the key you ask for doesn't exist, you'll get an error.

Let's see what happens when you ask for the point value of an alien that doesn't have a point value set:

alien_no_points.py

```
alien_0 = {'color': 'green', 'speed': 'slow'}
print(alien_0['points'])
```

This results in a traceback, showing a `KeyError`:

```
Traceback (most recent call last):
  File "alien_no_points.py", line 2, in <module>
    print(alien_0['points'])
          ~~~~~~~^^^^^^^^^^
KeyError: 'points'
```

You'll learn more about how to handle errors like this in general in Chapter 10. For dictionaries specifically, you can use the `get()` method to set a default value that will be returned if the requested key doesn't exist.

The `get()` method requires a key as a first argument. As a second optional argument, you can pass the value to be returned if the key doesn't exist:

```
alien_0 = {'color': 'green', 'speed': 'slow'}

point_value = alien_0.get('points', 'No point value assigned.')
print(point_value)
```

If the key `'points'` exists in the dictionary, you'll get the corresponding value. If it doesn't, you get the default value. In this case, `points` doesn't exist, and we get a clean message instead of an error:

```
No point value assigned.
```

If there's a chance the key you're asking for might not exist, consider using the `get()` method instead of the square bracket notation.

NOTE *If you leave out the second argument in the call to `get()` and the key doesn't exist, Python will return the value `None`. The special value `None` means "no value exists." This is not an error: it's a special value meant to indicate the absence of a value. You'll see more uses for `None` in Chapter 8.*

TRY IT YOURSELF

6-1. Person: Use a dictionary to store information about a person you know. Store their first name, last name, age, and the city in which they live. You should have keys such as `first_name`, `last_name`, `age`, and `city`. Print each piece of information stored in your dictionary.

6-2. Favorite Numbers: Use a dictionary to store people's favorite numbers. Think of five names, and use them as keys in your dictionary. Think of a favorite number for each person, and store each as a value in your dictionary. Print each person's name and their favorite number. For even more fun, poll a few friends and get some actual data for your program.

6-3. Glossary: A Python dictionary can be used to model an actual dictionary. However, to avoid confusion, let's call it a glossary.

- Think of five programming words you've learned about in the previous chapters. Use these words as the keys in your glossary, and store their meanings as values.
- Print each word and its meaning as neatly formatted output. You might print the word followed by a colon and then its meaning, or print the word on one line and then print its meaning indented on a second line. Use the newline character (\n) to insert a blank line between each word-meaning pair in your output.

Looping Through a Dictionary

A single Python dictionary can contain just a few key-value pairs or millions of pairs. Because a dictionary can contain large amounts of data, Python lets you loop through a dictionary. Dictionaries can be used to store information in a variety of ways; therefore, several different ways exist to loop through them. You can loop through all of a dictionary's key-value pairs, through its keys, or through its values.

Looping Through All Key-Value Pairs

Before we explore the different approaches to looping, let's consider a new dictionary designed to store information about a user on a website. The following dictionary would store one person's username, first name, and last name:

user.py

```
user_0 = {
    'username': 'efermi',
    'first': 'enrico',
    'last': 'fermi',
    }
```

You can access any single piece of information about `user_0` based on what you've already learned in this chapter. But what if you wanted to see everything stored in this user's dictionary? To do so, you could loop through the dictionary using a `for` loop:

```
user_0 = {
    'username': 'efermi',
    'first': 'enrico',
    'last': 'fermi',
    }
```

```
for key, value in user_0.items():
    print(f"\nKey: {key}")
    print(f"Value: {value}")
```

To write a for loop for a dictionary, you create names for the two variables that will hold the key and value in each key-value pair. You can choose any names you want for these two variables. This code would work just as well if you had used abbreviations for the variable names, like this:

```
for k, v in user_0.items()
```

The second half of the for statement includes the name of the dictionary followed by the method items(), which returns a sequence of key-value pairs. The for loop then assigns each of these pairs to the two variables provided. In the preceding example, we use the variables to print each key, followed by the associated value. The "\n" in the first print() call ensures that a blank line is inserted before each key-value pair in the output:

```
Key: username
Value: efermi

Key: first
Value: enrico

Key: last
Value: fermi
```

Looping through all key-value pairs works particularly well for dictionaries like the *favorite_languages.py* example on page 96, which stores the same kind of information for many different keys. If you loop through the favorite_languages dictionary, you get the name of each person in the dictionary and their favorite programming language. Because the keys always refer to a person's name and the value is always a language, we'll use the variables name and language in the loop instead of key and value. This will make it easier to follow what's happening inside the loop:

favorite_languages.py

```
favorite_languages = {
    'jen': 'python',
    'sarah': 'c',
    'edward': 'rust',
    'phil': 'python',
    }

for name, language in favorite_languages.items():
    print(f"{name.title()}'s favorite language is {language.title()}.")
```

This code tells Python to loop through each key-value pair in the dictionary. As it works through each pair the key is assigned to the variable name, and the value is assigned to the variable language. These descriptive names make it much easier to see what the print() call is doing.

Now, in just a few lines of code, we can display all of the information from the poll:

```
Jen's favorite language is Python.
Sarah's favorite language is C.
Edward's favorite language is Rust.
Phil's favorite language is Python.
```

This type of looping would work just as well if our dictionary stored the results from polling a thousand or even a million people.

Looping Through All the Keys in a Dictionary

The `keys()` method is useful when you don't need to work with all of the values in a dictionary. Let's loop through the `favorite_languages` dictionary and print the names of everyone who took the poll:

```
favorite_languages = {
    'jen': 'python',
    'sarah': 'c',
    'edward': 'rust',
    'phil': 'python',
    }

for name in favorite_languages.keys():
    print(name.title())
```

This `for` loop tells Python to pull all the keys from the dictionary `favorite_languages` and assign them one at a time to the variable `name`. The output shows the names of everyone who took the poll:

```
Jen
Sarah
Edward
Phil
```

Looping through the keys is actually the default behavior when looping through a dictionary, so this code would have exactly the same output if you wrote:

```
for name in favorite_languages:
```

rather than:

```
for name in favorite_languages.keys():
```

You can choose to use the `keys()` method explicitly if it makes your code easier to read, or you can omit it if you wish.

You can access the value associated with any key you care about inside the loop, by using the current key. Let's print a message to a couple of friends about the languages they chose. We'll loop through the names in the dictionary as we did previously, but when the name matches one of our friends, we'll display a message about their favorite language:

```
favorite_languages = {
    --snip--
    }

friends = ['phil', 'sarah']
for name in favorite_languages.keys():
    print(f"Hi {name.title()}.")

❶   if name in friends:
❷       language = favorite_languages[name].title()
        print(f"\t{name.title()}, I see you love {language}!")
```

First, we make a list of friends that we want to print a message to. Inside the loop, we print each person's name. Then we check whether the `name` we're working with is in the list `friends` ❶. If it is, we determine the person's favorite language using the name of the dictionary and the current value of `name` as the key ❷. We then print a special greeting, including a reference to their language of choice.

Everyone's name is printed, but our friends receive a special message:

```
Hi Jen.
Hi Sarah.
    Sarah, I see you love C!
Hi Edward.
Hi Phil.
    Phil, I see you love Python!
```

You can also use the `keys()` method to find out if a particular person was polled. This time, let's find out if Erin took the poll:

```
favorite_languages = {
    --snip--
    }

if 'erin' not in favorite_languages.keys():
    print("Erin, please take our poll!")
```

The `keys()` method isn't just for looping: it actually returns a sequence of all the keys, and the `if` statement simply checks if `'erin'` is in this sequence. Because she's not, a message is printed inviting her to take the poll:

```
Erin, please take our poll!
```

Looping Through a Dictionary's Keys in a Particular Order

Looping through a dictionary returns the items in the same order they were inserted. Sometimes, though, you'll want to loop through a dictionary in a different order.

One way to do this is to sort the keys as they're returned in the `for` loop. You can use the `sorted()` function to get a copy of the keys in order:

```
favorite_languages = {
    'jen': 'python',
    'sarah': 'c',
    'edward': 'rust',
    'phil': 'python',
    }

for name in sorted(favorite_languages.keys()):
    print(f"{name.title()}, thank you for taking the poll.")
```

This `for` statement is like other `for` statements, except that we've wrapped the `sorted()` function around the `dictionary.keys()` method. This tells Python to get all the keys in the dictionary and sort them before starting the loop. The output shows everyone who took the poll, with the names displayed in order:

```
Edward, thank you for taking the poll.
Jen, thank you for taking the poll.
Phil, thank you for taking the poll.
Sarah, thank you for taking the poll.
```

Looping Through All Values in a Dictionary

If you are primarily interested in the values that a dictionary contains, you can use the `values()` method to return a sequence of values without any keys. For example, say we simply want a list of all languages chosen in our programming language poll, without the name of the person who chose each language:

```
favorite_languages = {
    'jen': 'python',
    'sarah': 'c',
    'edward': 'rust',
    'phil': 'python',
    }

print("The following languages have been mentioned:")
for language in favorite_languages.values():
    print(language.title())
```

The `for` statement here pulls each value from the dictionary and assigns it to the variable `language`. When these values are printed, we get a list of all chosen languages:

```
The following languages have been mentioned:
Python
C
Rust
Python
```

This approach pulls all the values from the dictionary without checking for repeats. This might work fine with a small number of values, but in a poll with a large number of respondents, it would result in a very repetitive list. To see each language chosen without repetition, we can use a set. A *set* is a collection in which each item must be unique:

```
favorite_languages = {
    --snip--
    }

print("The following languages have been mentioned:")
for language in set(favorite_languages.values()):
    print(language.title())
```

When you wrap `set()` around a collection of values that contains duplicate items, Python identifies the unique items in the collection and builds a set from those items. Here we use `set()` to pull out the unique languages in `favorite_languages.values()`.

The result is a nonrepetitive list of languages that have been mentioned by people taking the poll:

```
The following languages have been mentioned:
Python
C
Rust
```

As you continue learning about Python, you'll often find a built-in feature of the language that helps you do exactly what you want with your data.

NOTE *You can build a set directly using braces and separating the elements with commas:*

```
>>> languages = {'python', 'rust', 'python', 'c'}
>>> languages
{'rust', 'python', 'c'}
```

It's easy to mistake sets for dictionaries because they're both wrapped in braces. When you see braces but no key-value pairs, you're probably looking at a set. Unlike lists and dictionaries, sets do not retain items in any specific order.

TRY IT YOURSELF

6-4. Glossary 2: Now that you know how to loop through a dictionary, clean up the code from Exercise 6-3 (page 99) by replacing your series of `print()` calls with a loop that runs through the dictionary's keys and values. When you're sure that your loop works, add five more Python terms to your glossary. When you run your program again, these new words and meanings should automatically be included in the output.

6-5. Rivers: Make a dictionary containing three major rivers and the country each river runs through. One key-value pair might be `'nile': 'egypt'`.

- Use a loop to print a sentence about each river, such as *The Nile runs through Egypt.*
- Use a loop to print the name of each river included in the dictionary.
- Use a loop to print the name of each country included in the dictionary.

6-6. Polling: Use the code in *favorite_languages.py* (page 96).

- Make a list of people who should take the favorite languages poll. Include some names that are already in the dictionary and some that are not.
- Loop through the list of people who should take the poll. If they have already taken the poll, print a message thanking them for responding. If they have not yet taken the poll, print a message inviting them to take the poll.

Nesting

Sometimes you'll want to store multiple dictionaries in a list, or a list of items as a value in a dictionary. This is called *nesting*. You can nest dictionaries inside a list, a list of items inside a dictionary, or even a dictionary inside another dictionary. Nesting is a powerful feature, as the following examples will demonstrate.

A List of Dictionaries

The `alien_0` dictionary contains a variety of information about one alien, but it has no room to store information about a second alien, much less a screen full of aliens. How can you manage a fleet of aliens? One way is to make a list of aliens in which each alien is a dictionary of information about that alien. For example, the following code builds a list of three aliens:

aliens.py

```
alien_0 = {'color': 'green', 'points': 5}
alien_1 = {'color': 'yellow', 'points': 10}
alien_2 = {'color': 'red', 'points': 15}

❶ aliens = [alien_0, alien_1, alien_2]

for alien in aliens:
    print(alien)
```

We first create three dictionaries, each representing a different alien. We store each of these dictionaries in a list called `aliens` ❶. Finally, we loop through the list and print out each alien:

```
{'color': 'green', 'points': 5}
{'color': 'yellow', 'points': 10}
{'color': 'red', 'points': 15}
```

A more realistic example would involve more than three aliens with code that automatically generates each alien. In the following example, we use `range()` to create a fleet of 30 aliens:

```
# Make an empty list for storing aliens.
aliens = []

# Make 30 green aliens.
❶ for alien_number in range(30):
❷     new_alien = {'color': 'green', 'points': 5, 'speed': 'slow'}
❸     aliens.append(new_alien)

# Show the first 5 aliens.
❹ for alien in aliens[:5]:
    print(alien)
print("...")

# Show how many aliens have been created.
print(f"Total number of aliens: {len(aliens)}")
```

This example begins with an empty list to hold all of the aliens that will be created. The `range()` function ❶ returns a series of numbers, which just tells Python how many times we want the loop to repeat. Each time the loop runs, we create a new alien ❷ and then append each new alien to the list `aliens` ❸. We use a slice to print the first five aliens ❹, and finally, we print the length of the list to prove we've actually generated the full fleet of 30 aliens:

```
{'color': 'green', 'points': 5, 'speed': 'slow'}
{'color': 'green', 'points': 5, 'speed': 'slow'}
{'color': 'green', 'points': 5, 'speed': 'slow'}
{'color': 'green', 'points': 5, 'speed': 'slow'}
{'color': 'green', 'points': 5, 'speed': 'slow'}
...

Total number of aliens: 30
```

These aliens all have the same characteristics, but Python considers each one a separate object, which allows us to modify each alien individually.

How might you work with a group of aliens like this? Imagine that one aspect of a game has some aliens changing color and moving faster as the game progresses. When it's time to change colors, we can use a `for` loop and an `if` statement to change the color of the aliens. For example, to change the first three aliens to yellow, medium-speed aliens worth 10 points each, we could do this:

```
# Make an empty list for storing aliens.
aliens = []

# Make 30 green aliens.
for alien_number in range (30):
    new_alien = {'color': 'green', 'points': 5, 'speed': 'slow'}
    aliens.append(new_alien)

for alien in aliens[:3]:
    if alien['color'] == 'green':
        alien['color'] = 'yellow'
        alien['speed'] = 'medium'
        alien['points'] = 10

# Show the first 5 aliens.
for alien in aliens[:5]:
    print(alien)
print("...")
```

Because we want to modify the first three aliens, we loop through a slice that includes only the first three aliens. All of the aliens are green now, but that won't always be the case, so we write an `if` statement to make sure we're only modifying green aliens. If the alien is green, we change the color to `'yellow'`, the speed to `'medium'`, and the point value to `10`, as shown in the following output:

```
{'color': 'yellow', 'points': 10, 'speed': 'medium'}
{'color': 'yellow', 'points': 10, 'speed': 'medium'}
{'color': 'yellow', 'points': 10, 'speed': 'medium'}
{'color': 'green', 'points': 5, 'speed': 'slow'}
{'color': 'green', 'points': 5, 'speed': 'slow'}
...
```

You could expand this loop by adding an `elif` block that turns yellow aliens into red, fast-moving ones worth 15 points each. Without showing the entire program again, that loop would look like this:

```
for alien in aliens[0:3]:
    if alien['color'] == 'green':
        alien['color'] = 'yellow'
        alien['speed'] = 'medium'
        alien['points'] = 10
    elif alien['color'] == 'yellow':
        alien['color'] = 'red'
        alien['speed'] = 'fast'
        alien['points'] = 15
```

It's common to store a number of dictionaries in a list when each dictionary contains many kinds of information about one object. For example, you might create a dictionary for each user on a website, as we did in *user.py* on page 99, and store the individual dictionaries in a list called users. All of the dictionaries in the list should have an identical structure, so you can loop through the list and work with each dictionary object in the same way.

A List in a Dictionary

Rather than putting a dictionary inside a list, it's sometimes useful to put a list inside a dictionary. For example, consider how you might describe a pizza that someone is ordering. If you were to use only a list, all you could really store is a list of the pizza's toppings. With a dictionary, a list of toppings can be just one aspect of the pizza you're describing.

In the following example, two kinds of information are stored for each pizza: a type of crust and a list of toppings. The list of toppings is a value associated with the key 'toppings'. To use the items in the list, we give the name of the dictionary and the key 'toppings', as we would any value in the dictionary. Instead of returning a single value, we get a list of toppings:

pizza.py

```
# Store information about a pizza being ordered.
pizza = {
    'crust': 'thick',
    'toppings': ['mushrooms', 'extra cheese'],
    }

# Summarize the order.
❶ print(f"You ordered a {pizza['crust']}-crust pizza "
    "with the following toppings:")

❷ for topping in pizza['toppings']:
    print(f"\t{topping}")
```

We begin with a dictionary that holds information about a pizza that has been ordered. One key in the dictionary is 'crust', and the associated value is the string 'thick'. The next key, 'toppings', has a list as its value that stores all requested toppings. We summarize the order before building the pizza ❶. When you need to break up a long line in a print() call, choose an appropriate point at which to break the line being printed, and end the line with a quotation mark. Indent the next line, add an opening quotation mark, and continue the string. Python will automatically combine all of the strings it finds inside the parentheses. To print the toppings, we write a for loop ❷. To access the list of toppings, we use the key 'toppings', and Python grabs the list of toppings from the dictionary.

The following output summarizes the pizza that we plan to build:

```
You ordered a thick-crust pizza with the following toppings:
    mushrooms
    extra cheese
```

You can nest a list inside a dictionary anytime you want more than one value to be associated with a single key in a dictionary. In the earlier example of favorite programming languages, if we were to store each person's responses in a list, people could choose more than one favorite language.

When we loop through the dictionary, the value associated with each person would be a list of languages rather than a single language. Inside the dictionary's for loop, we use another for loop to run through the list of languages associated with each person:

favorite_languages.py

```
favorite_languages = {
    'jen': ['python', 'rust'],
    'sarah': ['c'],
    'edward': ['rust', 'go'],
    'phil': ['python', 'haskell'],
    }

❶ for name, languages in favorite_languages.items():
    print(f"\n{name.title()}'s favorite languages are:")
❷     for language in languages:
        print(f"\t{language.title()}")
```

The value associated with each name in favorite_languages is now a list. Note that some people have one favorite language and others have multiple favorites. When we loop through the dictionary ❶, we use the variable name languages to hold each value from the dictionary, because we know that each value will be a list. Inside the main dictionary loop, we use another for loop ❷ to run through each person's list of favorite languages. Now each person can list as many favorite languages as they like:

```
Jen's favorite languages are:
    Python
    Rust

Sarah's favorite languages are:
    C

Edward's favorite languages are:
    Rust
    Go

Phil's favorite languages are:
    Python
    Haskell
```

To refine this program even further, you could include an if statement at the beginning of the dictionary's for loop to see whether each person has more than one favorite language by examining the value of len(languages). If a person has more than one favorite, the output would stay the same. If the person has only one favorite language, you could change the wording to reflect that. For example, you could say, "Sarah's favorite language is C."

NOTE *You should not nest lists and dictionaries too deeply. If you're nesting items much deeper than what you see in the preceding examples, or if you're working with someone else's code with significant levels of nesting, there's most likely a simpler way to solve the problem.*

A Dictionary in a Dictionary

You can nest a dictionary inside another dictionary, but your code can get complicated quickly when you do. For example, if you have several users for a website, each with a unique username, you can use the usernames as the keys in a dictionary. You can then store information about each user by using a dictionary as the value associated with their username. In the following listing, we store three pieces of information about each user: their first name, last name, and location. We'll access this information by looping through the usernames and the dictionary of information associated with each username:

many_users.py

```
users = {
    'aeinstein': {
        'first': 'albert',
        'last': 'einstein',
        'location': 'princeton',
        },

    'mcurie': {
        'first': 'marie',
        'last': 'curie',
        'location': 'paris',
        },

    }

❶ for username, user_info in users.items():
❷     print(f"\nUsername: {username}")
❸     full_name = f"{user_info['first']} {user_info['last']}"
      location = user_info['location']

❹     print(f"\tFull name: {full_name.title()}")
      print(f"\tLocation: {location.title()}")
```

We first define a dictionary called `users` with two keys: one each for the usernames `'aeinstein'` and `'mcurie'`. The value associated with each key is a dictionary that includes each user's first name, last name, and location. Then, we loop through the `users` dictionary ❶. Python assigns each key to the variable `username`, and the dictionary associated with each username is assigned to the variable `user_info`. Once inside the main dictionary loop, we print the username ❷.

Then, we start accessing the inner dictionary ❸. The variable `user_info`, which contains the dictionary of user information, has three keys: `'first'`, `'last'`, and `'location'`. We use each key to generate a neatly formatted full name and location for each person, and then print a summary of what we know about each user ❹:

```
Username: aeinstein
    Full name: Albert Einstein
    Location: Princeton
```

```
Username: mcurie
    Full name: Marie Curie
    Location: Paris
```

Notice that the structure of each user's dictionary is identical. Although not required by Python, this structure makes nested dictionaries easier to work with. If each user's dictionary had different keys, the code inside the `for` loop would be more complicated.

TRY IT YOURSELF

6-7. People: Start with the program you wrote for Exercise 6-1 (page 98). Make two new dictionaries representing different people, and store all three dictionaries in a list called `people`. Loop through your list of people. As you loop through the list, print everything you know about each person.

6-8. Pets: Make several dictionaries, where each dictionary represents a different pet. In each dictionary, include the kind of animal and the owner's name. Store these dictionaries in a list called `pets`. Next, loop through your list and as you do, print everything you know about each pet.

6-9. Favorite Places: Make a dictionary called `favorite_places`. Think of three names to use as keys in the dictionary, and store one to three favorite places for each person. To make this exercise a bit more interesting, ask some friends to name a few of their favorite places. Loop through the dictionary, and print each person's name and their favorite places.

6-10. Favorite Numbers: Modify your program from Exercise 6-2 (page 98) so each person can have more than one favorite number. Then print each person's name along with their favorite numbers.

6-11. Cities: Make a dictionary called `cities`. Use the names of three cities as keys in your dictionary. Create a dictionary of information about each city and include the country that the city is in, its approximate population, and one fact about that city. The keys for each city's dictionary should be something like `country`, `population`, and `fact`. Print the name of each city and all of the information you have stored about it.

6-12. Extensions: We're now working with examples that are complex enough that they can be extended in any number of ways. Use one of the example programs from this chapter, and extend it by adding new keys and values, changing the context of the program, or improving the formatting of the output.

Summary

In this chapter, you learned how to define a dictionary and how to work with the information stored in a dictionary. You learned how to access and modify individual elements in a dictionary, and how to loop through all of the information in a dictionary. You learned to loop through a dictionary's key-value pairs, its keys, and its values. You also learned how to nest multiple dictionaries in a list, nest lists in a dictionary, and nest a dictionary inside a dictionary.

In the next chapter you'll learn about `while` loops and how to accept input from people who are using your programs. This will be an exciting chapter, because you'll learn to make all of your programs interactive: they'll be able to respond to user input.

7

USER INPUT AND WHILE LOOPS

Most programs are written to solve an end user's problem. To do so, you usually need to get some information from the user. For example, say someone wants to find out whether they're old enough to vote. If you write a program to answer this question, you need to know the user's age before you can provide an answer. The program will need to ask the user to enter, or *input*, their age; once the program has this input, it can compare it to the voting age to determine if the user is old enough and then report the result.

In this chapter you'll learn how to accept user input so your program can then work with it. When your program needs a name, you'll be able to prompt the user for a name. When your program needs a list of names, you'll be able to prompt the user for a series of names. To do this, you'll use the `input()` function.

You'll also learn how to keep programs running as long as users want them to, so they can enter as much information as they need to; then, your program can work with that information. You'll use Python's `while` loop to keep programs running as long as certain conditions remain true.

With the ability to work with user input and the ability to control how long your programs run, you'll be able to write fully interactive programs.

How the input() Function Works

The `input()` function pauses your program and waits for the user to enter some text. Once Python receives the user's input, it assigns that input to a variable to make it convenient for you to work with.

For example, the following program asks the user to enter some text, then displays that message back to the user:

parrot.py

```
message = input("Tell me something, and I will repeat it back to you: ")
print(message)
```

The `input()` function takes one argument: the *prompt* that we want to display to the user, so they know what kind of information to enter. In this example, when Python runs the first line, the user sees the prompt `Tell me something, and I will repeat it back to you: `. The program waits while the user enters their response and continues after the user presses ENTER. The response is assigned to the variable `message`, then `print(message)` displays the input back to the user:

```
Tell me something, and I will repeat it back to you: Hello everyone!
Hello everyone!
```

NOTE *Some text editors won't run programs that prompt the user for input. You can use these editors to write programs that prompt for input, but you'll need to run these programs from a terminal.* *See "Running Python Programs from a Terminal" on page 11.*

Writing Clear Prompts

Each time you use the `input()` function, you should include a clear, easy-to-follow prompt that tells the user exactly what kind of information you're looking for. Any statement that tells the user what to enter should work. For example:

greeter.py

```
name = input("Please enter your name: ")
print(f"\nHello, {name}!")
```

Add a space at the end of your prompts (after the colon in the preceding example) to separate the prompt from the user's response and to make it clear to your user where to enter their text. For example:

```
Please enter your name: Eric
Hello, Eric!
```

Sometimes you'll want to write a prompt that's longer than one line. For example, you might want to tell the user why you're asking for certain input. You can assign your prompt to a variable and pass that variable to the input() function. This allows you to build your prompt over several lines, then write a clean input() statement.

greeter.py

```
prompt = "If you share your name, we can personalize the messages you see."
prompt += "\nWhat is your first name? "

name = input(prompt)
print(f"\nHello, {name}!")
```

This example shows one way to build a multiline string. The first line assigns the first part of the message to the variable prompt. In the second line, the operator += takes the string that was assigned to prompt and adds the new string onto the end.

The prompt now spans two lines, again with space after the question mark for clarity:

```
If you share your name, we can personalize the messages you see.
What is your first name? Eric

Hello, Eric!
```

Using int() to Accept Numerical Input

When you use the input() function, Python interprets everything the user enters as a string. Consider the following interpreter session, which asks for the user's age:

```
>>> age = input("How old are you? ")
How old are you? 21
>>> age
'21'
```

The user enters the number 21, but when we ask Python for the value of age, it returns '21', the string representation of the numerical value entered. We know Python interpreted the input as a string because the number is now enclosed in quotes. If all you want to do is print the input, this works well. But if you try to use the input as a number, you'll get an error:

```
  >>> age = input("How old are you? ")
  How old are you? 21
❶ >>> age >= 18
  Traceback (most recent call last):
    File "<stdin>", line 1, in <module>
❷ TypeError: '>=' not supported between instances of 'str' and 'int'
```

When you try to use the input to do a numerical comparison ❶, Python produces an error because it can't compare a string to an integer: the string '21' that's assigned to age can't be compared to the numerical value 18 ❷.

We can resolve this issue by using the int() function, which converts the input string to a numerical value. This allows the comparison to run successfully:

```
>>> age = input("How old are you? ")
How old are you? 21
❶ >>> age = int(age)
>>> age >= 18
True
```

In this example, when we enter 21 at the prompt, Python interprets the number as a string, but the value is then converted to a numerical representation by int() ❶. Now Python can run the conditional test: it compares age (which now represents the numerical value 21) and 18 to see if age is greater than or equal to 18. This test evaluates to True.

How do you use the int() function in an actual program? Consider a program that determines whether people are tall enough to ride a roller coaster:

rollercoaster.py

```
height = input("How tall are you, in inches? ")
height = int(height)

if height >= 48:
    print("\nYou're tall enough to ride!")
else:
    print("\nYou'll be able to ride when you're a little older.")
```

The program can compare height to 48 because height = int(height) converts the input value to a numerical representation before the comparison is made. If the number entered is greater than or equal to 48, we tell the user that they're tall enough:

```
How tall are you, in inches? 71

You're tall enough to ride!
```

When you use numerical input to do calculations and comparisons, be sure to convert the input value to a numerical representation first.

The Modulo Operator

A useful tool for working with numerical information is the *modulo operator* (%), which divides one number by another number and returns the remainder:

```
>>> 4 % 3
1
>>> 5 % 3
2
>>> 6 % 3
0
>>> 7 % 3
1
```

The modulo operator doesn't tell you how many times one number fits into another; it only tells you what the remainder is.

When one number is divisible by another number, the remainder is 0, so the modulo operator always returns 0. You can use this fact to determine if a number is even or odd:

even_or_odd.py

```
number = input("Enter a number, and I'll tell you if it's even or odd: ")
number = int(number)

if number % 2 == 0:
    print(f"\nThe number {number} is even.")
else:
    print(f"\nThe number {number} is odd.")
```

Even numbers are always divisible by two, so if the modulo of a number and two is zero (here, if `number % 2 == 0`) the number is even. Otherwise, it's odd.

```
Enter a number, and I'll tell you if it's even or odd: 42

The number 42 is even.
```

TRY IT YOURSELF

7-1. Rental Car: Write a program that asks the user what kind of rental car they would like. Print a message about that car, such as "Let me see if I can find you a Subaru."

7-2. Restaurant Seating: Write a program that asks the user how many people are in their dinner group. If the answer is more than eight, print a message saying they'll have to wait for a table. Otherwise, report that their table is ready.

7-3. Multiples of Ten: Ask the user for a number, and then report whether the number is a multiple of 10 or not.

Introducing while Loops

The `for` loop takes a collection of items and executes a block of code once for each item in the collection. In contrast, the `while` loop runs as long as, or *while*, a certain condition is true.

The while Loop in Action

You can use a `while` loop to count up through a series of numbers. For example, the following `while` loop counts from 1 to 5:

counting.py

```
current_number = 1
while current_number <= 5:
```

```
    print(current_number)
    current_number += 1
```

In the first line, we start counting from 1 by assigning current_number the value 1. The while loop is then set to keep running as long as the value of current_number is less than or equal to 5. The code inside the loop prints the value of current_number and then adds 1 to that value with current_number += 1. (The += operator is shorthand for current_number = current_number + 1.)

Python repeats the loop as long as the condition current_number <= 5 is true. Because 1 is less than 5, Python prints 1 and then adds 1, making the current number 2. Because 2 is less than 5, Python prints 2 and adds 1 again, making the current number 3, and so on. Once the value of current_number is greater than 5, the loop stops running and the program ends:

```
1
2
3
4
5
```

The programs you use every day most likely contain while loops. For example, a game needs a while loop to keep running as long as you want to keep playing, and so it can stop running as soon as you ask it to quit. Programs wouldn't be fun to use if they stopped running before we told them to or kept running even after we wanted to quit, so while loops are quite useful.

Letting the User Choose When to Quit

We can make the *parrot.py* program run as long as the user wants by putting most of the program inside a while loop. We'll define a *quit value* and then keep the program running as long as the user has not entered the quit value:

parrot.py

```
prompt = "\nTell me something, and I will repeat it back to you:"
prompt += "\nEnter 'quit' to end the program. "

message = ""
while message != 'quit':
    message = input(prompt)
    print(message)
```

We first define a prompt that tells the user their two options: entering a message or entering the quit value (in this case, 'quit'). Then we set up a variable message to keep track of whatever value the user enters. We define message as an empty string, "", so Python has something to check the first time it reaches the while line. The first time the program runs and Python reaches the while statement, it needs to compare the value of message to 'quit', but no user input has been entered yet. If Python has nothing to compare, it won't be able to continue running the program. To solve this problem, we make sure to give `message` an initial value. Although it's just an empty string, it will make sense to Python and allow it to perform the comparison that makes the `while` loop work. This `while` loop runs as long as the value of `message` is not `'quit'`.

The first time through the loop, `message` is just an empty string, so Python enters the loop. At `message = input(prompt)`, Python displays the prompt and waits for the user to enter their input. Whatever they enter is assigned to `message` and printed; then, Python reevaluates the condition in the `while` statement. As long as the user has not entered the word `'quit'`, the prompt is displayed again and Python waits for more input. When the user finally enters `'quit'`, Python stops executing the `while` loop and the program ends:

```
Tell me something, and I will repeat it back to you:
Enter 'quit' to end the program. Hello everyone!
Hello everyone!

Tell me something, and I will repeat it back to you:
Enter 'quit' to end the program. Hello again.
Hello again.

Tell me something, and I will repeat it back to you:
Enter 'quit' to end the program. quit
quit
```

This program works well, except that it prints the word `'quit'` as if it were an actual message. A simple `if` test fixes this:

```
prompt = "\nTell me something, and I will repeat it back to you:"
prompt += "\nEnter 'quit' to end the program. "

message = ""
while message != 'quit':
    message = input(prompt)

    if message != 'quit':
        print(message)
```

Now the program makes a quick check before displaying the message and only prints the message if it does not match the quit value:

```
Tell me something, and I will repeat it back to you:
Enter 'quit' to end the program. Hello everyone!
Hello everyone!

Tell me something, and I will repeat it back to you:
Enter 'quit' to end the program. Hello again.
Hello again.

Tell me something, and I will repeat it back to you:
Enter 'quit' to end the program. quit
```

Using a Flag

In the previous example, we had the program perform certain tasks while a given condition was true. But what about more complicated programs in which many different events could cause the program to stop running?

For example, in a game, several different events can end the game. When the player runs out of ships, their time runs out, or the cities they were supposed to protect are all destroyed, the game should end. It needs to end if any one of these events happens. If many possible events might occur to stop the program, trying to test all these conditions in one `while` statement becomes complicated and difficult.

For a program that should run only as long as many conditions are true, you can define one variable that determines whether or not the entire program is active. This variable, called a *flag*, acts as a signal to the program. We can write our programs so they run while the flag is set to `True` and stop running when any of several events sets the value of the flag to `False`. As a result, our overall `while` statement needs to check only one condition: whether the flag is currently `True`. Then, all our other tests (to see if an event has occurred that should set the flag to `False`) can be neatly organized in the rest of the program.

Let's add a flag to *parrot.py* from the previous section. This flag, which we'll call `active` (though you can call it anything), will monitor whether or not the program should continue running:

```
prompt = "\nTell me something, and I will repeat it back to you:"
prompt += "\nEnter 'quit' to end the program. "

active = True
❶ while active:
    message = input(prompt)

    if message == 'quit':
        active = False
    else:
        print(message)
```

We set the variable `active` to `True` so the program starts in an active state. Doing so makes the `while` statement simpler because no comparison is made in the `while` statement itself; the logic is taken care of in other parts of the program. As long as the `active` variable remains `True`, the loop will continue running ❶.

In the `if` statement inside the `while` loop, we check the value of `message` once the user enters their input. If the user enters `'quit'`, we set `active` to `False`, and the `while` loop stops. If the user enters anything other than `'quit'`, we print their input as a message.

This program has the same output as the previous example where we placed the conditional test directly in the `while` statement. But now that we have a flag to indicate whether the overall program is in an active state, it would be easy to add more tests (such as `elif` statements) for events that should cause `active` to become `False`. This is useful in complicated programs like games, in which there may be many events that should each make the program stop running. When any of these events causes the active flag to become `False`, the main game loop will exit, a *Game Over* message can be displayed, and the player can be given the option to play again.

Using break to Exit a Loop

To exit a `while` loop immediately without running any remaining code in the loop, regardless of the results of any conditional test, use the `break` statement. The `break` statement directs the flow of your program; you can use it to control which lines of code are executed and which aren't, so the program only executes code that you want it to, when you want it to.

For example, consider a program that asks the user about places they've visited. We can stop the `while` loop in this program by calling `break` as soon as the user enters the `'quit'` value:

cities.py

```
prompt = "\nPlease enter the name of a city you have visited:"
prompt += "\n(Enter 'quit' when you are finished.) "

❶ while True:
    city = input(prompt)

    if city == 'quit':
        break
    else:
        print(f"I'd love to go to {city.title()}!")
```

A loop that starts with `while True` ❶ will run forever unless it reaches a `break` statement. The loop in this program continues asking the user to enter the names of cities they've been to until they enter `'quit'`. When they enter `'quit'`, the `break` statement runs, causing Python to exit the loop:

```
Please enter the name of a city you have visited:
(Enter 'quit' when you are finished.) New York
I'd love to go to New York!

Please enter the name of a city you have visited:
(Enter 'quit' when you are finished.) San Francisco
I'd love to go to San Francisco!

Please enter the name of a city you have visited:
(Enter 'quit' when you are finished.) quit
```

NOTE *You can use the `break` statement in any of Python's loops. For example, you could use `break` to quit a `for` loop that's working through a list or a dictionary.*

Using continue in a Loop

Rather than breaking out of a loop entirely without executing the rest of its code, you can use the `continue` statement to return to the beginning of the loop, based on the result of a conditional test. For example, consider a loop that counts from 1 to 10 but prints only the odd numbers in that range:

counting.py

```
current_number = 0
while current_number < 10:
❶    current_number += 1
    if current_number % 2 == 0:
        continue

    print(current_number)
```

First, we set `current_number` to 0. Because it's less than 10, Python enters the `while` loop. Once inside the loop, we increment the count by 1 ❶, so `current_number` is 1. The `if` statement then checks the modulo of `current_number` and 2. If the modulo is 0 (which means `current_number` is divisible by 2), the `continue` statement tells Python to ignore the rest of the loop and return to the beginning. If the current number is not divisible by 2, the rest of the loop is executed and Python prints the current number:

```
1
3
5
7
9
```

Avoiding Infinite Loops

Every `while` loop needs a way to stop running so it won't continue to run forever. For example, this counting loop should count from 1 to 5:

counting.py

```
x = 1
while x <= 5:
    print(x)
    x += 1
```

However, if you accidentally omit the line `x += 1`, the loop will run forever:

```
# This loop runs forever!
x = 1
while x <= 5:
    print(x)
```

Now the value of `x` will start at `1` but never change. As a result, the conditional test `x <= 5` will always evaluate to `True` and the `while` loop will run forever, printing a series of 1s, like this:

```
1
1
1
1
--snip--
```

Every programmer accidentally writes an infinite `while` loop from time to time, especially when a program's loops have subtle exit conditions. If your program gets stuck in an infinite loop, press CTRL-C or just close the terminal window displaying your program's output.

To avoid writing infinite loops, test every `while` loop and make sure the loop stops when you expect it to. If you want your program to end when the user enters a certain input value, run the program and enter that value. If the program doesn't end, scrutinize the way your program handles the value that should cause the loop to exit. Make sure at least one part of the program can make the loop's condition `False` or cause it to reach a `break` statement.

NOTE *VS Code, like many editors, displays output in an embedded terminal window. To cancel an infinite loop, make sure you click in the output area of the editor before pressing CTRL-C.*

TRY IT YOURSELF

7-4. Pizza Toppings: Write a loop that prompts the user to enter a series of pizza toppings until they enter a `'quit'` value. As they enter each topping, print a message saying you'll add that topping to their pizza.

7-5. Movie Tickets: A movie theater charges different ticket prices depending on a person's age. If a person is under the age of 3, the ticket is free; if they are between 3 and 12, the ticket is $10; and if they are over age 12, the ticket is $15. Write a loop in which you ask users their age, and then tell them the cost of their movie ticket.

7-6. Three Exits: Write different versions of either Exercise 7-4 or 7-5 that do each of the following at least once:

- Use a conditional test in the `while` statement to stop the loop.
- Use an `active` variable to control how long the loop runs.
- Use a `break` statement to exit the loop when the user enters a `'quit'` value.

7-7. Infinity: Write a loop that never ends, and run it. (To end the loop, press CTRL-C or close the window displaying the output.)

Using a while Loop with Lists and Dictionaries

So far, we've worked with only one piece of user information at a time. We received the user's input and then printed the input or a response to it. The next time through the `while` loop, we'd receive another input value and respond to that. But to keep track of many users and pieces of information, we'll need to use lists and dictionaries with our `while` loops.

A `for` loop is effective for looping through a list, but you shouldn't modify a list inside a `for` loop because Python will have trouble keeping track of the items in the list. To modify a list as you work through it, use a `while` loop. Using `while` loops with lists and dictionaries allows you to collect, store, and organize lots of input to examine and report on later.

Moving Items from One List to Another

Consider a list of newly registered but unverified users of a website. After we verify these users, how can we move them to a separate list of confirmed users? One way would be to use a `while` loop to pull users from the list of unconfirmed users as we verify them and then add them to a separate list of confirmed users. Here's what that code might look like:

confirmed_users.py

```
  # Start with users that need to be verified,
  #  and an empty list to hold confirmed users.
❶ unconfirmed_users = ['alice', 'brian', 'candace']
  confirmed_users = []

  # Verify each user until there are no more unconfirmed users.
  #  Move each verified user into the list of confirmed users.
❷ while unconfirmed_users:
❸     current_user = unconfirmed_users.pop()

      print(f"Verifying user: {current_user.title()}")
❹     confirmed_users.append(current_user)

  # Display all confirmed users.
  print("\nThe following users have been confirmed:")
  for confirmed_user in confirmed_users:
      print(confirmed_user.title())
```

We begin with a list of unconfirmed users ❶ (Alice, Brian, and Candace) and an empty list to hold confirmed users. The `while` loop runs as long as the list `unconfirmed_users` is not empty ❷. Within this loop, the `pop()` method removes unverified users one at a time from the end of `unconfirmed_users` ❸. Because Candace is last in the `unconfirmed_users` list, her name will be the first to be removed, assigned to `current_user`, and added to the `confirmed_users` list ❹. Next is Brian, then Alice.

We simulate confirming each user by printing a verification message and then adding them to the list of confirmed users. As the list of unconfirmed users shrinks, the list of confirmed users grows. When the list of unconfirmed users is empty, the loop stops and the list of confirmed users is printed:

```
Verifying user: Candace
Verifying user: Brian
Verifying user: Alice

The following users have been confirmed:
Candace
Brian
Alice
```

Removing All Instances of Specific Values from a List

In Chapter 3, we used `remove()` to remove a specific value from a list. The `remove()` function worked because the value we were interested in appeared only once in the list. But what if you want to remove all instances of a value from a list?

Say you have a list of pets with the value `'cat'` repeated several times. To remove all instances of that value, you can run a `while` loop until `'cat'` is no longer in the list, as shown here:

pets.py

```
pets = ['dog', 'cat', 'dog', 'goldfish', 'cat', 'rabbit', 'cat']
print(pets)

while 'cat' in pets:
    pets.remove('cat')

print(pets)
```

We start with a list containing multiple instances of `'cat'`. After printing the list, Python enters the `while` loop because it finds the value `'cat'` in the list at least once. Once inside the loop, Python removes the first instance of `'cat'`, returns to the `while` line, and then reenters the loop when it finds that `'cat'` is still in the list. It removes each instance of `'cat'` until the value is no longer in the list, at which point Python exits the loop and prints the list again:

```
['dog', 'cat', 'dog', 'goldfish', 'cat', 'rabbit', 'cat']
['dog', 'dog', 'goldfish', 'rabbit']
```

Filling a Dictionary with User Input

You can prompt for as much input as you need in each pass through a `while` loop. Let's make a polling program in which each pass through the loop prompts for the participant's name and response. We'll store the data we gather in a dictionary, because we want to connect each response with a particular user:

mountain_poll.py

```
responses = {}
# Set a flag to indicate that polling is active.
polling_active = True
```

```
while polling_active:
    # Prompt for the person's name and response.
❶   name = input("\nWhat is your name? ")
    response = input("Which mountain would you like to climb someday? ")

    # Store the response in the dictionary.
❷   responses[name] = response

    # Find out if anyone else is going to take the poll.
❸   repeat = input("Would you like to let another person respond? (yes/ no) ")
    if repeat == 'no':
        polling_active = False

# Polling is complete. Show the results.
print("\n--- Poll Results ---")
❹ for name, response in responses.items():
    print(f"{name} would like to climb {response}.")
```

The program first defines an empty dictionary (responses) and sets a flag (polling_active) to indicate that polling is active. As long as polling_active is True, Python will run the code in the while loop.

Within the loop, the user is prompted to enter their name and a mountain they'd like to climb ❶. That information is stored in the responses dictionary ❷, and the user is asked whether or not to keep the poll running ❸. If they enter yes, the program enters the while loop again. If they enter no, the polling_active flag is set to False, the while loop stops running, and the final code block ❹ displays the results of the poll.

If you run this program and enter sample responses, you should see output like this:

```
What is your name? Eric
Which mountain would you like to climb someday? Denali
Would you like to let another person respond? (yes/ no) yes

What is your name? Lynn
Which mountain would you like to climb someday? Devil's Thumb
Would you like to let another person respond? (yes/ no) no

--- Poll Results ---
Eric would like to climb Denali.
Lynn would like to climb Devil's Thumb.
```

TRY IT YOURSELF

7-8. Deli: Make a list called `sandwich_orders` and fill it with the names of various sandwiches. Then make an empty list called `finished_sandwiches`. Loop through the list of sandwich orders and print a message for each order, such as `I made your tuna sandwich`. As each sandwich is made, move it to the list of finished sandwiches. After all the sandwiches have been made, print a message listing each sandwich that was made.

7-9. No Pastrami: Using the list `sandwich_orders` from Exercise 7-8, make sure the sandwich `'pastrami'` appears in the list at least three times. Add code near the beginning of your program to print a message saying the deli has run out of pastrami, and then use a `while` loop to remove all occurrences of `'pastrami'` from `sandwich_orders`. Make sure no pastrami sandwiches end up in `finished_sandwiches`.

7-10. Dream Vacation: Write a program that polls users about their dream vacation. Write a prompt similar to *If you could visit one place in the world, where would you go?* Include a block of code that prints the results of the poll.

Summary

In this chapter, you learned how to use `input()` to allow users to provide their own information in your programs. You learned to work with both text and numerical input and how to use `while` loops to make your programs run as long as your users want them to. You saw several ways to control the flow of a `while` loop by setting an `active` flag, using the `break` statement, and using the `continue` statement. You learned how to use a `while` loop to move items from one list to another and how to remove all instances of a value from a list. You also learned how `while` loops can be used with dictionaries.

In Chapter 8 you'll learn about functions. *Functions* allow you to break your programs into small parts, each of which does one specific job. You can call a function as many times as you want, and you can store your functions in separate files. By using functions, you'll be able to write more efficient code that's easier to troubleshoot and maintain and that can be reused in many different programs.

8

FUNCTIONS

In this chapter you'll learn to write *functions*, which are named blocks of code designed to do one specific job. When you want to perform a particular task that you've defined in a function, you *call* the function responsible for it. If you need to perform that task multiple times throughout your program, you don't need to type all the code for the same task again and again; you just call the function dedicated to handling that task, and the call tells Python to run the code inside the function. You'll find that using functions makes your programs easier to write, read, test, and fix.

In this chapter you'll also learn a variety of ways to pass information to functions. You'll learn how to write certain functions whose primary job is to display information and other functions designed to process data and return a value or set of values. Finally, you'll learn to store functions in separate files called *modules* to help organize your main program files.

Defining a Function

Here's a simple function named greet_user() that prints a greeting:

greeter.py

```
def greet_user():
    """Display a simple greeting."""
    print("Hello!")

greet_user()
```

This example shows the simplest structure of a function. The first line uses the keyword def to inform Python that you're defining a function. This is the *function definition*, which tells Python the name of the function and, if applicable, what kind of information the function needs to do its job. The parentheses hold that information. In this case, the name of the function is greet_user(), and it needs no information to do its job, so its parentheses are empty. (Even so, the parentheses are required.) Finally, the definition ends in a colon.

Any indented lines that follow def greet_user(): make up the *body* of the function. The text on the second line is a comment called a *docstring*, which describes what the function does. When Python generates documentation for the functions in your programs, it looks for a string immediately after the function's definition. These strings are usually enclosed in triple quotes, which lets you write multiple lines.

The line print("Hello!") is the only line of actual code in the body of this function, so greet_user() has just one job: print("Hello!").

When you want to use this function, you have to call it. A *function call* tells Python to execute the code in the function. To *call* a function, you write the name of the function, followed by any necessary information in parentheses. Because no information is needed here, calling our function is as simple as entering greet_user(). As expected, it prints Hello!:

```
Hello!
```

Passing Information to a Function

If you modify the function greet_user() slightly, it can greet the user by name. For the function to do this, you enter username in the parentheses of the function's definition at def greet_user(). By adding username here, you allow the function to accept any value of username you specify. The function now expects you to provide a value for username each time you call it. When you call greet_user(), you can pass it a name, such as 'jesse', inside the parentheses:

```
def greet_user(username):
    """Display a simple greeting."""
    print(f"Hello, {username.title()}!")

greet_user('jesse')
```

Entering greet_user('jesse') calls greet_user() and gives the function the information it needs to execute the print() call. The function accepts the name you passed it and displays the greeting for that name:

```
Hello, Jesse!
```

Likewise, entering greet_user('sarah') calls greet_user(), passes it 'sarah', and prints Hello, Sarah! You can call greet_user() as often as you want and pass it any name you want to produce a predictable output every time.

Arguments and Parameters

In the preceding greet_user() function, we defined greet_user() to require a value for the variable username. Once we called the function and gave it the information (a person's name), it printed the right greeting.

The variable username in the definition of greet_user() is an example of a *parameter*, a piece of information the function needs to do its job. The value 'jesse' in greet_user('jesse') is an example of an argument. An *argument* is a piece of information that's passed from a function call to a function. When we call the function, we place the value we want the function to work with in parentheses. In this case the argument 'jesse' was passed to the function greet_user(), and the value was assigned to the parameter username.

NOTE *People sometimes speak of arguments and parameters interchangeably. Don't be surprised if you see the variables in a function definition referred to as arguments or the variables in a function call referred to as parameters.*

TRY IT YOURSELF

8-1. Message: Write a function called display_message() that prints one sentence telling everyone what you are learning about in this chapter. Call the function, and make sure the message displays correctly.

8-2. Favorite Book: Write a function called favorite_book() that accepts one parameter, title. The function should print a message, such as One of my favorite books is Alice in Wonderland. Call the function, making sure to include a book title as an argument in the function call.

Passing Arguments

Because a function definition can have multiple parameters, a function call may need multiple arguments. You can pass arguments to your functions in a number of ways. You can use *positional arguments*, which need to be in the same order the parameters were written; *keyword arguments*, where each argument consists of a variable name and a value; and lists and dictionaries of values. Let's look at each of these in turn.

Positional Arguments

When you call a function, Python must match each argument in the function call with a parameter in the function definition. The simplest way to do this is based on the order of the arguments provided. Values matched up this way are called *positional arguments*.

To see how this works, consider a function that displays information about pets. The function tells us what kind of animal each pet is and the pet's name, as shown here:

pets.py

```
❶ def describe_pet(animal_type, pet_name):
      """Display information about a pet."""
      print(f"\nI have a {animal_type}.")
      print(f"My {animal_type}'s name is {pet_name.title()}.")

❷ describe_pet('hamster', 'harry')
```

The definition shows that this function needs a type of animal and the animal's name ❶. When we call `describe_pet()`, we need to provide an animal type and a name, in that order. For example, in the function call, the argument `'hamster'` is assigned to the parameter `animal_type` and the argument `'harry'` is assigned to the parameter `pet_name` ❷. In the function body, these two parameters are used to display information about the pet being described.

The output describes a hamster named Harry:

```
I have a hamster.
My hamster's name is Harry.
```

Multiple Function Calls

You can call a function as many times as needed. Describing a second, different pet requires just one more call to `describe_pet()`:

```
def describe_pet(animal_type, pet_name):
    """Display information about a pet."""
    print(f"\nI have a {animal_type}.")
    print(f"My {animal_type}'s name is {pet_name.title()}.")

describe_pet('hamster', 'harry')
describe_pet('dog', 'willie')
```

In this second function call, we pass `describe_pet()` the arguments `'dog'` and `'willie'`. As with the previous set of arguments we used, Python matches `'dog'` with the parameter `animal_type` and `'willie'` with the parameter `pet_name`. As before, the function does its job, but this time it prints values for a dog named Willie. Now we have a hamster named Harry and a dog named Willie:

```
I have a hamster.
My hamster's name is Harry.
```

```
I have a dog.
My dog's name is Willie.
```

Calling a function multiple times is a very efficient way to work. The code describing a pet is written once in the function. Then, anytime you want to describe a new pet, you call the function with the new pet's information. Even if the code for describing a pet were to expand to 10 lines, you could still describe a new pet in just one line by calling the function again.

Order Matters in Positional Arguments

You can get unexpected results if you mix up the order of the arguments in a function call when using positional arguments:

```
def describe_pet(animal_type, pet_name):
    """Display information about a pet."""
    print(f"\nI have a {animal_type}.")
    print(f"My {animal_type}'s name is {pet_name.title()}.")

describe_pet('harry', 'hamster')
```

In this function call, we list the name first and the type of animal second. Because the argument `'harry'` is listed first this time, that value is assigned to the parameter `animal_type`. Likewise, `'hamster'` is assigned to `pet_name`. Now we have a "harry" named "Hamster":

```
I have a harry.
My harry's name is Hamster.
```

If you get funny results like this, check to make sure the order of the arguments in your function call matches the order of the parameters in the function's definition.

Keyword Arguments

A *keyword argument* is a name-value pair that you pass to a function. You directly associate the name and the value within the argument, so when you pass the argument to the function, there's no confusion (you won't end up with a harry named Hamster). Keyword arguments free you from having to worry about correctly ordering your arguments in the function call, and they clarify the role of each value in the function call.

Let's rewrite *pets.py* using keyword arguments to call `describe_pet()`:

```
def describe_pet(animal_type, pet_name):
    """Display information about a pet."""
    print(f"\nI have a {animal_type}.")
    print(f"My {animal_type}'s name is {pet_name.title()}.")

describe_pet(animal_type='hamster', pet_name='harry')
```

The function describe_pet() hasn't changed. But when we call the function, we explicitly tell Python which parameter each argument should be matched with. When Python reads the function call, it knows to assign the argument 'hamster' to the parameter animal_type and the argument 'harry' to pet_name. The output correctly shows that we have a hamster named Harry.

The order of keyword arguments doesn't matter because Python knows where each value should go. The following two function calls are equivalent:

```
describe_pet(animal_type='hamster', pet_name='harry')
describe_pet(pet_name='harry', animal_type='hamster')
```

NOTE *When you use keyword arguments, be sure to use the exact names of the parameters in the function's definition.*

Default Values

When writing a function, you can define a *default value* for each parameter. If an argument for a parameter is provided in the function call, Python uses the argument value. If not, it uses the parameter's default value. So when you define a default value for a parameter, you can exclude the corresponding argument you'd usually write in the function call. Using default values can simplify your function calls and clarify the ways your functions are typically used.

For example, if you notice that most of the calls to describe_pet() are being used to describe dogs, you can set the default value of animal_type to 'dog'. Now anyone calling describe_pet() for a dog can omit that information:

```
def describe_pet(pet_name, animal_type='dog'):
    """Display information about a pet."""
    print(f"\nI have a {animal_type}.")
    print(f"My {animal_type}'s name is {pet_name.title()}.")

describe_pet(pet_name='willie')
```

We changed the definition of describe_pet() to include a default value, 'dog', for animal_type. Now when the function is called with no animal_type specified, Python knows to use the value 'dog' for this parameter:

```
I have a dog.
My dog's name is Willie.
```

Note that the order of the parameters in the function definition had to be changed. Because the default value makes it unnecessary to specify a type of animal as an argument, the only argument left in the function call is the pet's name. Python still interprets this as a positional argument, so if the function is called with just a pet's name, that argument will match up with the first parameter listed in the function's definition. This is the reason the first parameter needs to be pet_name.

The simplest way to use this function now is to provide just a dog's name in the function call:

```
describe_pet('willie')
```

This function call would have the same output as the previous example. The only argument provided is `'willie'`, so it is matched up with the first parameter in the definition, `pet_name`. Because no argument is provided for `animal_type`, Python uses the default value `'dog'`.

To describe an animal other than a dog, you could use a function call like this:

```
describe_pet(pet_name='harry', animal_type='hamster')
```

Because an explicit argument for `animal_type` is provided, Python will ignore the parameter's default value.

NOTE *When you use default values, any parameter with a default value needs to be listed after all the parameters that don't have default values. This allows Python to continue interpreting positional arguments correctly.*

Equivalent Function Calls

Because positional arguments, keyword arguments, and default values can all be used together, you'll often have several equivalent ways to call a function. Consider the following definition for `describe_pet()` with one default value provided:

```
def describe_pet(pet_name, animal_type='dog'):
```

With this definition, an argument always needs to be provided for `pet_name`, and this value can be provided using the positional or keyword format. If the animal being described is not a dog, an argument for `animal_type` must be included in the call, and this argument can also be specified using the positional or keyword format.

All of the following calls would work for this function:

```
# A dog named Willie.
describe_pet('willie')
describe_pet(pet_name='willie')

# A hamster named Harry.
describe_pet('harry', 'hamster')
describe_pet(pet_name='harry', animal_type='hamster')
describe_pet(animal_type='hamster', pet_name='harry')
```

Each of these function calls would have the same output as the previous examples.

It doesn't really matter which calling style you use. As long as your function calls produce the output you want, just use the style you find easiest to understand.

Avoiding Argument Errors

When you start to use functions, don't be surprised if you encounter errors about unmatched arguments. Unmatched arguments occur when you provide fewer or more arguments than a function needs to do its work. For example, here's what happens if we try to call `describe_pet()` with no arguments:

```
def describe_pet(animal_type, pet_name):
    """Display information about a pet."""
    print(f"\nI have a {animal_type}.")
    print(f"My {animal_type}'s name is {pet_name.title()}.")

describe_pet()
```

Python recognizes that some information is missing from the function call, and the traceback tells us that:

```
Traceback (most recent call last):
❶   File "pets.py", line 6, in <module>
❷     describe_pet()
      ^^^^^^^^^^^^^^
❸ TypeError: describe_pet() missing 2 required positional arguments:
    'animal_type' and 'pet_name'
```

The traceback first tells us the location of the problem ❶, allowing us to look back and see that something went wrong in our function call. Next, the offending function call is written out for us to see ❷. Last, the traceback tells us the call is missing two arguments and reports the names of the missing arguments ❸. If this function were in a separate file, we could probably rewrite the call correctly without having to open that file and read the function code.

Python is helpful in that it reads the function's code for us and tells us the names of the arguments we need to provide. This is another motivation for giving your variables and functions descriptive names. If you do, Python's error messages will be more useful to you and anyone else who might use your code.

If you provide too many arguments, you should get a similar traceback that can help you correctly match your function call to the function definition.

TRY IT YOURSELF

8-3. T-Shirt: Write a function called `make_shirt()` that accepts a size and the text of a message that should be printed on the shirt. The function should print a sentence summarizing the size of the shirt and the message printed on it.

Call the function once using positional arguments to make a shirt. Call the function a second time using keyword arguments.

8-4. Large Shirts: Modify the `make_shirt()` function so that shirts are large by default with a message that reads *I love Python*. Make a large shirt and a medium shirt with the default message, and a shirt of any size with a different message.

8-5. Cities: Write a function called `describe_city()` that accepts the name of a city and its country. The function should print a simple sentence, such as `Reykjavik is in Iceland`. Give the parameter for the country a default value. Call your function for three different cities, at least one of which is not in the default country.

Return Values

A function doesn't always have to display its output directly. Instead, it can process some data and then return a value or set of values. The value the function returns is called a *return value*. The `return` statement takes a value from inside a function and sends it back to the line that called the function. Return values allow you to move much of your program's grunt work into functions, which can simplify the body of your program.

Returning a Simple Value

Let's look at a function that takes a first and last name, and returns a neatly formatted full name:

formatted_name.py

```
def get_formatted_name(first_name, last_name):
    """Return a full name, neatly formatted."""
❶   full_name = f"{first_name} {last_name}"
❷   return full_name.title()

❸ musician = get_formatted_name('jimi', 'hendrix')
print(musician)
```

The definition of `get_formatted_name()` takes as parameters a first and last name. The function combines these two names, adds a space between them, and assigns the result to `full_name` ❶. The value of `full_name` is converted to title case, and then returned to the calling line ❷.

When you call a function that returns a value, you need to provide a variable that the return value can be assigned to. In this case, the returned value is assigned to the variable `musician` ❸. The output shows a neatly formatted name made up of the parts of a person's name:

```
Jimi Hendrix
```

This might seem like a lot of work to get a neatly formatted name when we could have just written:

```
print("Jimi Hendrix")
```

However, when you consider working with a large program that needs to store many first and last names separately, functions like `get_formatted_name()` become very useful. You store first and last names separately and then call this function whenever you want to display a full name.

Making an Argument Optional

Sometimes it makes sense to make an argument optional, so that people using the function can choose to provide extra information only if they want to. You can use default values to make an argument optional.

For example, say we want to expand `get_formatted_name()` to handle middle names as well. A first attempt to include middle names might look like this:

```
def get_formatted_name(first_name, middle_name, last_name):
    """Return a full name, neatly formatted."""
    full_name = f"{first_name} {middle_name} {last_name}"
    return full_name.title()

musician = get_formatted_name('john', 'lee', 'hooker')
print(musician)
```

This function works when given a first, middle, and last name. The function takes in all three parts of a name and then builds a string out of them. The function adds spaces where appropriate and converts the full name to title case:

```
John Lee Hooker
```

But middle names aren't always needed, and this function as written would not work if you tried to call it with only a first name and a last name. To make the middle name optional, we can give the `middle_name` argument an empty default value and ignore the argument unless the user provides a value. To make `get_formatted_name()` work without a middle name, we set the default value of `middle_name` to an empty string and move it to the end of the list of parameters:

```
  def get_formatted_name(first_name, last_name, middle_name=''):
      """Return a full name, neatly formatted."""
❶     if middle_name:
          full_name = f"{first_name} {middle_name} {last_name}"
❷     else:
          full_name = f"{first_name} {last_name}"
      return full_name.title()

  musician = get_formatted_name('jimi', 'hendrix')
  print(musician)

❸ musician = get_formatted_name('john', 'hooker', 'lee')
  print(musician)
```

In this example, the name is built from three possible parts. Because there's always a first and last name, these parameters are listed first in the function's definition. The middle name is optional, so it's listed last in the definition, and its default value is an empty string.

In the body of the function, we check to see if a middle name has been provided. Python interprets non-empty strings as `True`, so the conditional test `if middle_name` evaluates to `True` if a middle name argument is in the function call ❶. If a middle name is provided, the first, middle, and last names are combined to form a full name. This name is then changed to title case and returned to the function call line, where it's assigned to the variable `musician` and printed. If no middle name is provided, the empty string fails the `if` test and the `else` block runs ❷. The full name is made with just a first and last name, and the formatted name is returned to the calling line where it's assigned to `musician` and printed.

Calling this function with a first and last name is straightforward. If we're using a middle name, however, we have to make sure the middle name is the last argument passed so Python will match up the positional arguments correctly ❸.

This modified version of our function works for people with just a first and last name, and it works for people who have a middle name as well:

```
Jimi Hendrix
John Lee Hooker
```

Optional values allow functions to handle a wide range of use cases while letting function calls remain as simple as possible.

Returning a Dictionary

A function can return any kind of value you need it to, including more complicated data structures like lists and dictionaries. For example, the following function takes in parts of a name and returns a dictionary representing a person:

person.py

```
def build_person(first_name, last_name):
    """Return a dictionary of information about a person."""
❶   person = {'first': first_name, 'last': last_name}
❷   return person

musician = build_person('jimi', 'hendrix')
❸ print(musician)
```

The function `build_person()` takes in a first and last name, and puts these values into a dictionary ❶. The value of `first_name` is stored with the key `'first'`, and the value of `last_name` is stored with the key `'last'`. Then, the entire dictionary representing the person is returned ❷. The return value is printed ❸ with the original two pieces of textual information now stored in a dictionary:

```
{'first': 'jimi', 'last': 'hendrix'}
```

This function takes in simple textual information and puts it into a more meaningful data structure that lets you work with the information beyond just printing it. The strings 'jimi' and 'hendrix' are now labeled as a first name and last name. You can easily extend this function to accept optional values like a middle name, an age, an occupation, or any other information you want to store about a person. For example, the following change allows you to store a person's age as well:

```
def build_person(first_name, last_name, age=None):
    """Return a dictionary of information about a person."""
    person = {'first': first_name, 'last': last_name}
    if age:
        person['age'] = age
    return person

musician = build_person('jimi', 'hendrix', age=27)
print(musician)
```

We add a new optional parameter age to the function definition and assign the parameter the special value None, which is used when a variable has no specific value assigned to it. You can think of None as a placeholder value. In conditional tests, None evaluates to False. If the function call includes a value for age, that value is stored in the dictionary. This function always stores a person's name, but it can also be modified to store any other information you want about a person.

Using a Function with a while Loop

You can use functions with all the Python structures you've learned about so far. For example, let's use the get_formatted_name() function with a while loop to greet users more formally. Here's a first attempt at greeting people using their first and last names:

greeter.py

```
def get_formatted_name(first_name, last_name):
    """Return a full name, neatly formatted."""
    full_name = f"{first_name} {last_name}"
    return full_name.title()

# This is an infinite loop!
while True:
❶   print("\nPlease tell me your name:")
    f_name = input("First name: ")
    l_name = input("Last name: ")

    formatted_name = get_formatted_name(f_name, l_name)
    print(f"\nHello, {formatted_name}!")
```

For this example, we use a simple version of get_formatted_name() that doesn't involve middle names. The while loop asks the user to enter their name, and we prompt for their first and last name separately ❶.

But there's one problem with this while loop: We haven't defined a quit condition. Where do you put a quit condition when you ask for a series of inputs? We want the user to be able to quit as easily as possible, so each prompt should offer a way to quit. The break statement offers a straightforward way to exit the loop at either prompt:

```
def get_formatted_name(first_name, last_name):
    """Return a full name, neatly formatted."""
    full_name = f"{first_name} {last_name}"
    return full_name.title()

while True:
    print("\nPlease tell me your name:")
    print("(enter 'q' at any time to quit)")

    f_name = input("First name: ")
    if f_name == 'q':
        break

    l_name = input("Last name: ")
    if l_name == 'q':
        break

    formatted_name = get_formatted_name(f_name, l_name)
    print(f"\nHello, {formatted_name}!")
```

We add a message that informs the user how to quit, and then we break out of the loop if the user enters the quit value at either prompt. Now the program will continue greeting people until someone enters q for either name:

```
Please tell me your name:
(enter 'q' at any time to quit)
First name: eric
Last name: matthes

Hello, Eric Matthes!

Please tell me your name:
(enter 'q' at any time to quit)
First name: q
```

TRY IT YOURSELF

8-6. City Names: Write a function called city_country() that takes in the name of a city and its country. The function should return a string formatted like this:

```
"Santiago, Chile"
```

Call your function with at least three city-country pairs, and print the values that are returned.

(continued)

8-7. Album: Write a function called make_album() that builds a dictionary describing a music album. The function should take in an artist name and an album title, and it should return a dictionary containing these two pieces of information. Use the function to make three dictionaries representing different albums. Print each return value to show that the dictionaries are storing the album information correctly.

Use None to add an optional parameter to make_album() that allows you to store the number of songs on an album. If the calling line includes a value for the number of songs, add that value to the album's dictionary. Make at least one new function call that includes the number of songs on an album.

8-8. User Albums: Start with your program from Exercise 8-7. Write a while loop that allows users to enter an album's artist and title. Once you have that information, call make_album() with the user's input and print the dictionary that's created. Be sure to include a quit value in the while loop.

Passing a List

You'll often find it useful to pass a list to a function, whether it's a list of names, numbers, or more complex objects, such as dictionaries. When you pass a list to a function, the function gets direct access to the contents of the list. Let's use functions to make working with lists more efficient.

Say we have a list of users and want to print a greeting to each. The following example sends a list of names to a function called greet_users(), which greets each person in the list individually:

greet_users.py

```
def greet_users(names):
    """Print a simple greeting to each user in the list."""
    for name in names:
        msg = f"Hello, {name.title()}!"
        print(msg)

usernames = ['hannah', 'ty', 'margot']
greet_users(usernames)
```

We define greet_users() so it expects a list of names, which it assigns to the parameter names. The function loops through the list it receives and prints a greeting to each user. Outside of the function, we define a list of users and then pass the list usernames to greet_users() in the function call:

```
Hello, Hannah!
Hello, Ty!
Hello, Margot!
```

This is the output we wanted. Every user sees a personalized greeting, and you can call the function anytime you want to greet a specific set of users.

Modifying a List in a Function

When you pass a list to a function, the function can modify the list. Any changes made to the list inside the function's body are permanent, allowing you to work efficiently even when you're dealing with large amounts of data.

Consider a company that creates 3D printed models of designs that users submit. Designs that need to be printed are stored in a list, and after being printed they're moved to a separate list. The following code does this without using functions:

printing_models.py

```
# Start with some designs that need to be printed.
unprinted_designs = ['phone case', 'robot pendant', 'dodecahedron']
completed_models = []

# Simulate printing each design, until none are left.
#  Move each design to completed_models after printing.
while unprinted_designs:
    current_design = unprinted_designs.pop()
    print(f"Printing model: {current_design}")
    completed_models.append(current_design)

# Display all completed models.
print("\nThe following models have been printed:")
for completed_model in completed_models:
    print(completed_model)
```

This program starts with a list of designs that need to be printed and an empty list called `completed_models` that each design will be moved to after it has been printed. As long as designs remain in `unprinted_designs`, the `while` loop simulates printing each design by removing a design from the end of the list, storing it in `current_design`, and displaying a message that the current design is being printed. It then adds the design to the list of completed models. When the loop is finished running, a list of the designs that have been printed is displayed:

```
Printing model: dodecahedron
Printing model: robot pendant
Printing model: phone case

The following models have been printed:
dodecahedron
robot pendant
phone case
```

We can reorganize this code by writing two functions, each of which does one specific job. Most of the code won't change; we're just structuring it more carefully. The first function will handle printing the designs, and the second will summarize the prints that have been made:

```
❶ def print_models(unprinted_designs, completed_models):
    """
    Simulate printing each design, until none are left.
```

```
    Move each design to completed_models after printing.
    """
    while unprinted_designs:
        current_design = unprinted_designs.pop()
        print(f"Printing model: {current_design}")
        completed_models.append(current_design)

❷ def show_completed_models(completed_models):
    """Show all the models that were printed."""
    print("\nThe following models have been printed:")
    for completed_model in completed_models:
        print(completed_model)

unprinted_designs = ['phone case', 'robot pendant', 'dodecahedron']
completed_models = []

print_models(unprinted_designs, completed_models)
show_completed_models(completed_models)
```

We define the function `print_models()` with two parameters: a list of designs that need to be printed and a list of completed models ❶. Given these two lists, the function simulates printing each design by emptying the list of unprinted designs and filling up the list of completed models. We then define the function `show_completed_models()` with one parameter: the list of completed models ❷. Given this list, `show_completed_models()` displays the name of each model that was printed.

This program has the same output as the version without functions, but the code is much more organized. The code that does most of the work has been moved to two separate functions, which makes the main part of the program easier to understand. Look at the body of the program and notice how easily you can follow what's happening:

```
unprinted_designs = ['phone case', 'robot pendant', 'dodecahedron']
completed_models = []

print_models(unprinted_designs, completed_models)
show_completed_models(completed_models)
```

We set up a list of unprinted designs and an empty list that will hold the completed models. Then, because we've already defined our two functions, all we have to do is call them and pass them the right arguments. We call `print_models()` and pass it the two lists it needs; as expected, `print_models()` simulates printing the designs. Then we call `show_completed_models()` and pass it the list of completed models so it can report the models that have been printed. The descriptive function names allow others to read this code and understand it, even without comments.

This program is easier to extend and maintain than the version without functions. If we need to print more designs later on, we can simply call `print_models()` again. If we realize the printing code needs to be modified, we can change the code once, and our changes will take place everywhere the function is called. This technique is more efficient than having to update code separately in several places in the program.

This example also demonstrates the idea that every function should have one specific job. The first function prints each design, and the second displays the completed models. This is more beneficial than using one function to do both jobs. If you're writing a function and notice the function is doing too many different tasks, try to split the code into two functions. Remember that you can always call a function from another function, which can be helpful when splitting a complex task into a series of steps.

Preventing a Function from Modifying a List

Sometimes you'll want to prevent a function from modifying a list. For example, say that you start with a list of unprinted designs and write a function to move them to a list of completed models, as in the previous example. You may decide that even though you've printed all the designs, you want to keep the original list of unprinted designs for your records. But because you moved all the design names out of `unprinted_designs`, the list is now empty, and the empty list is the only version you have; the original is gone. In this case, you can address this issue by passing the function a copy of the list, not the original. Any changes the function makes to the list will affect only the copy, leaving the original list intact.

You can send a copy of a list to a function like this:

```
function_name(list_name[:])
```

The slice notation `[:]` makes a copy of the list to send to the function. If we didn't want to empty the list of unprinted designs in *printing_models.py*, we could call `print_models()` like this:

```
print_models(unprinted_designs[:], completed_models)
```

The function `print_models()` can do its work because it still receives the names of all unprinted designs. But this time it uses a copy of the original unprinted designs list, not the actual `unprinted_designs` list. The list `completed_models` will fill up with the names of printed models like it did before, but the original list of unprinted designs will be unaffected by the function.

Even though you can preserve the contents of a list by passing a copy of it to your functions, you should pass the original list to functions unless you have a specific reason to pass a copy. It's more efficient for a function to work with an existing list, because this avoids using the time and memory needed to make a separate copy. This is especially true when working with large lists.

TRY IT YOURSELF

8-9. Messages: Make a list containing a series of short text messages. Pass the list to a function called `show_messages()`, which prints each text message.

8-10. Sending Messages: Start with a copy of your program from Exercise 8-9. Write a function called `send_messages()` that prints each text message and moves each message to a new list called `sent_messages` as it's printed. After calling the function, print both of your lists to make sure the messages were moved correctly.

8-11. Archived Messages: Start with your work from Exercise 8-10. Call the function `send_messages()` with a copy of the list of messages. After calling the function, print both of your lists to show that the original list has retained its messages.

Passing an Arbitrary Number of Arguments

Sometimes you won't know ahead of time how many arguments a function needs to accept. Fortunately, Python allows a function to collect an arbitrary number of arguments from the calling statement.

For example, consider a function that builds a pizza. It needs to accept a number of toppings, but you can't know ahead of time how many toppings a person will want. The function in the following example has one parameter, `*toppings`, but this parameter collects as many arguments as the calling line provides:

pizza.py

```
def make_pizza(*toppings):
    """Print the list of toppings that have been requested."""
    print(toppings)

make_pizza('pepperoni')
make_pizza('mushrooms', 'green peppers', 'extra cheese')
```

The asterisk in the parameter name `*toppings` tells Python to make a tuple called `toppings`, containing all the values this function receives. The `print()` call in the function body produces output showing that Python can handle a function call with one value and a call with three values. It treats the different calls similarly. Note that Python packs the arguments into a tuple, even if the function receives only one value:

```
('pepperoni',)
('mushrooms', 'green peppers', 'extra cheese')
```

Now we can replace the `print()` call with a loop that runs through the list of toppings and describes the pizza being ordered:

```
def make_pizza(*toppings):
    """Summarize the pizza we are about to make."""
```

```
    print("\nMaking a pizza with the following toppings:")
    for topping in toppings:
        print(f"- {topping}")

make_pizza('pepperoni')
make_pizza('mushrooms', 'green peppers', 'extra cheese')
```

The function responds appropriately, whether it receives one value or three values:

```
Making a pizza with the following toppings:
- pepperoni

Making a pizza with the following toppings:
- mushrooms
- green peppers
- extra cheese
```

This syntax works no matter how many arguments the function receives.

Mixing Positional and Arbitrary Arguments

If you want a function to accept several different kinds of arguments, the parameter that accepts an arbitrary number of arguments must be placed last in the function definition. Python matches positional and keyword arguments first and then collects any remaining arguments in the final parameter.

For example, if the function needs to take in a size for the pizza, that parameter must come before the parameter `*toppings`:

```
def make_pizza(size, *toppings):
    """Summarize the pizza we are about to make."""
    print(f"\nMaking a {size}-inch pizza with the following toppings:")
    for topping in toppings:
        print(f"- {topping}")

make_pizza(16, 'pepperoni')
make_pizza(12, 'mushrooms', 'green peppers', 'extra cheese')
```

In the function definition, Python assigns the first value it receives to the parameter `size`. All other values that come after are stored in the tuple `toppings`. The function calls include an argument for the size first, followed by as many toppings as needed.

Now each pizza has a size and a number of toppings, and each piece of information is printed in the proper place, showing size first and toppings after:

```
Making a 16-inch pizza with the following toppings:
- pepperoni
```

```
Making a 12-inch pizza with the following toppings:
- mushrooms
- green peppers
- extra cheese
```

You'll often see the generic parameter name `*args`, which collects arbitrary positional arguments like this.

Using Arbitrary Keyword Arguments

Sometimes you'll want to accept an arbitrary number of arguments, but you won't know ahead of time what kind of information will be passed to the function. In this case, you can write functions that accept as many key-value pairs as the calling statement provides. One example involves building user profiles: you know you'll get information about a user, but you're not sure what kind of information you'll receive. The function `build_profile()` in the following example always takes in a first and last name, but it accepts an arbitrary number of keyword arguments as well:

user_profile.py

```
def build_profile(first, last, **user_info):
    """Build a dictionary containing everything we know about a user."""
❶   user_info['first_name'] = first
    user_info['last_name'] = last
    return user_info

user_profile = build_profile('albert', 'einstein',
                             location='princeton',
                             field='physics')
print(user_profile)
```

The definition of `build_profile()` expects a first and last name, and then it allows the user to pass in as many name-value pairs as they want. The double asterisks before the parameter `**user_info` cause Python to create a dictionary called `user_info` containing all the extra name-value pairs the function receives. Within the function, you can access the key-value pairs in `user_info` just as you would for any dictionary.

In the body of `build_profile()`, we add the first and last names to the `user_info` dictionary because we'll always receive these two pieces of information from the user ❶, and they haven't been placed into the dictionary yet. Then we return the `user_info` dictionary to the function call line.

We call `build_profile()`, passing it the first name `'albert'`, the last name `'einstein'`, and the two key-value pairs `location='princeton'` and `field='physics'`. We assign the returned `profile` to `user_profile` and print `user_profile`:

```
{'location': 'princeton', 'field': 'physics',
'first_name': 'albert', 'last_name': 'einstein'}
```

The returned dictionary contains the user's first and last names and, in this case, the location and field of study as well. The function will work no matter how many additional key-value pairs are provided in the function call.

You can mix positional, keyword, and arbitrary values in many different ways when writing your own functions. It's useful to know that all these argument types exist because you'll see them often when you start reading other people's code. It takes practice to use the different types correctly and to know when to use each type. For now, remember to use the simplest approach that gets the job done. As you progress, you'll learn to use the most efficient approach each time.

NOTE *You'll often see the parameter name* `**kwargs` *used to collect nonspecific keyword arguments.*

TRY IT YOURSELF

8-12. Sandwiches: Write a function that accepts a list of items a person wants on a sandwich. The function should have one parameter that collects as many items as the function call provides, and it should print a summary of the sandwich that's being ordered. Call the function three times, using a different number of arguments each time.

8-13. User Profile: Start with a copy of *user_profile.py* from page 148. Build a profile of yourself by calling `build_profile()`, using your first and last names and three other key-value pairs that describe you.

8-14. Cars: Write a function that stores information about a car in a dictionary. The function should always receive a manufacturer and a model name. It should then accept an arbitrary number of keyword arguments. Call the function with the required information and two other name-value pairs, such as a color or an optional feature. Your function should work for a call like this one:

```
car = make_car('subaru', 'outback', color='blue', tow_package=True)
```

Print the dictionary that's returned to make sure all the information was stored correctly.

Storing Your Functions in Modules

One advantage of functions is the way they separate blocks of code from your main program. When you use descriptive names for your functions, your programs become much easier to follow. You can go a step further by storing your functions in a separate file called a *module* and then *importing* that module into your main program. An `import` statement tells Python to make the code in a module available in the currently running program file.

Storing your functions in a separate file allows you to hide the details of your program's code and focus on its higher-level logic. It also allows you to reuse functions in many different programs. When you store your functions in separate files, you can share those files with other programmers without having to share your entire program. Knowing how to import functions also allows you to use libraries of functions that other programmers have written.

There are several ways to import a module, and I'll show you each of these briefly.

Importing an Entire Module

To start importing functions, we first need to create a module. A *module* is a file ending in *.py* that contains the code you want to import into your program. Let's make a module that contains the function `make_pizza()`. To make this module, we'll remove everything from the file *pizza.py* except the function `make_pizza()`:

pizza.py

```
def make_pizza(size, *toppings):
    """Summarize the pizza we are about to make."""
    print(f"\nMaking a {size}-inch pizza with the following toppings:")
    for topping in toppings:
        print(f"- {topping}")
```

Now we'll make a separate file called *making_pizzas.py* in the same directory as *pizza.py*. This file imports the module we just created and then makes two calls to `make_pizza()`:

making_pizzas.py

```
import pizza

❶ pizza.make_pizza(16, 'pepperoni')
pizza.make_pizza(12, 'mushrooms', 'green peppers', 'extra cheese')
```

When Python reads this file, the line `import pizza` tells Python to open the file *pizza.py* and copy all the functions from it into this program. You don't actually see code being copied between files because Python copies the code behind the scenes, just before the program runs. All you need to know is that any function defined in *pizza.py* will now be available in *making_pizzas.py*.

To call a function from an imported module, enter the name of the module you imported, `pizza`, followed by the name of the function, `make_pizza()`, separated by a dot ❶. This code produces the same output as the original program that didn't import a module:

```
Making a 16-inch pizza with the following toppings:
- pepperoni

Making a 12-inch pizza with the following toppings:
- mushrooms
- green peppers
- extra cheese
```

This first approach to importing, in which you simply write `import` followed by the name of the module, makes every function from the module available in your program. If you use this kind of import statement to import an entire module named *module_name.py*, each function in the module is available through the following syntax:

```
module_name.function_name()
```

Importing Specific Functions

You can also import a specific function from a module. Here's the general syntax for this approach:

```
from module_name import function_name
```

You can import as many functions as you want from a module by separating each function's name with a comma:

```
from module_name import function_0, function_1, function_2
```

The *making_pizzas.py* example would look like this if we want to import just the function we're going to use:

```
from pizza import make_pizza

make_pizza(16, 'pepperoni')
make_pizza(12, 'mushrooms', 'green peppers', 'extra cheese')
```

With this syntax, you don't need to use the dot notation when you call a function. Because we've explicitly imported the function `make_pizza()` in the `import` statement, we can call it by name when we use the function.

Using as to Give a Function an Alias

If the name of a function you're importing might conflict with an existing name in your program, or if the function name is long, you can use a short, unique *alias*—an alternate name similar to a nickname for the function. You'll give the function this special nickname when you import the function.

Here we give the function `make_pizza()` an alias, `mp()`, by importing `make_pizza as mp`. The `as` keyword renames a function using the alias you provide:

```
from pizza import make_pizza as mp

mp(16, 'pepperoni')
mp(12, 'mushrooms', 'green peppers', 'extra cheese')
```

The `import` statement shown here renames the function `make_pizza()` to `mp()` in this program. Anytime we want to call `make_pizza()` we can simply write `mp()` instead, and Python will run the code in `make_pizza()` while avoiding any confusion with another `make_pizza()` function you might have written in this program file.

The general syntax for providing an alias is:

```
from module_name import function_name as fn
```

Using as to Give a Module an Alias

You can also provide an alias for a module name. Giving a module a short alias, like `p` for `pizza`, allows you to call the module's functions more quickly. Calling `p.make_pizza()` is more concise than calling `pizza.make_pizza()`:

```
import pizza as p

p.make_pizza(16, 'pepperoni')
p.make_pizza(12, 'mushrooms', 'green peppers', 'extra cheese')
```

The module `pizza` is given the alias `p` in the `import` statement, but all of the module's functions retain their original names. Calling the functions by writing `p.make_pizza()` is not only more concise than `pizza.make_pizza()`, but it also redirects your attention from the module name and allows you to focus on the descriptive names of its functions. These function names, which clearly tell you what each function does, are more important to the readability of your code than using the full module name.

The general syntax for this approach is:

```
import module_name as mn
```

Importing All Functions in a Module

You can tell Python to import every function in a module by using the asterisk (`*`) operator:

```
from pizza import *

make_pizza(16, 'pepperoni')
make_pizza(12, 'mushrooms', 'green peppers', 'extra cheese')
```

The asterisk in the `import` statement tells Python to copy every function from the module `pizza` into this program file. Because every function is imported, you can call each function by name without using the dot notation. However, it's best not to use this approach when you're working with larger modules that you didn't write: if the module has a function name that matches an existing name in your project, you can get unexpected results. Python may see several functions or variables with the same name, and instead of importing all the functions separately, it will overwrite the functions.

The best approach is to import the function or functions you want, or import the entire module and use the dot notation. This leads to clear code that's easy to read and understand. I include this section so you'll recognize `import` statements like the following when you see them in other people's code:

```
from module_name import *
```

Styling Functions

You need to keep a few details in mind when you're styling functions. Functions should have descriptive names, and these names should use lowercase letters and underscores. Descriptive names help you and others understand what your code is trying to do. Module names should use these conventions as well.

Every function should have a comment that explains concisely what the function does. This comment should appear immediately after the function definition and use the docstring format. In a well-documented function, other programmers can use the function by reading only the description in the docstring. They should be able to trust that the code works as described, and as long as they know the name of the function, the arguments it needs, and the kind of value it returns, they should be able to use it in their programs.

If you specify a default value for a parameter, no spaces should be used on either side of the equal sign:

```
def function_name(parameter_0, parameter_1='default value')
```

The same convention should be used for keyword arguments in function calls:

```
function_name(value_0, parameter_1='value')
```

PEP 8 (*https://www.python.org/dev/peps/pep-0008*) recommends that you limit lines of code to 79 characters so every line is visible in a reasonably sized editor window. If a set of parameters causes a function's definition to be longer than 79 characters, press ENTER after the opening parenthesis on the definition line. On the next line, press the TAB key twice to separate the list of arguments from the body of the function, which will only be indented one level.

Most editors automatically line up any additional lines of arguments to match the indentation you have established on the first line:

```
def function_name(
        parameter_0, parameter_1, parameter_2,
        parameter_3, parameter_4, parameter_5):
    function body...
```

If your program or module has more than one function, you can separate each by two blank lines to make it easier to see where one function ends and the next one begins.

All `import` statements should be written at the beginning of a file. The only exception is if you use comments at the beginning of your file to describe the overall program.

TRY IT YOURSELF

8-15. Printing Models: Put the functions for the example *printing_models.py* in a separate file called *printing_functions.py*. Write an import statement at the top of *printing_models.py*, and modify the file to use the imported functions.

8-16. Imports: Using a program you wrote that has one function in it, store that function in a separate file. Import the function into your main program file, and call the function using each of these approaches:

```
import module_name
from module_name import function_name
from module_name import function_name as fn
import module_name as mn
from module_name import *
```

8-17. Styling Functions: Choose any three programs you wrote for this chapter, and make sure they follow the styling guidelines described in this section.

Summary

In this chapter, you learned how to write functions and to pass arguments so that your functions have access to the information they need to do their work. You learned how to use positional and keyword arguments, and also how to accept an arbitrary number of arguments. You saw functions that display output and functions that return values. You learned how to use functions with lists, dictionaries, `if` statements, and `while` loops. You also saw how to store your functions in separate files called *modules*, so your program files will be simpler and easier to understand. Finally, you learned to style your functions so your programs will continue to be well-structured and as easy as possible for you and others to read.

One of your goals as a programmer should be to write simple code that does what you want it to, and functions help you do this. They allow you to write blocks of code and leave them alone once you know they work. When you know a function does its job correctly, you can trust that it will continue to work and move on to your next coding task.

Functions allow you to write code once and then reuse that code as many times as you want. When you need to run the code in a function, all you need to do is write a one-line call and the function does its job. When you need to modify a function's behavior, you only have to modify one block of code, and your change takes effect everywhere you've made a call to that function.

Using functions makes your programs easier to read, and good function names summarize what each part of a program does. Reading a series of function calls gives you a much quicker sense of what a program does than reading a long series of code blocks.

Functions also make your code easier to test and debug. When the bulk of your program's work is done by a set of functions, each of which has a specific job, it's much easier to test and maintain the code you've written. You can write a separate program that calls each function and tests whether each function works in all the situations it may encounter. When you do this, you can be confident that your functions will work properly each time you call them.

In Chapter 9, you'll learn to write classes. *Classes* combine functions and data into one neat package that can be used in flexible and efficient ways.

9

CLASSES

Object-oriented programming (OOP) is one of the most effective approaches to writing software. In object-oriented programming, you write *classes* that represent real-world things and situations, and you create *objects* based on these classes. When you write a class, you define the general behavior that a whole category of objects can have.

When you create individual objects from the class, each object is automatically equipped with the general behavior; you can then give each object whatever unique traits you desire. You'll be amazed how well real-world situations can be modeled with object-oriented programming.

Making an object from a class is called *instantiation*, and you work with *instances* of a class. In this chapter you'll write classes and create instances of those classes. You'll specify the kind of information that can be stored in instances, and you'll define actions that can be taken with these instances. You'll also write classes that extend the functionality of existing classes, so similar classes can share common functionality, and you can do more with less code. You'll store your classes in modules and import classes written by other programmers into your own program files.

Learning about object-oriented programming will help you see the world as a programmer does. It'll help you understand your code—not just what's happening line by line, but also the bigger concepts behind it. Knowing the logic behind classes will train you to think logically, so you can write programs that effectively address almost any problem you encounter.

Classes also make life easier for you and the other programmers you'll work with as you take on increasingly complex challenges. When you and other programmers write code based on the same kind of logic, you'll be able to understand each other's work. Your programs will make sense to the people you work with, allowing everyone to accomplish more.

Creating and Using a Class

You can model almost anything using classes. Let's start by writing a simple class, Dog, that represents a dog—not one dog in particular, but any dog. What do we know about most pet dogs? Well, they all have a name and an age. We also know that most dogs sit and roll over. Those two pieces of information (name and age) and those two behaviors (sit and roll over) will go in our Dog class because they're common to most dogs. This class will tell Python how to make an object representing a dog. After our class is written, we'll use it to make individual instances, each of which represents one specific dog.

Creating the Dog Class

Each instance created from the Dog class will store a name and an age, and we'll give each dog the ability to sit() and roll_over():

dog.py

```
❶ class Dog:
      """A simple attempt to model a dog."""

❷     def __init__(self, name, age):
          """Initialize name and age attributes."""
❸         self.name = name
          self.age = age

❹     def sit(self):
          """Simulate a dog sitting in response to a command."""
          print(f"{self.name} is now sitting.")

      def roll_over(self):
          """Simulate rolling over in response to a command."""
          print(f"{self.name} rolled over!")
```

There's a lot to notice here, but don't worry. You'll see this structure throughout this chapter and have lots of time to get used to it. We first define a class called Dog ❶. By convention, capitalized names refer to classes in Python. There are no parentheses in the class definition because we're creating this class from scratch. We then write a docstring describing what this class does.

The __init__() Method

A function that's part of a class is a *method*. Everything you learned about functions applies to methods as well; the only practical difference for now is the way we'll call methods. The `__init__()` method ❷ is a special method that Python runs automatically whenever we create a new instance based on the `Dog` class. This method has two leading underscores and two trailing underscores, a convention that helps prevent Python's default method names from conflicting with your method names. Make sure to use two underscores on each side of `__init__()`. If you use just one on each side, the method won't be called automatically when you use your class, which can result in errors that are difficult to identify.

We define the `__init__()` method to have three parameters: `self`, `name`, and `age`. The `self` parameter is required in the method definition, and it must come first, before the other parameters. It must be included in the definition because when Python calls this method later (to create an instance of `Dog`), the method call will automatically pass the `self` argument. Every method call associated with an instance automatically passes `self`, which is a reference to the instance itself; it gives the individual instance access to the attributes and methods in the class. When we make an instance of `Dog`, Python will call the `__init__()` method from the `Dog` class. We'll pass `Dog()` a name and an age as arguments; `self` is passed automatically, so we don't need to pass it. Whenever we want to make an instance from the `Dog` class, we'll provide values for only the last two parameters, `name` and `age`.

The two variables defined in the body of the `__init__()` method each have the prefix `self` ❸. Any variable prefixed with `self` is available to every method in the class, and we'll also be able to access these variables through any instance created from the class. The line `self.name = name` takes the value associated with the parameter `name` and assigns it to the variable `name`, which is then attached to the instance being created. The same process happens with `self.age = age`. Variables that are accessible through instances like this are called *attributes*.

The `Dog` class has two other methods defined: `sit()` and `roll_over()` ❹. Because these methods don't need additional information to run, we just define them to have one parameter, `self`. The instances we create later will have access to these methods. In other words, they'll be able to sit and roll over. For now, `sit()` and `roll_over()` don't do much. They simply print a message saying the dog is sitting or rolling over. But the concept can be extended to realistic situations: if this class were part of a computer game, these methods would contain code to make an animated dog sit and roll over. If this class was written to control a robot, these methods would direct movements that cause a robotic dog to sit and roll over.

Making an Instance from a Class

Think of a class as a set of instructions for how to make an instance. The `Dog` class is a set of instructions that tells Python how to make individual instances representing specific dogs.

Let's make an instance representing a specific dog:

```
class Dog:
    --snip--

❶ my_dog = Dog('Willie', 6)

❷ print(f"My dog's name is {my_dog.name}.")
❸ print(f"My dog is {my_dog.age} years old.")
```

The Dog class we're using here is the one we just wrote in the previous example. Here, we tell Python to create a dog whose name is 'Willie' and whose age is 6 ❶. When Python reads this line, it calls the __init__() method in Dog with the arguments 'Willie' and 6. The __init__() method creates an instance representing this particular dog and sets the name and age attributes using the values we provided. Python then returns an instance representing this dog. We assign that instance to the variable my_dog. The naming convention is helpful here; we can usually assume that a capitalized name like Dog refers to a class, and a lowercase name like my_dog refers to a single instance created from a class.

Accessing Attributes

To access the attributes of an instance, you use dot notation. We access the value of my_dog's attribute name ❷ by writing:

```
my_dog.name
```

Dot notation is used often in Python. This syntax demonstrates how Python finds an attribute's value. Here, Python looks at the instance my_dog and then finds the attribute name associated with my_dog. This is the same attribute referred to as self.name in the class Dog. We use the same approach to work with the attribute age ❸.

The output is a summary of what we know about my_dog:

```
My dog's name is Willie.
My dog is 6 years old.
```

Calling Methods

After we create an instance from the class Dog, we can use dot notation to call any method defined in Dog. Let's make our dog sit and roll over:

```
class Dog:
    --snip--

my_dog = Dog('Willie', 6)
my_dog.sit()
my_dog.roll_over()
```

To call a method, give the name of the instance (in this case, my_dog) and the method you want to call, separated by a dot. When Python reads my_dog.sit(), it looks for the method sit() in the class Dog and runs that code. Python interprets the line my_dog.roll_over() in the same way.

Now Willie does what we tell him to:

```
Willie is now sitting.
Willie rolled over!
```

This syntax is quite useful. When attributes and methods have been given appropriately descriptive names like name, age, sit(), and roll_over(), we can easily infer what a block of code, even one we've never seen before, is supposed to do.

Creating Multiple Instances

You can create as many instances from a class as you need. Let's create a second dog called your_dog:

```
class Dog:
    --snip--

my_dog = Dog('Willie', 6)
your_dog = Dog('Lucy', 3)

print(f"My dog's name is {my_dog.name}.")
print(f"My dog is {my_dog.age} years old.")
my_dog.sit()

print(f"\nYour dog's name is {your_dog.name}.")
print(f"Your dog is {your_dog.age} years old.")
your_dog.sit()
```

In this example we create a dog named Willie and a dog named Lucy. Each dog is a separate instance with its own set of attributes, capable of the same set of actions:

```
My dog's name is Willie.
My dog is 6 years old.
Willie is now sitting.

Your dog's name is Lucy.
Your dog is 3 years old.
Lucy is now sitting.
```

Even if we used the same name and age for the second dog, Python would still create a separate instance from the Dog class. You can make as many instances from one class as you need, as long as you give each instance a unique variable name or it occupies a unique spot in a list or dictionary.

TRY IT YOURSELF

9-1. Restaurant: Make a class called Restaurant. The __init__() method for Restaurant should store two attributes: a restaurant_name and a cuisine_type. Make a method called describe_restaurant() that prints these two pieces of information, and a method called open_restaurant() that prints a message indicating that the restaurant is open.

Make an instance called restaurant from your class. Print the two attributes individually, and then call both methods.

9-2. Three Restaurants: Start with your class from Exercise 9-1. Create three different instances from the class, and call describe_restaurant() for each instance.

9-3. Users: Make a class called User. Create two attributes called first_name and last_name, and then create several other attributes that are typically stored in a user profile. Make a method called describe_user() that prints a summary of the user's information. Make another method called greet_user() that prints a personalized greeting to the user.

Create several instances representing different users, and call both methods for each user.

Working with Classes and Instances

You can use classes to represent many real-world situations. Once you write a class, you'll spend most of your time working with instances created from that class. One of the first tasks you'll want to do is modify the attributes associated with a particular instance. You can modify the attributes of an instance directly or write methods that update attributes in specific ways.

The Car Class

Let's write a new class representing a car. Our class will store information about the kind of car we're working with, and it will have a method that summarizes this information:

car.py

```
class Car:
    """A simple attempt to represent a car."""

❶   def __init__(self, make, model, year):
        """Initialize attributes to describe a car."""
        self.make = make
        self.model = model
        self.year = year

❷   def get_descriptive_name(self):
        """Return a neatly formatted descriptive name."""
        long_name = f"{self.year} {self.make} {self.model}"
```

```
        return long_name.title()

❸ my_new_car = Car('audi', 'a4', 2024)
print(my_new_car.get_descriptive_name())
```

In the Car class, we define the `__init__()` method with the self parameter first ❶, just like we did with the Dog class. We also give it three other parameters: make, model, and year. The `__init__()` method takes in these parameters and assigns them to the attributes that will be associated with instances made from this class. When we make a new Car instance, we'll need to specify a make, model, and year for our instance.

We define a method called `get_descriptive_name()` ❷ that puts a car's year, make, and model into one string neatly describing the car. This will spare us from having to print each attribute's value individually. To work with the attribute values in this method, we use self.make, self.model, and self.year. Outside of the class, we make an instance from the Car class and assign it to the variable my_new_car ❸. Then we call `get_descriptive_name()` to show what kind of car we have:

```
2024 Audi A4
```

To make the class more interesting, let's add an attribute that changes over time. We'll add an attribute that stores the car's overall mileage.

Setting a Default Value for an Attribute

When an instance is created, attributes can be defined without being passed in as parameters. These attributes can be defined in the `__init__()` method, where they are assigned a default value.

Let's add an attribute called odometer_reading that always starts with a value of 0. We'll also add a method `read_odometer()` that helps us read each car's odometer:

```
class Car:

    def __init__(self, make, model, year):
        """Initialize attributes to describe a car."""
        self.make = make
        self.model = model
        self.year = year
❶       self.odometer_reading = 0

    def get_descriptive_name(self):
        --snip--

❷   def read_odometer(self):
        """Print a statement showing the car's mileage."""
        print(f"This car has {self.odometer_reading} miles on it.")

my_new_car = Car('audi', 'a4', 2024)
print(my_new_car.get_descriptive_name())
my_new_car.read_odometer()
```

This time, when Python calls the `__init__()` method to create a new instance, it stores the make, model, and year values as attributes, like it did in the previous example. Then Python creates a new attribute called `odometer_reading` and sets its initial value to 0 ❶. We also have a new method called `read_odometer()` ❷ that makes it easy to read a car's mileage.

Our car starts with a mileage of 0:

```
2024 Audi A4
This car has 0 miles on it.
```

Not many cars are sold with exactly 0 miles on the odometer, so we need a way to change the value of this attribute.

Modifying Attribute Values

You can change an attribute's value in three ways: you can change the value directly through an instance, set the value through a method, or increment the value (add a certain amount to it) through a method. Let's look at each of these approaches.

Modifying an Attribute's Value Directly

The simplest way to modify the value of an attribute is to access the attribute directly through an instance. Here we set the odometer reading to 23 directly:

```
class Car:
    --snip--

my_new_car = Car('audi', 'a4', 2024)
print(my_new_car.get_descriptive_name())

my_new_car.odometer_reading = 23
my_new_car.read_odometer()
```

We use dot notation to access the car's `odometer_reading` attribute, and set its value directly. This line tells Python to take the instance `my_new_car`, find the attribute `odometer_reading` associated with it, and set the value of that attribute to 23:

```
2024 Audi A4
This car has 23 miles on it.
```

Sometimes you'll want to access attributes directly like this, but other times you'll want to write a method that updates the value for you.

Modifying an Attribute's Value Through a Method

It can be helpful to have methods that update certain attributes for you. Instead of accessing the attribute directly, you pass the new value to a method that handles the updating internally.

Here's an example showing a method called `update_odometer()`:

```
class Car:
    --snip--

    def update_odometer(self, mileage):
        """Set the odometer reading to the given value."""
        self.odometer_reading = mileage

my_new_car = Car('audi', 'a4', 2024)
print(my_new_car.get_descriptive_name())

❶ my_new_car.update_odometer(23)
my_new_car.read_odometer()
```

The only modification to `Car` is the addition of `update_odometer()`. This method takes in a mileage value and assigns it to `self.odometer_reading`. Using the `my_new_car` instance, we call `update_odometer()` with `23` as an argument ❶. This sets the odometer reading to 23, and `read_odometer()` prints the reading:

```
2024 Audi A4
This car has 23 miles on it.
```

We can extend the method `update_odometer()` to do additional work every time the odometer reading is modified. Let's add a little logic to make sure no one tries to roll back the odometer reading:

```
class Car:
    --snip--

    def update_odometer(self, mileage):
        """
        Set the odometer reading to the given value.
        Reject the change if it attempts to roll the odometer back.
        """
❶       if mileage >= self.odometer_reading:
            self.odometer_reading = mileage
        else:
❷           print("You can't roll back an odometer!")
```

Now `update_odometer()` checks that the new reading makes sense before modifying the attribute. If the value provided for `mileage` is greater than or equal to the existing mileage, `self.odometer_reading`, you can update the odometer reading to the new mileage ❶. If the new mileage is less than the existing mileage, you'll get a warning that you can't roll back an odometer ❷.

Incrementing an Attribute's Value Through a Method

Sometimes you'll want to increment an attribute's value by a certain amount, rather than set an entirely new value. Say we buy a used car and put 100 miles on it between the time we buy it and the time we register it. Here's a method that allows us to pass this incremental amount and add that value to the odometer reading:

```
class Car:
    --snip--

    def update_odometer(self, mileage):
        --snip--

    def increment_odometer(self, miles):
        """Add the given amount to the odometer reading."""
        self.odometer_reading += miles

❶ my_used_car = Car('subaru', 'outback', 2019)
print(my_used_car.get_descriptive_name())

❷ my_used_car.update_odometer(23_500)
my_used_car.read_odometer()

my_used_car.increment_odometer(100)
my_used_car.read_odometer()
```

The new method `increment_odometer()` takes in a number of miles, and adds this value to `self.odometer_reading`. First, we create a used car, `my_used_car` ❶. We set its odometer to 23,500 by calling `update_odometer()` and passing it `23_500` ❷. Finally, we call `increment_odometer()` and pass it `100` to add the 100 miles that we drove between buying the car and registering it:

```
2019 Subaru Outback
This car has 23500 miles on it.
This car has 23600 miles on it.
```

You can modify this method to reject negative increments so no one uses this function to roll back an odometer as well.

NOTE *You can use methods like this to control how users of your program update values such as an odometer reading, but anyone with access to the program can set the odometer reading to any value by accessing the attribute directly. Effective security takes extreme attention to detail in addition to basic checks like those shown here.*

TRY IT YOURSELF

9-4. Number Served: Start with your program from Exercise 9-1 (page 162). Add an attribute called `number_served` with a default value of 0. Create an instance called `restaurant` from this class. Print the number of customers the restaurant has served, and then change this value and print it again.

Add a method called `set_number_served()` that lets you set the number of customers that have been served. Call this method with a new number and print the value again.

Add a method called `increment_number_served()` that lets you increment the number of customers who've been served. Call this method with any number you like that could represent how many customers were served in, say, a day of business.

9-5. Login Attempts: Add an attribute called `login_attempts` to your `User` class from Exercise 9-3 (page 162). Write a method called `increment_login_attempts()` that increments the value of `login_attempts` by 1. Write another method called `reset_login_attempts()` that resets the value of `login_attempts` to 0.

Make an instance of the `User` class and call `increment_login_attempts()` several times. Print the value of `login_attempts` to make sure it was incremented properly, and then call `reset_login_attempts()`. Print `login_attempts` again to make sure it was reset to 0.

Inheritance

You don't always have to start from scratch when writing a class. If the class you're writing is a specialized version of another class you wrote, you can use *inheritance*. When one class *inherits* from another, it takes on the attributes and methods of the first class. The original class is called the *parent class*, and the new class is the *child class*. The child class can inherit any or all of the attributes and methods of its parent class, but it's also free to define new attributes and methods of its own.

The __init__() Method for a Child Class

When you're writing a new class based on an existing class, you'll often want to call the `__init__()` method from the parent class. This will initialize any attributes that were defined in the parent `__init__()` method and make them available in the child class.

As an example, let's model an electric car. An electric car is just a specific kind of car, so we can base our new `ElectricCar` class on the `Car` class we wrote earlier. Then we'll only have to write code for the attributes and behaviors specific to electric cars.

Let's start by making a simple version of the `ElectricCar` class, which does everything the `Car` class does:

electric_car.py ❶

```
class Car:
    """A simple attempt to represent a car."""

    def __init__(self, make, model, year):
        """Initialize attributes to describe a car."""
        self.make = make
        self.model = model
```

```
        self.year = year
        self.odometer_reading = 0

    def get_descriptive_name(self):
        """Return a neatly formatted descriptive name."""
        long_name = f"{self.year} {self.make} {self.model}"
        return long_name.title()

    def read_odometer(self):
        """Print a statement showing the car's mileage."""
        print(f"This car has {self.odometer_reading} miles on it.")

    def update_odometer(self, mileage):
        """Set the odometer reading to the given value."""
        if mileage >= self.odometer_reading:
            self.odometer_reading = mileage
        else:
            print("You can't roll back an odometer!")

    def increment_odometer(self, miles):
        """Add the given amount to the odometer reading."""
        self.odometer_reading += miles

❷ class ElectricCar(Car):
    """Represent aspects of a car, specific to electric vehicles."""

❸   def __init__(self, make, model, year):
        """Initialize attributes of the parent class."""
❹       super().__init__(make, model, year)


❺ my_leaf = ElectricCar('nissan', 'leaf', 2024)
  print(my_leaf.get_descriptive_name())
```

We start with `Car` ❶. When you create a child class, the parent class must be part of the current file and must appear before the child class in the file. We then define the child class, `ElectricCar` ❷. The name of the parent class must be included in parentheses in the definition of a child class. The `__init__()` method takes in the information required to make a `Car` instance ❸.

The `super()` function ❹ is a special function that allows you to call a method from the parent class. This line tells Python to call the `__init__()` method from `Car`, which gives an `ElectricCar` instance all the attributes defined in that method. The name *super* comes from a convention of calling the parent class a *superclass* and the child class a *subclass*.

We test whether inheritance is working properly by trying to create an electric car with the same kind of information we'd provide when making a regular car. We make an instance of the `ElectricCar` class and assign it to `my_leaf` ❺. This line calls the `__init__()` method defined in `ElectricCar`, which in turn tells Python to call the `__init__()` method defined in the parent class `Car`. We provide the arguments `'nissan'`, `'leaf'`, and `2024`.

Aside from `__init__()`, there are no attributes or methods yet that are particular to an electric car. At this point we're just making sure the electric car has the appropriate `Car` behaviors:

```
2024 Nissan Leaf
```

The `ElectricCar` instance works just like an instance of `Car`, so now we can begin defining attributes and methods specific to electric cars.

Defining Attributes and Methods for the Child Class

Once you have a child class that inherits from a parent class, you can add any new attributes and methods necessary to differentiate the child class from the parent class.

Let's add an attribute that's specific to electric cars (a battery, for example) and a method to report on this attribute. We'll store the battery size and write a method that prints a description of the battery:

```
class Car:
    --snip--

class ElectricCar(Car):
    """Represent aspects of a car, specific to electric vehicles."""

    def __init__(self, make, model, year):
        """
        Initialize attributes of the parent class.
        Then initialize attributes specific to an electric car.
        """
        super().__init__(make, model, year)
❶       self.battery_size = 40

❷   def describe_battery(self):
        """Print a statement describing the battery size."""
        print(f"This car has a {self.battery_size}-kWh battery.")

my_leaf = ElectricCar('nissan', 'leaf', 2024)
print(my_leaf.get_descriptive_name())
my_leaf.describe_battery()
```

We add a new attribute `self.battery_size` and set its initial value to `40` ❶. This attribute will be associated with all instances created from the `ElectricCar` class but won't be associated with any instances of `Car`. We also add a method called `describe_battery()` that prints information about the battery ❷. When we call this method, we get a description that is clearly specific to an electric car:

```
2024 Nissan Leaf
This car has a 40-kWh battery.
```

There's no limit to how much you can specialize the `ElectricCar` class. You can add as many attributes and methods as you need to model an electric car to whatever degree of accuracy you need. An attribute or method that could belong to any car, rather than one that's specific to an electric car, should be added to the Car class instead of the ElectricCar class. Then anyone who uses the Car class will have that functionality available as well, and the ElectricCar class will only contain code for the information and behavior specific to electric vehicles.

Overriding Methods from the Parent Class

You can override any method from the parent class that doesn't fit what you're trying to model with the child class. To do this, you define a method in the child class with the same name as the method you want to override in the parent class. Python will disregard the parent class method and only pay attention to the method you define in the child class.

Say the class Car had a method called fill_gas_tank(). This method is meaningless for an all-electric vehicle, so you might want to override this method. Here's one way to do that:

```
class ElectricCar(Car):
    --snip--

    def fill_gas_tank(self):
        """Electric cars don't have gas tanks."""
        print("This car doesn't have a gas tank!")
```

Now if someone tries to call fill_gas_tank() with an electric car, Python will ignore the method fill_gas_tank() in Car and run this code instead. When you use inheritance, you can make your child classes retain what you need and override anything you don't need from the parent class.

Instances as Attributes

When modeling something from the real world in code, you may find that you're adding more and more detail to a class. You'll find that you have a growing list of attributes and methods and that your files are becoming lengthy. In these situations, you might recognize that part of one class can be written as a separate class. You can break your large class into smaller classes that work together; this approach is called *composition*.

For example, if we continue adding detail to the ElectricCar class, we might notice that we're adding many attributes and methods specific to the car's battery. When we see this happening, we can stop and move those attributes and methods to a separate class called Battery. Then we can use a Battery instance as an attribute in the ElectricCar class:

```
class Car:
    --snip--

class Battery:
    """A simple attempt to model a battery for an electric car."""

❶   def __init__(self, battery_size=40):
```

```
        """Initialize the battery's attributes."""
        self.battery_size = battery_size

❷   def describe_battery(self):
        """Print a statement describing the battery size."""
        print(f"This car has a {self.battery_size}-kWh battery.")


class ElectricCar(Car):
    """Represent aspects of a car, specific to electric vehicles."""

    def __init__(self, make, model, year):
        """
        Initialize attributes of the parent class.
        Then initialize attributes specific to an electric car.
        """
        super().__init__(make, model, year)
❸       self.battery = Battery()

my_leaf = ElectricCar('nissan', 'leaf', 2024)
print(my_leaf.get_descriptive_name())
my_leaf.battery.describe_battery()
```

We define a new class called Battery that doesn't inherit from any other class. The __init__() method ❶ has one parameter, battery_size, in addition to self. This is an optional parameter that sets the battery's size to 40 if no value is provided. The method describe_battery() has been moved to this class as well ❷.

In the ElectricCar class, we now add an attribute called self.battery ❸. This line tells Python to create a new instance of Battery (with a default size of 40, because we're not specifying a value) and assign that instance to the attribute self.battery. This will happen every time the __init__() method is called; any ElectricCar instance will now have a Battery instance created automatically.

We create an electric car and assign it to the variable my_leaf. When we want to describe the battery, we need to work through the car's battery attribute:

```
my_leaf.battery.describe_battery()
```

This line tells Python to look at the instance my_leaf, find its battery attribute, and call the method describe_battery() that's associated with the Battery instance assigned to the attribute.

The output is identical to what we saw previously:

```
2024 Nissan Leaf
This car has a 40-kWh battery.
```

This looks like a lot of extra work, but now we can describe the battery in as much detail as we want without cluttering the ElectricCar class. Let's add another method to Battery that reports the range of the car based on the battery size:

```
class Car:
    --snip--

class Battery:
    --snip--

    def get_range(self):
        """Print a statement about the range this battery provides."""
        if self.battery_size == 40:
            range = 150
        elif self.battery_size == 65:
            range = 225

        print(f"This car can go about {range} miles on a full charge.")

class ElectricCar(Car):
    --snip--

my_leaf = ElectricCar('nissan', 'leaf', 2024)
print(my_leaf.get_descriptive_name())
my_leaf.battery.describe_battery()
❶ my_leaf.battery.get_range()
```

The new method get_range() performs some simple analysis. If the battery's capacity is 40 kWh, get_range() sets the range to 150 miles, and if the capacity is 65 kWh, it sets the range to 225 miles. It then reports this value. When we want to use this method, we again have to call it through the car's battery attribute ❶.

The output tells us the range of the car based on its battery size:

```
2024 Nissan Leaf
This car has a 40-kWh battery.
This car can go about 150 miles on a full charge.
```

Modeling Real-World Objects

As you begin to model more complicated things like electric cars, you'll wrestle with interesting questions. Is the range of an electric car a property of the battery or of the car? If we're only describing one car, it's probably fine to maintain the association of the method get_range() with the Battery class. But if we're describing a manufacturer's entire line of cars, we probably want to move get_range() to the ElectricCar class. The get_range() method would still check the battery size before determining the range, but it would report a range specific to the kind of car it's associated with. Alternatively, we could maintain the association of the get_range() method with the battery but pass it a parameter such as car_model. The get_range() method would then report a range based on the battery size and car model.

This brings you to an interesting point in your growth as a programmer. When you wrestle with questions like these, you're thinking at a higher logical level rather than a syntax-focused level. You're thinking not about Python, but about how to represent the real world in code. When you reach this point, you'll realize there are often no right or wrong approaches to modeling real-world situations. Some approaches are more efficient than others, but it takes practice to find the most efficient representations. If your code is working as you want it to, you're doing well! Don't be discouraged if you find you're ripping apart your classes and rewriting them several times using different approaches. In the quest to write accurate, efficient code, everyone goes through this process.

TRY IT YOURSELF

9-6. Ice Cream Stand: An ice cream stand is a specific kind of restaurant. Write a class called `IceCreamStand` that inherits from the `Restaurant` class you wrote in Exercise 9-1 (page 162) or Exercise 9-4 (page 166). Either version of the class will work; just pick the one you like better. Add an attribute called `flavors` that stores a list of ice cream flavors. Write a method that displays these flavors. Create an instance of `IceCreamStand`, and call this method.

9-7. Admin: An administrator is a special kind of user. Write a class called `Admin` that inherits from the `User` class you wrote in Exercise 9-3 (page 162) or Exercise 9-5 (page 167). Add an attribute, `privileges`, that stores a list of strings like `"can add post"`, `"can delete post"`, `"can ban user"`, and so on. Write a method called `show_privileges()` that lists the administrator's set of privileges. Create an instance of `Admin`, and call your method.

9-8. Privileges: Write a separate `Privileges` class. The class should have one attribute, `privileges`, that stores a list of strings as described in Exercise 9-7. Move the `show_privileges()` method to this class. Make a `Privileges` instance as an attribute in the `Admin` class. Create a new instance of `Admin` and use your method to show its privileges.

9-9. Battery Upgrade: Use the final version of *electric_car.py* from this section. Add a method to the `Battery` class called `upgrade_battery()`. This method should check the battery size and set the capacity to 65 if it isn't already. Make an electric car with a default battery size, call `get_range()` once, and then call `get_range()` a second time after upgrading the battery. You should see an increase in the car's range.

Importing Classes

As you add more functionality to your classes, your files can get long, even when you use inheritance and composition properly. In keeping with the overall philosophy of Python, you'll want to keep your files as uncluttered as possible. To help, Python lets you store classes in modules and then import the classes you need into your main program.

Importing a Single Class

Let's create a module containing just the Car class. This brings up a subtle naming issue: we already have a file named *car.py* in this chapter, but this module should be named *car.py* because it contains code representing a car. We'll resolve this naming issue by storing the Car class in a module named *car.py*, replacing the *car.py* file we were previously using. From now on, any program that uses this module will need a more specific filename, such as *my_car.py*. Here's *car.py* with just the code from the class Car:

car.py

```
❶ """A class that can be used to represent a car."""

class Car:
    """A simple attempt to represent a car."""

    def __init__(self, make, model, year):
        """Initialize attributes to describe a car."""
        self.make = make
        self.model = model
        self.year = year
        self.odometer_reading = 0

    def get_descriptive_name(self):
        """Return a neatly formatted descriptive name."""
        long_name = f"{self.year} {self.make} {self.model}"
        return long_name.title()

    def read_odometer(self):
        """Print a statement showing the car's mileage."""
        print(f"This car has {self.odometer_reading} miles on it.")

    def update_odometer(self, mileage):
        """
        Set the odometer reading to the given value.
        Reject the change if it attempts to roll the odometer back.
        """
        if mileage >= self.odometer_reading:
            self.odometer_reading = mileage
        else:
            print("You can't roll back an odometer!")

    def increment_odometer(self, miles):
        """Add the given amount to the odometer reading."""
        self.odometer_reading += miles
```

We include a module-level docstring that briefly describes the contents of this module ❶. You should write a docstring for each module you create.

Now we make a separate file called *my_car.py*. This file will import the Car class and then create an instance from that class:

my_car.py

```
❶ from car import Car

my_new_car = Car('audi', 'a4', 2024)
print(my_new_car.get_descriptive_name())
```

```
my_new_car.odometer_reading = 23
my_new_car.read_odometer()
```

The import statement ❶ tells Python to open the car module and import the class Car. Now we can use the Car class as if it were defined in this file. The output is the same as we saw earlier:

```
2024 Audi A4
This car has 23 miles on it.
```

Importing classes is an effective way to program. Picture how long this program file would be if the entire Car class were included. When you instead move the class to a module and import the module, you still get all the same functionality, but you keep your main program file clean and easy to read. You also store most of the logic in separate files; once your classes work as you want them to, you can leave those files alone and focus on the higher-level logic of your main program.

Storing Multiple Classes in a Module

You can store as many classes as you need in a single module, although each class in a module should be related somehow. The classes Battery and ElectricCar both help represent cars, so let's add them to the module *car.py*.

car.py

```
"""A set of classes used to represent gas and electric cars."""

class Car:
    --snip--

class Battery:
    """A simple attempt to model a battery for an electric car."""

    def __init__(self, battery_size=40):
        """Initialize the battery's attributes."""
        self.battery_size = battery_size

    def describe_battery(self):
        """Print a statement describing the battery size."""
        print(f"This car has a {self.battery_size}-kWh battery.")

    def get_range(self):
        """Print a statement about the range this battery provides."""
        if self.battery_size == 40:
            range = 150
        elif self.battery_size == 65:
            range = 225

        print(f"This car can go about {range} miles on a full charge.")

class ElectricCar(Car):
    """Models aspects of a car, specific to electric vehicles."""

    def __init__(self, make, model, year):
        """
```

```
        Initialize attributes of the parent class.
        Then initialize attributes specific to an electric car.
        """
        super().__init__(make, model, year)
        self.battery = Battery()
```

Now we can make a new file called *my_electric_car.py*, import the `ElectricCar` class, and make an electric car:

my_electric_car.py

```
from car import ElectricCar

my_leaf = ElectricCar('nissan', 'leaf', 2024)
print(my_leaf.get_descriptive_name())
my_leaf.battery.describe_battery()
my_leaf.battery.get_range()
```

This has the same output we saw earlier, even though most of the logic is hidden away in a module:

```
2024 Nissan Leaf
This car has a 40-kWh battery.
This car can go about 150 miles on a full charge.
```

Importing Multiple Classes from a Module

You can import as many classes as you need into a program file. If we want to make a regular car and an electric car in the same file, we need to import both classes, `Car` and `ElectricCar`:

my_cars.py

```
❶ from car import Car, ElectricCar

❷ my_mustang = Car('ford', 'mustang', 2024)
  print(my_mustang.get_descriptive_name())
❸ my_leaf = ElectricCar('nissan', 'leaf', 2024)
  print(my_leaf.get_descriptive_name())
```

You import multiple classes from a module by separating each class with a comma ❶. Once you've imported the necessary classes, you're free to make as many instances of each class as you need.

In this example we make a gas-powered Ford Mustang ❷ and then an electric Nissan Leaf ❸:

```
2024 Ford Mustang
2024 Nissan Leaf
```

Importing an Entire Module

You can also import an entire module and then access the classes you need using dot notation. This approach is simple and results in code that is easy to read. Because every call that creates an instance of a class includes the module name, you won't have naming conflicts with any names used in the current file.

Here's what it looks like to import the entire car module and then create a regular car and an electric car:

my_cars.py

```
❶ import car

❷ my_mustang = car.Car('ford', 'mustang', 2024)
  print(my_mustang.get_descriptive_name())

❸ my_leaf = car.ElectricCar('nissan', 'leaf', 2024)
  print(my_leaf.get_descriptive_name())
```

First we import the entire car module ❶. We then access the classes we need through the *module_name.ClassName* syntax. We again create a Ford Mustang ❷, and a Nissan Leaf ❸.

Importing All Classes from a Module

You can import every class from a module using the following syntax:

```
from module_name import *
```

This method is not recommended for two reasons. First, it's helpful to be able to read the import statements at the top of a file and get a clear sense of which classes a program uses. With this approach it's unclear which classes you're using from the module. This approach can also lead to confusion with names in the file. If you accidentally import a class with the same name as something else in your program file, you can create errors that are hard to diagnose. I show this here because even though it's not a recommended approach, you're likely to see it in other people's code at some point.

If you need to import many classes from a module, you're better off importing the entire module and using the *module_name.ClassName* syntax. You won't see all the classes used at the top of the file, but you'll see clearly where the module is used in the program. You'll also avoid the potential naming conflicts that can arise when you import every class in a module.

Importing a Module into a Module

Sometimes you'll want to spread out your classes over several modules to keep any one file from growing too large and avoid storing unrelated classes in the same module. When you store your classes in several modules, you may find that a class in one module depends on a class in another module. When this happens, you can import the required class into the first module.

For example, let's store the Car class in one module and the ElectricCar and Battery classes in a separate module. We'll make a new module called *electric_car.py*—replacing the *electric_car.py* file we created earlier—and copy just the Battery and ElectricCar classes into this file:

electric_car.py

```
"""A set of classes that can be used to represent electric cars."""

from car import Car
```

```
class Battery:
    --snip--

class ElectricCar(Car):
    --snip--
```

The class `ElectricCar` needs access to its parent class `Car`, so we import `Car` directly into the module. If we forget this line, Python will raise an error when we try to import the `electric_car` module. We also need to update the `Car` module so it contains only the `Car` class:

car.py

```
"""A class that can be used to represent a car."""

class Car:
    --snip--
```

Now we can import from each module separately and create whatever kind of car we need:

my_cars.py

```
from car import Car
from electric_car import ElectricCar

my_mustang = Car('ford', 'mustang', 2024)
print(my_mustang.get_descriptive_name())

my_leaf = ElectricCar('nissan', 'leaf', 2024)
print(my_leaf.get_descriptive_name())
```

We import `Car` from its module, and `ElectricCar` from its module. We then create one regular car and one electric car. Both cars are created correctly:

```
2024 Ford Mustang
2024 Nissan Leaf
```

Using Aliases

As you saw in Chapter 8, aliases can be quite helpful when using modules to organize your projects' code. You can use aliases when importing classes as well.

As an example, consider a program where you want to make a bunch of electric cars. It might get tedious to type (and read) `ElectricCar` over and over again. You can give `ElectricCar` an alias in the import statement:

```
from electric_car import ElectricCar as EC
```

Now you can use this alias whenever you want to make an electric car:

```
my_leaf = EC('nissan', 'leaf', 2024)
```

You can also give a module an alias. Here's how to import the entire electric_car module using an alias:

```
import electric_car as ec
```

Now you can use this module alias with the full class name:

```
my_leaf = ec.ElectricCar('nissan', 'leaf', 2024)
```

Finding Your Own Workflow

As you can see, Python gives you many options for how to structure code in a large project. It's important to know all these possibilities so you can determine the best ways to organize your projects as well as understand other people's projects.

When you're starting out, keep your code structure simple. Try doing everything in one file and moving your classes to separate modules once everything is working. If you like how modules and files interact, try storing your classes in modules when you start a project. Find an approach that lets you write code that works, and go from there.

TRY IT YOURSELF

9-10. Imported Restaurant: Using your latest Restaurant class, store it in a module. Make a separate file that imports Restaurant. Make a Restaurant instance, and call one of Restaurant's methods to show that the import statement is working properly.

9-11. Imported Admin: Start with your work from Exercise 9-8 (page 173). Store the classes User, Privileges, and Admin in one module. Create a separate file, make an Admin instance, and call show_privileges() to show that everything is working correctly.

9-12. Multiple Modules: Store the User class in one module, and store the Privileges and Admin classes in a separate module. In a separate file, create an Admin instance and call show_privileges() to show that everything is still working correctly.

The Python Standard Library

The *Python standard library* is a set of modules included with every Python installation. Now that you have a basic understanding of how functions and classes work, you can start to use modules like these that other programmers have written. You can use any function or class in the standard library by including a simple import statement at the top of your file. Let's look at one module, random, which can be useful in modeling many real-world situations.

One interesting function from the random module is `randint()`. This function takes two integer arguments and returns a randomly selected integer between (and including) those numbers.

Here's how to generate a random number between 1 and 6:

```
>>> from random import randint
>>> randint(1, 6)
3
```

Another useful function is `choice()`. This function takes in a list or tuple and returns a randomly chosen element:

```
>>> from random import choice
>>> players = ['charles', 'martina', 'michael', 'florence', 'eli']
>>> first_up = choice(players)
>>> first_up
'florence'
```

The `random` module shouldn't be used when building security-related applications, but it works well for many fun and interesting projects.

NOTE *You can also download modules from external sources. You'll see a number of these examples in Part II, where we'll need external modules to complete each project.*

TRY IT YOURSELF

9-13. Dice: Make a class `Die` with one attribute called `sides`, which has a default value of 6. Write a method called `roll_die()` that prints a random number between 1 and the number of sides the die has. Make a 6-sided die and roll it 10 times.

Make a 10-sided die and a 20-sided die. Roll each die 10 times.

9-14. Lottery: Make a list or tuple containing a series of 10 numbers and 5 letters. Randomly select 4 numbers or letters from the list and print a message saying that any ticket matching these 4 numbers or letters wins a prize.

9-15. Lottery Analysis: You can use a loop to see how hard it might be to win the kind of lottery you just modeled. Make a list or tuple called `my_ticket`. Write a loop that keeps pulling numbers until your ticket wins. Print a message reporting how many times the loop had to run to give you a winning ticket.

9-16. Python Module of the Week: One excellent resource for exploring the Python standard library is a site called *Python Module of the Week*. Go to *https://pymotw.com* and look at the table of contents. Find a module that looks interesting to you and read about it, perhaps starting with the `random` module.

Styling Classes

A few styling issues related to classes are worth clarifying, especially as your programs become more complicated.

Class names should be written in *CamelCase.* To do this, capitalize the first letter of each word in the name, and don't use underscores. Instance and module names should be written in lowercase, with underscores between words.

Every class should have a docstring immediately following the class definition. The docstring should be a brief description of what the class does, and you should follow the same formatting conventions you used for writing docstrings in functions. Each module should also have a docstring describing what the classes in a module can be used for.

You can use blank lines to organize code, but don't use them excessively. Within a class you can use one blank line between methods, and within a module you can use two blank lines to separate classes.

If you need to import a module from the standard library and a module that you wrote, place the import statement for the standard library module first. Then add a blank line and the import statement for the module you wrote. In programs with multiple import statements, this convention makes it easier to see where the different modules used in the program come from.

Summary

In this chapter, you learned how to write your own classes. You learned how to store information in a class using attributes and how to write methods that give your classes the behavior they need. You learned to write `__init__()` methods that create instances from your classes with exactly the attributes you want. You saw how to modify the attributes of an instance directly and through methods. You learned that inheritance can simplify the creation of classes that are related to each other, and you learned to use instances of one class as attributes in another class to keep each class simple.

You saw how storing classes in modules and importing classes you need into the files where they'll be used can keep your projects organized. You started learning about the Python standard library, and you saw an example based on the `random` module. Finally, you learned to style your classes using Python conventions.

In Chapter 10, you'll learn to work with files so you can save the work you've done in a program and the work you've allowed users to do. You'll also learn about *exceptions*, a special Python class designed to help you respond to errors when they arise.

10

FILES AND EXCEPTIONS

Now that you've mastered the basic skills you need to write organized programs that are easy to use, it's time to think about making your programs even more relevant and usable. In this chapter, you'll learn to work with files so your programs can quickly analyze lots of data.

You'll learn to handle errors so your programs don't crash when they encounter unexpected situations. You'll learn about *exceptions*, which are special objects Python creates to manage errors that arise while a program is running. You'll also learn about the `json` module, which allows you to save user data so it isn't lost when your program stops running.

Learning to work with files and save data will make your programs easier for people to use. Users will be able to choose what data to enter and when to enter it. People will be able to run your program, do some work, and then close the program and pick up where they left off. Learning to handle exceptions will help you deal with situations in which files don't exist and deal with other problems that can cause your programs to crash. This will make your programs more robust when they encounter bad data, whether it comes from innocent mistakes or from malicious attempts to break your programs. With the skills you'll learn in this chapter, you'll make your programs more applicable, usable, and stable.

Reading from a File

An incredible amount of data is available in text files. Text files can contain weather data, traffic data, socioeconomic data, literary works, and more. Reading from a file is particularly useful in data analysis applications, but it's also applicable to any situation in which you want to analyze or modify information stored in a file. For example, you can write a program that reads in the contents of a text file and rewrites the file with formatting that allows a browser to display it.

When you want to work with the information in a text file, the first step is to read the file into memory. You can then work through all of the file's contents at once or work through the contents line by line.

Reading the Contents of a File

To begin, we need a file with a few lines of text in it. Let's start with a file that contains *pi* to 30 decimal places, with 10 decimal places per line:

pi_digits.txt

```
3.1415926535
  8979323846
  2643383279
```

To try the following examples yourself, you can enter these lines in an editor and save the file as *pi_digits.txt*, or you can download the file from the book's resources through *https://ehmatthes.github.io/pcc_3e*. Save the file in the same directory where you'll store this chapter's programs.

Here's a program that opens this file, reads it, and prints the contents of the file to the screen:

file_reader.py

```
from pathlib import Path

❶ path = Path('pi_digits.txt')
❷ contents = path.read_text()
print(contents)
```

To work with the contents of a file, we need to tell Python the path to the file. A *path* is the exact location of a file or folder on a system. Python provides a module called `pathlib` that makes it easier to work with files and directories, no matter which operating system you or your program's users are working with. A module that provides specific functionality like this is often called a *library*, hence the name `pathlib`.

We start by importing the `Path` class from `pathlib`. There's a lot you can do with a `Path` object that points to a file. For example, you can check that the file exists before working with it, read the file's contents, or write new data to the file. Here, we build a `Path` object representing the file *pi_digits.txt*, which we assign to the variable `path` ❶. Since this file is saved in the same directory as the *.py* file we're writing, the filename is all that Path needs to access the file.

NOTE *VS Code looks for files in the folder that was most recently opened. If you're using VS Code, start by opening the folder where you're storing this chapter's programs. For example, if you're saving your program files in a folder called* chapter_10, *press CTRL-O (⌘-O on macOS), and open that folder.*

Once we have a Path object representing *pi_digits.txt*, we use the read_text() method to read the entire contents of the file ❷. The contents of the file are returned as a single string, which we assign to the variable contents. When we print the value of contents, we see the entire contents of the text file:

```
3.1415926535
  8979323846
  2643383279

```

The only difference between this output and the original file is the extra blank line at the end of the output. The blank line appears because read_text() returns an empty string when it reaches the end of the file; this empty string shows up as a blank line.

We can remove the extra blank line by using rstrip() on the contents string:

```
from pathlib import Path

path = Path('pi_digits.txt')
contents = path.read_text()
contents = contents.rstrip()
print(contents)
```

Recall from Chapter 2 that Python's rstrip() method removes, or strips, any whitespace characters from the right side of a string. Now the output matches the contents of the original file exactly:

```
3.1415926535
  8979323846
  2643383279
```

We can strip the trailing newline character when we read the contents of the file, by applying the rstrip() method immediately after calling read_text():

```
contents = path.read_text().rstrip()
```

This line tells Python to call the read_text() method on the file we're working with. Then it applies the rstrip() method to the string that read_text() returns. The cleaned-up string is then assigned to the variable contents. This approach is called *method chaining*, and you'll see it used often in programming.

Relative and Absolute File Paths

When you pass a simple filename like *pi_digits.txt* to `Path`, Python looks in the directory where the file that's currently being executed (that is, your *.py* program file) is stored.

Sometimes, depending on how you organize your work, the file you want to open won't be in the same directory as your program file. For example, you might store your program files in a folder called *python_work*; inside *python_work*, you might have another folder called *text_files* to distinguish your program files from the text files they're manipulating. Even though *text_files* is in *python_work*, just passing `Path` the name of a file in *text_files* won't work, because Python will only look in *python_work* and stop there; it won't go on and look in *text_files*. To get Python to open files from a directory other than the one where your program file is stored, you need to provide the correct path.

There are two main ways to specify paths in programming. A *relative file path* tells Python to look for a given location relative to the directory where the currently running program file is stored. Since *text_files* is inside *python_work*, we need to build a path that starts with the directory *text_files*, and ends with the filename. Here's how to build this path:

```
path = Path('text_files/filename.txt')
```

You can also tell Python exactly where the file is on your computer, regardless of where the program that's being executed is stored. This is called an *absolute file path*. You can use an absolute path if a relative path doesn't work. For instance, if you've put *text_files* in some folder other than *python_work*, then just passing `Path` the path `'text_files/filename.txt'` won't work because Python will only look for that location inside *python_work*. You'll need to write out an absolute path to clarify where you want Python to look.

Absolute paths are usually longer than relative paths, because they start at your system's root folder:

```
path = Path('/home/eric/data_files/text_files/filename.txt')
```

Using absolute paths, you can read files from any location on your system. For now it's easiest to store files in the same directory as your program files, or in a folder such as *text_files* within the directory that stores your program files.

NOTE *Windows systems use a backslash (\) instead of a forward slash (/) when displaying file paths, but you should use forward slashes in your code, even on Windows. The* `pathlib` *library will automatically use the correct representation of the path when it interacts with your system, or any user's system.*

Accessing a File's Lines

When you're working with a file, you'll often want to examine each line of the file. You might be looking for certain information in the file, or you might want to modify the text in the file in some way. For example, you might want to read through a file of weather data and work with any line that includes the word *sunny* in the description of that day's weather. In a news report, you might look for any line with the tag `<headline>` and rewrite that line with a specific kind of formatting.

You can use the `splitlines()` method to turn a long string into a set of lines, and then use a `for` loop to examine each line from a file, one at a time:

file_reader.py

```
from pathlib import Path

path = Path('pi_digits.txt')
❶ contents = path.read_text()

❷ lines = contents.splitlines()
for line in lines:
    print(line)
```

We start out by reading the entire contents of the file, as we did earlier ❶. If you're planning to work with the individual lines in a file, you don't need to strip any whitespace when reading the file. The `splitlines()` method returns a list of all lines in the file, and we assign this list to the variable `lines` ❷. We then loop over these lines and print each one:

```
3.1415926535
  8979323846
  2643383279
```

Since we haven't modified any of the lines, the output matches the original text file exactly.

Working with a File's Contents

After you've read the contents of a file into memory, you can do whatever you want with that data, so let's briefly explore the digits of *pi*. First, we'll attempt to build a single string containing all the digits in the file with no whitespace in it:

pi_string.py

```
from pathlib import Path

path = Path('pi_digits.txt')
contents = path.read_text()

lines = contents.splitlines()
pi_string = ''
❶ for line in lines:
    pi_string += line

print(pi_string)
print(len(pi_string))
```

We start by reading the file and storing each line of digits in a list, just as we did in the previous example. We then create a variable, `pi_string`, to hold the digits of *pi*. We write a loop that adds each line of digits to `pi_string` ❶. We print this string, and also show how long the string is:

```
3.1415926535  8979323846  2643383279
36
```

The variable `pi_string` contains the whitespace that was on the left side of the digits in each line, but we can get rid of that by using `lstrip()` on each line:

```
--snip--
for line in lines:
    pi_string += line.lstrip()

print(pi_string)
print(len(pi_string))
```

Now we have a string containing *pi* to 30 decimal places. The string is 32 characters long because it also includes the leading `3` and a decimal point:

```
3.141592653589793238462643383279
32
```

NOTE *When Python reads from a text file, it interprets all text in the file as a string. If you read in a number and want to work with that value in a numerical context, you'll have to convert it to an integer using the* `int()` *function or a float using the* `float()` *function.*

Large Files: One Million Digits

So far, we've focused on analyzing a text file that contains only three lines, but the code in these examples would work just as well on much larger files. If we start with a text file that contains *pi* to 1,000,000 decimal places, instead of just 30, we can create a single string containing all these digits. We don't need to change our program at all, except to pass it a different file. We'll also print just the first 50 decimal places, so we don't have to watch a million digits scroll by in the terminal:

pi_string.py

```
from pathlib import Path

path = Path('pi_million_digits.txt')
contents = path.read_text()

lines = contents.splitlines()
pi_string = ''
for line in lines:
    pi_string += line.lstrip()

print(f"{pi_string[:52]}...")
print(len(pi_string))
```

The output shows that we do indeed have a string containing *pi* to 1,000,000 decimal places:

```
3.14159265358979323846264338327950288419716939937510...
1000002
```

Python has no inherent limit to how much data you can work with; you can work with as much data as your system's memory can handle.

NOTE *To run this program (and many of the examples that follow), you'll need to download the resources available at* https://ehmatthes.github.io/pcc_3e.

Is Your Birthday Contained in Pi?

I've always been curious to know if my birthday appears anywhere in the digits of *pi*. Let's use the program we just wrote to find out if someone's birthday appears anywhere in the first million digits of *pi*. We can do this by expressing each birthday as a string of digits and seeing if that string appears anywhere in `pi_string`:

pi_birthday.py

```
--snip--
for line in lines:
    pi_string += line.strip()

birthday = input("Enter your birthday, in the form mmddyy: ")
if birthday in pi_string:
    print("Your birthday appears in the first million digits of pi!")
else:
    print("Your birthday does not appear in the first million digits of pi.")
```

We first prompt for the user's birthday, and then check if that string is in `pi_string`. Let's try it:

```
Enter your birthdate, in the form mmddyy: 120372
Your birthday appears in the first million digits of pi!
```

My birthday does appear in the digits of *pi*! Once you've read from a file, you can analyze its contents in just about any way you can imagine.

TRY IT YOURSELF

10-1. Learning Python: Open a blank file in your text editor and write a few lines summarizing what you've learned about Python so far. Start each line with the phrase *In Python you can. . . .* Save the file as *learning_python.txt* in the same directory as your exercises from this chapter. Write a program that reads the file and prints what you wrote two times: print the contents once by reading in the entire file, and once by storing the lines in a list and then looping over each line.

(continued)

10-2. Learning C: You can use the `replace()` method to replace any word in a string with a different word. Here's a quick example showing how to replace `'dog'` with `'cat'` in a sentence:

```
>>> message = "I really like dogs."
>>> message.replace('dog', 'cat')
'I really like cats.'
```

Read in each line from the file you just created, *learning_python.txt*, and replace the word *Python* with the name of another language, such as *C*. Print each modified line to the screen.

10-3. Simpler Code: The program *file_reader.py* in this section uses a temporary variable, `lines`, to show how `splitlines()` works. You can skip the temporary variable and loop directly over the list that `splitlines()` returns:

```
for line in contents.splitlines():
```

Remove the temporary variable from each of the programs in this section, to make them more concise.

Writing to a File

One of the simplest ways to save data is to write it to a file. When you write text to a file, the output will still be available after you close the terminal containing your program's output. You can examine output after a program finishes running, and you can share the output files with others as well. You can also write programs that read the text back into memory and work with it again later.

Writing a Single Line

Once you have a path defined, you can write to a file using the `write_text()` method. To see how this works, let's write a simple message and store it in a file instead of printing it to the screen:

write_message.py

```
from pathlib import Path

path = Path('programming.txt')
path.write_text("I love programming.")
```

The `write_text()` method takes a single argument: the string that you want to write to the file. This program has no terminal output, but if you open the file *programming.txt*, you'll see one line:

programming.txt

```
I love programming.
```

This file behaves like any other file on your computer. You can open it, write new text in it, copy from it, paste to it, and so forth.

NOTE *Python can only write strings to a text file. If you want to store numerical data in a text file, you'll have to convert the data to string format first using the* `str()` *function.*

Writing Multiple Lines

The `write_text()` method does a few things behind the scenes. If the file that `path` points to doesn't exist, it creates that file. Also, after writing the string to the file, it makes sure the file is closed properly. Files that aren't closed properly can lead to missing or corrupted data.

To write more than one line to a file, you need to build a string containing the entire contents of the file, and then call `write_text()` with that string. Let's write several lines to the *programming.txt* file:

```
from pathlib import Path

contents = "I love programming.\n"
contents += "I love creating new games.\n"
contents += "I also love working with data.\n"

path = Path('programming.txt')
path.write_text(contents)
```

We define a variable called `contents` that will hold the entire contents of the file. On the next line, we use the `+=` operator to add to this string. You can do this as many times as you need, to build strings of any length. In this case we include newline characters at the end of each line, to make sure each statement appears on its own line.

If you run this and then open *programming.txt*, you'll see each of these lines in the text file:

```
I love programming.
I love creating new games.
I also love working with data.
```

You can also use spaces, tab characters, and blank lines to format your output, just as you've been doing with terminal-based output. There's no limit to the length of your strings, and this is how many computer-generated documents are created.

NOTE *Be careful when calling* `write_text()` *on a path object. If the file already exists,* `write_text()` *will erase the current contents of the file and write new contents to the file. Later in this chapter, you'll learn to check whether a file exists using* `pathlib`*.*

> **TRY IT YOURSELF**
>
> **10-4. Guest:** Write a program that prompts the user for their name. When they respond, write their name to a file called *guest.txt*.
>
> **10-5. Guest Book:** Write a `while` loop that prompts users for their name. Collect all the names that are entered, and then write these names to a file called *guest_book.txt*. Make sure each entry appears on a new line in the file.

Exceptions

Python uses special objects called *exceptions* to manage errors that arise during a program's execution. Whenever an error occurs that makes Python unsure of what to do next, it creates an exception object. If you write code that handles the exception, the program will continue running. If you don't handle the exception, the program will halt and show a *traceback*, which includes a report of the exception that was raised.

Exceptions are handled with `try-except` blocks. A *try-except* block asks Python to do something, but it also tells Python what to do if an exception is raised. When you use `try-except` blocks, your programs will continue running even if things start to go wrong. Instead of tracebacks, which can be confusing for users to read, users will see friendly error messages that you've written.

Handling the ZeroDivisionError Exception

Let's look at a simple error that causes Python to raise an exception. You probably know that it's impossible to divide a number by zero, but let's ask Python to do it anyway:

division_calculator.py

```
print(5/0)
```

Python can't do this, so we get a traceback:

```
Traceback (most recent call last):
  File "division_calculator.py", line 1, in <module>
    print(5/0)
          ~^~
❶ ZeroDivisionError: division by zero
```

The error reported in the traceback, `ZeroDivisionError`, is an exception object ❶. Python creates this kind of object in response to a situation where it can't do what we ask it to. When this happens, Python stops the program and tells us the kind of exception that was raised. We can use this information to modify our program. We'll tell Python what to do when this kind of exception occurs; that way, if it happens again, we'll be prepared.

Using try-except Blocks

When you think an error may occur, you can write a try-except block to handle the exception that might be raised. You tell Python to try running some code, and you tell it what to do if the code results in a particular kind of exception.

Here's what a try-except block for handling the ZeroDivisionError exception looks like:

```
try:
    print(5/0)
except ZeroDivisionError:
    print("You can't divide by zero!")
```

We put print(5/0), the line that caused the error, inside a try block. If the code in a try block works, Python skips over the except block. If the code in the try block causes an error, Python looks for an except block whose error matches the one that was raised, and runs the code in that block.

In this example, the code in the try block produces a ZeroDivisionError, so Python looks for an except block telling it how to respond. Python then runs the code in that block, and the user sees a friendly error message instead of a traceback:

```
You can't divide by zero!
```

If more code followed the try-except block, the program would continue running because we told Python how to handle the error. Let's look at an example where catching an error can allow a program to continue running.

Using Exceptions to Prevent Crashes

Handling errors correctly is especially important when the program has more work to do after the error occurs. This happens often in programs that prompt users for input. If the program responds to invalid input appropriately, it can prompt for more valid input instead of crashing.

Let's create a simple calculator that does only division:

division_calculator.py

```
print("Give me two numbers, and I'll divide them.")
print("Enter 'q' to quit.")

while True:
❶    first_number = input("\nFirst number: ")
    if first_number == 'q':
        break
❷    second_number = input("Second number: ")
    if second_number == 'q':
        break
❸    answer = int(first_number) / int(second_number)
    print(answer)
```

This program prompts the user to input a `first_number` ❶ and, if the user does not enter `q` to quit, a `second_number` ❷. We then divide these two numbers to get an `answer` ❸. This program does nothing to handle errors, so asking it to divide by zero causes it to crash:

```
Give me two numbers, and I'll divide them.
Enter 'q' to quit.

First number: 5
Second number: 0
Traceback (most recent call last):
  File "division_calculator.py", line 11, in <module>
    answer = int(first_number) / int(second_number)
             ~~~~~~~~~~~~~~~~~~^~~~~~~~~~~~~~~~~~~~
ZeroDivisionError: division by zero
```

It's bad that the program crashed, but it's also not a good idea to let users see tracebacks. Nontechnical users will be confused by them, and in a malicious setting, attackers will learn more than you want them to. For example, they'll know the name of your program file, and they'll see a part of your code that isn't working properly. A skilled attacker can sometimes use this information to determine which kind of attacks to use against your code.

The else Block

We can make this program more error resistant by wrapping the line that might produce errors in a `try-except` block. The error occurs on the line that performs the division, so that's where we'll put the `try-except` block. This example also includes an `else` block. Any code that depends on the `try` block executing successfully goes in the `else` block:

```
--snip--
while True:
    --snip--
    if second_number == 'q':
        break
❶   try:
        answer = int(first_number) / int(second_number)
❷   except ZeroDivisionError:
        print("You can't divide by 0!")
❸   else:
        print(answer)
```

We ask Python to try to complete the division operation in a `try` block ❶, which includes only the code that might cause an error. Any code that depends on the `try` block succeeding is added to the `else` block. In this case, if the division operation is successful, we use the `else` block to print the result ❸.

The `except` block tells Python how to respond when a `ZeroDivisionError` arises ❷. If the `try` block doesn't succeed because of a division-by-zero error, we print a friendly message telling the user how to avoid this kind of error. The program continues to run, and the user never sees a traceback:

```
Give me two numbers, and I'll divide them.
Enter 'q' to quit.

First number: 5
Second number: 0
You can't divide by 0!

First number: 5
Second number: 2
2.5

First number: q
```

The only code that should go in a try block is code that might cause an exception to be raised. Sometimes you'll have additional code that should run only if the try block was successful; this code goes in the else block. The except block tells Python what to do in case a certain exception arises when it tries to run the code in the try block.

By anticipating likely sources of errors, you can write robust programs that continue to run even when they encounter invalid data and missing resources. Your code will be resistant to innocent user mistakes and malicious attacks.

Handling the FileNotFoundError Exception

One common issue when working with files is handling missing files. The file you're looking for might be in a different location, the filename might be misspelled, or the file might not exist at all. You can handle all of these situations with a try-except block.

Let's try to read a file that doesn't exist. The following program tries to read in the contents of *Alice in Wonderland*, but I haven't saved the file *alice.txt* in the same directory as *alice.py*:

alice.py

```
from pathlib import Path

path = Path('alice.txt')
contents = path.read_text(encoding='utf-8')
```

Note that we're using read_text() in a slightly different way here than what you saw earlier. The encoding argument is needed when your system's default encoding doesn't match the encoding of the file that's being read. This is most likely to happen when reading from a file that wasn't created on your system.

Python can't read from a missing file, so it raises an exception:

```
Traceback (most recent call last):
❶ File "alice.py", line 4, in <module>
❷   contents = path.read_text(encoding='utf-8')
               ^^^^^^^^^^^^^^^^^^^^^^^^^^^^^^^^^
```

```
  File "/.../pathlib.py", line 1056, in read_text
    with self.open(mode='r', encoding=encoding, errors=errors) as f:
         ^^^^^^^^^^^^^^^^^^^^^^^^^^^^^^^^^^^^^^^^^^^^^^^^^^^^^^^^^^^
  File "/.../pathlib.py", line 1042, in open
    return io.open(self, mode, buffering, encoding, errors, newline)
           ^^^^^^^^^^^^^^^^^^^^^^^^^^^^^^^^^^^^^^^^^^^^^^^^^^^^^^^^^^^^^
❸ FileNotFoundError: [Errno 2] No such file or directory: 'alice.txt'
```

This is a longer traceback than the ones we've seen previously, so let's look at how you can make sense of more complex tracebacks. It's often best to start at the very end of the traceback. On the last line, we can see that a `FileNotFoundError` exception was raised ❸. This is important because it tells us what kind of exception to use in the `except` block that we'll write.

Looking back near the beginning of the traceback ❶, we can see that the error occurred at line 4 in the file *alice.py*. The next line shows the line of code that caused the error ❷. The rest of the traceback shows some code from the libraries that are involved in opening and reading from files. You don't usually need to read through or understand all of these lines in a traceback.

To handle the error that's being raised, the `try` block will begin with the line that was identified as problematic in the traceback. In our example, this is the line that contains `read_text()`:

```
from pathlib import Path

path = Path('alice.txt')
try:
    contents = path.read_text(encoding='utf-8')
❶ except FileNotFoundError:
    print(f"Sorry, the file {path} does not exist.")
```

In this example, the code in the `try` block produces a `FileNotFoundError`, so we write an `except` block that matches that error ❶. Python then runs the code in that block when the file can't be found, and the result is a friendly error message instead of a traceback:

```
Sorry, the file alice.txt does not exist.
```

The program has nothing more to do if the file doesn't exist, so this is all the output we see. Let's build on this example and see how exception handling can help when you're working with more than one file.

Analyzing Text

You can analyze text files containing entire books. Many classic works of literature are available as simple text files because they are in the public domain. The texts used in this section come from Project Gutenberg (*https://gutenberg.org*). Project Gutenberg maintains a collection of literary works that are available in the public domain, and it's a great resource if you're interested in working with literary texts in your programming projects.

Let's pull in the text of *Alice in Wonderland* and try to count the number of words in the text. To do this, we'll use the string method `split()`, which by default splits a string wherever it finds any whitespace:

```
from pathlib import Path

path = Path('alice.txt')
try:
    contents = path.read_text(encoding='utf-8')
except FileNotFoundError:
    print(f"Sorry, the file {path} does not exist.")
else:
    # Count the approximate number of words in the file:
❶   words = contents.split()
❷   num_words = len(words)
    print(f"The file {path} has about {num_words} words.")
```

I moved the file *alice.txt* to the correct directory, so the `try` block will work this time. We take the string `contents`, which now contains the entire text of *Alice in Wonderland* as one long string, and use `split()` to produce a list of all the words in the book ❶. Using `len()` on this list ❷ gives us a good approximation of the number of words in the original text. Lastly, we print a statement that reports how many words were found in the file. This code is placed in the `else` block because it only works if the code in the `try` block was executed successfully.

The output tells us how many words are in *alice.txt*:

```
The file alice.txt has about 29594 words.
```

The count is a little high because extra information is provided by the publisher in the text file used here, but it's a good approximation of the length of *Alice in Wonderland*.

Working with Multiple Files

Let's add more books to analyze, but before we do, let's move the bulk of this program to a function called `count_words()`. This will make it easier to run the analysis for multiple books:

word_count.py

```
from pathlib import Path

def count_words(path):
❶   """Count the approximate number of words in a file."""
    try:
        contents = path.read_text(encoding='utf-8')
    except FileNotFoundError:
        print(f"Sorry, the file {path} does not exist.")
    else:
        # Count the approximate number of words in the file:
        words = contents.split()
        num_words = len(words)
        print(f"The file {path} has about {num_words} words.")
```

```
path = Path('alice.txt')
count_words(path)
```

Most of this code is unchanged. It's only been indented, and moved into the body of `count_words()`. It's a good habit to keep comments up to date when you're modifying a program, so the comment has also been changed to a docstring and reworded slightly ❶.

Now we can write a short loop to count the words in any text we want to analyze. We do this by storing the names of the files we want to analyze in a list, and then we call `count_words()` for each file in the list. We'll try to count the words for *Alice in Wonderland*, *Siddhartha*, *Moby Dick*, and *Little Women*, which are all available in the public domain. I've intentionally left *siddhartha.txt* out of the directory containing *word_count.py*, so we can see how well our program handles a missing file:

```
from pathlib import Path

def count_words(filename):
    --snip--

filenames = ['alice.txt', 'siddhartha.txt', 'moby_dick.txt',
        'little_women.txt']
for filename in filenames:
❶   path = Path(filename)
    count_words(path)
```

The names of the files are stored as simple strings. Each string is then converted to a `Path` object ❶, before the call to `count_words()`. The missing *siddhartha.txt* file has no effect on the rest of the program's execution:

```
The file alice.txt has about 29594 words.
Sorry, the file siddhartha.txt does not exist.
The file moby_dick.txt has about 215864 words.
The file little_women.txt has about 189142 words.
```

Using the `try-except` block in this example provides two significant advantages. We prevent our users from seeing a traceback, and we let the program continue analyzing the texts it's able to find. If we don't catch the `FileNotFoundError` that *siddhartha.txt* raises, the user would see a full traceback, and the program would stop running after trying to analyze *Siddhartha*. It would never analyze *Moby Dick* or *Little Women*.

Failing Silently

In the previous example, we informed our users that one of the files was unavailable. But you don't need to report every exception you catch. Sometimes, you'll want the program to fail silently when an exception occurs and continue on as if nothing happened. To make a program fail silently, you write a `try` block as usual, but you explicitly tell Python to do nothing in the except block. Python has a pass statement that tells it to do nothing in a block:

```
def count_words(path):
    """Count the approximate number of words in a file."""
    try:
        --snip--
    except FileNotFoundError:
        pass
    else:
        --snip--
```

The only difference between this listing and the previous one is the pass statement in the except block. Now when a FileNotFoundError is raised, the code in the except block runs, but nothing happens. No traceback is produced, and there's no output in response to the error that was raised. Users see the word counts for each file that exists, but they don't see any indication that a file wasn't found:

```
The file alice.txt has about 29594 words.
The file moby_dick.txt has about 215864 words.
The file little_women.txt has about 189142 words.
```

The pass statement also acts as a placeholder. It's a reminder that you're choosing to do nothing at a specific point in your program's execution and that you might want to do something there later. For example, in this program we might decide to write any missing filenames to a file called *missing_files.txt*. Our users wouldn't see this file, but we'd be able to read the file and deal with any missing texts.

Deciding Which Errors to Report

How do you know when to report an error to your users and when to let your program fail silently? If users know which texts are supposed to be analyzed, they might appreciate a message informing them why some texts were not analyzed. If users expect to see some results but don't know which books are supposed to be analyzed, they might not need to know that some texts were unavailable. Giving users information they aren't looking for can decrease the usability of your program. Python's error-handling structures give you fine-grained control over how much to share with users when things go wrong; it's up to you to decide how much information to share.

Well-written, properly tested code is not very prone to internal errors, such as syntax or logical errors. But every time your program depends on something external such as user input, the existence of a file, or the availability of a network connection, there is a possibility of an exception being raised. A little experience will help you know where to include exception-handling blocks in your program and how much to report to users about errors that arise.

TRY IT YOURSELF

10-6. Addition: One common problem when prompting for numerical input occurs when people provide text instead of numbers. When you try to convert the input to an `int`, you'll get a `ValueError`. Write a program that prompts for two numbers. Add them together and print the result. Catch the `ValueError` if either input value is not a number, and print a friendly error message. Test your program by entering two numbers and then by entering some text instead of a number.

10-7. Addition Calculator: Wrap your code from Exercise 10-6 in a `while` loop so the user can continue entering numbers, even if they make a mistake and enter text instead of a number.

10-8. Cats and Dogs: Make two files, *cats.txt* and *dogs.txt*. Store at least three names of cats in the first file and three names of dogs in the second file. Write a program that tries to read these files and print the contents of the file to the screen. Wrap your code in a `try-except` block to catch the `FileNotFound` error, and print a friendly message if a file is missing. Move one of the files to a different location on your system, and make sure the code in the `except` block executes properly.

10-9. Silent Cats and Dogs: Modify your `except` block in Exercise 10-8 to fail silently if either file is missing.

10-10. Common Words: Visit Project Gutenberg (*https://gutenberg.org*) and find a few texts you'd like to analyze. Download the text files for these works, or copy the raw text from your browser into a text file on your computer.

You can use the `count()` method to find out how many times a word or phrase appears in a string. For example, the following code counts the number of times `'row'` appears in a string:

```
>>> line = "Row, row, row your boat"
>>> line.count('row')
2
>>> line.lower().count('row')
3
```

Notice that converting the string to lowercase using `lower()` catches all appearances of the word you're looking for, regardless of how it's formatted.

Write a program that reads the files you found at Project Gutenberg and determines how many times the word `'the'` appears in each text. This will be an approximation because it will also count words such as `'then'` and `'there'`. Try counting `'the '`, with a space in the string, and see how much lower your count is.

Storing Data

Many of your programs will ask users to input certain kinds of information. You might allow users to store preferences in a game or provide data for a visualization. Whatever the focus of your program is, you'll store the information users provide in data structures such as lists and dictionaries. When users close a program, you'll almost always want to save the information they entered. A simple way to do this involves storing your data using the `json` module.

The `json` module allows you to convert simple Python data structures into JSON-formatted strings, and then load the data from that file the next time the program runs. You can also use `json` to share data between different Python programs. Even better, the JSON data format is not specific to Python, so you can share data you store in the JSON format with people who work in many other programming languages. It's a useful and portable format, and it's easy to learn.

NOTE *The* JSON (JavaScript Object Notation) *format was originally developed for JavaScript. However, it has since become a common format used by many languages, including Python.*

Using json.dumps() and json.loads()

Let's write a short program that stores a set of numbers and another program that reads these numbers back into memory. The first program will use `json.dumps()` to store the set of numbers, and the second program will use `json.loads()`.

The `json.dumps()` function takes one argument: a piece of data that should be converted to the JSON format. The function returns a string, which we can then write to a data file:

number_writer.py

```
from pathlib import Path
import json

numbers = [2, 3, 5, 7, 11, 13]

❶ path = Path('numbers.json')
❷ contents = json.dumps(numbers)
path.write_text(contents)
```

We first import the `json` module, and then create a list of numbers to work with. Then we choose a filename in which to store the list of numbers ❶. It's customary to use the file extension *.json* to indicate that the data in the file is stored in the JSON format. Next, we use the `json.dumps()` ❷ function to generate a string containing the JSON representation of the data we're working with. Once we have this string, we write it to the file using the same `write_text()` method we used earlier.

This program has no output, but let's open the file *numbers.json* and look at it. The data is stored in a format that looks just like Python:

```
[2, 3, 5, 7, 11, 13]
```

Now we'll write a separate program that uses `json.loads()` to read the list back into memory:

number_reader.py

```
from pathlib import Path
import json

❶ path = Path('numbers.json')
❷ contents = path.read_text()
❸ numbers = json.loads(contents)

print(numbers)
```

We make sure to read from the same file we wrote to ❶. Since the data file is just a text file with specific formatting, we can read it with the `read_text()` method ❷. We then pass the contents of the file to `json.loads()` ❸. This function takes in a JSON-formatted string and returns a Python object (in this case, a list), which we assign to `numbers`. Finally, we print the recovered list of numbers and see that it's the same list created in *number_writer.py*:

```
[2, 3, 5, 7, 11, 13]
```

This is a simple way to share data between two programs.

Saving and Reading User-Generated Data

Saving data with `json` is useful when you're working with user-generated data, because if you don't store your user's information somehow, you'll lose it when the program stops running. Let's look at an example where we prompt the user for their name the first time they run a program and then remember their name when they run the program again.

Let's start by storing the user's name:

remember_me.py

```
from pathlib import Path
import json

❶ username = input("What is your name? ")

❷ path = Path('username.json')
contents = json.dumps(username)
path.write_text(contents)

❸ print(f"We'll remember you when you come back, {username}!")
```

We first prompt for a username to store ❶. Next, we write the data we just collected to a file called *username.json* ❷. Then we print a message informing the user that we've stored their information ❸:

```
What is your name? Eric
We'll remember you when you come back, Eric!
```

Now let's write a new program that greets a user whose name has already been stored:

greet_user.py

```
from pathlib import Path
import json

❶ path = Path('username.json')
contents = path.read_text()
❷ username = json.loads(contents)

print(f"Welcome back, {username}!")
```

We read the contents of the data file ❶ and then use `json.loads()` to assign the recovered data to the variable `username` ❷. Since we've recovered the username, we can welcome the user back with a personalized greeting:

```
Welcome back, Eric!
```

We need to combine these two programs into one file. When someone runs *remember_me.py*, we want to retrieve their username from memory if possible; if not, we'll prompt for a username and store it in *username.json* for next time. We could write a `try-except` block here to respond appropriately if *username.json* doesn't exist, but instead we'll use a handy method from the `pathlib` module:

remember_me.py

```
from pathlib import Path
import json

path = Path('username.json')
❶ if path.exists():
    contents = path.read_text()
    username = json.loads(contents)
    print(f"Welcome back, {username}!")
❷ else:
    username = input("What is your name? ")
    contents = json.dumps(username)
    path.write_text(contents)
    print(f"We'll remember you when you come back, {username}!")
```

There are many helpful methods you can use with `Path` objects. The `exists()` method returns `True` if a file or folder exists and `False` if it doesn't. Here we use `path.exists()` to find out if a username has already been stored ❶. If *username.json* exists, we load the username and print a personalized greeting to the user.

If the file *username.json* doesn't exist ❷, we prompt for a username and store the value that the user enters. We also print the familiar message that we'll remember them when they come back.

Whichever block executes, the result is a username and an appropriate greeting. If this is the first time the program runs, this is the output:

```
What is your name? Eric
We'll remember you when you come back, Eric!
```

Otherwise:

```
Welcome back, Eric!
```

This is the output you see if the program was already run at least once. Even though the data in this section is just a single string, the program would work just as well with any data that can be converted to a JSON-formatted string.

Refactoring

Often, you'll come to a point where your code will work, but you'll recognize that you could improve the code by breaking it up into a series of functions that have specific jobs. This process is called *refactoring*. Refactoring makes your code cleaner, easier to understand, and easier to extend.

We can refactor *remember_me.py* by moving the bulk of its logic into one or more functions. The focus of *remember_me.py* is on greeting the user, so let's move all of our existing code into a function called greet_user():

remember_me.py

```
from pathlib import Path
import json

def greet_user():
❶    """Greet the user by name."""
    path = Path('username.json')
    if path.exists():
        contents = path.read_text()
        username = json.loads(contents)
        print(f"Welcome back, {username}!")
    else:
        username = input("What is your name? ")
        contents = json.dumps(username)
        path.write_text(contents)
        print(f"We'll remember you when you come back, {username}!")

greet_user()
```

Because we're using a function now, we rewrite the comments as a docstring that reflects how the program currently works ❶. This file is a little cleaner, but the function greet_user() is doing more than just greeting the user—it's also retrieving a stored username if one exists and prompting for a new username if one doesn't.

Let's refactor greet_user() so it's not doing so many different tasks. We'll start by moving the code for retrieving a stored username to a separate function:

```
from pathlib import Path
import json

def get_stored_username(path):
❶    """Get stored username if available."""
```

```
    if path.exists():
        contents = path.read_text()
        username = json.loads(contents)
        return username
    else:
❷       return None

def greet_user():
    """Greet the user by name."""
    path = Path('username.json')
    username = get_stored_username(path)
❸   if username:
        print(f"Welcome back, {username}!")
    else:
        username = input("What is your name? ")
        contents = json.dumps(username)
        path.write_text(contents)
        print(f"We'll remember you when you come back, {username}!")

greet_user()
```

The new function get_stored_username() ❶ has a clear purpose, as stated in the docstring. This function retrieves a stored username and returns the username if it finds one. If the path that's passed to get_stored_username() doesn't exist, the function returns None ❷. This is good practice: a function should either return the value you're expecting, or it should return None. This allows us to perform a simple test with the return value of the function. We print a welcome back message to the user if the attempt to retrieve a username is successful ❸, and if it isn't, we prompt for a new username.

We should factor one more block of code out of greet_user(). If the username doesn't exist, we should move the code that prompts for a new username to a function dedicated to that purpose:

```
from pathlib import Path
import json

def get_stored_username(path):
    """Get stored username if available."""
    --snip--

def get_new_username(path):
    """Prompt for a new username."""
    username = input("What is your name? ")
    contents = json.dumps(username)
    path.write_text(contents)
    return username

def greet_user():
    """Greet the user by name."""
    path = Path('username.json')
❶   username = get_stored_username(path)
    if username:
        print(f"Welcome back, {username}!")
```

```
    else:
❷       username = get_new_username(path)
        print(f"We'll remember you when you come back, {username}!")

greet_user()
```

Each function in this final version of *remember_me.py* has a single, clear purpose. We call greet_user(), and that function prints an appropriate message: it either welcomes back an existing user or greets a new user. It does this by calling get_stored_username() ❶, which is responsible only for retrieving a stored username if one exists. Finally, if necessary, greet_user() calls get_new_username() ❷, which is responsible only for getting a new username and storing it. This compartmentalization of work is an essential part of writing clear code that will be easy to maintain and extend.

TRY IT YOURSELF

10-11. Favorite Number: Write a program that prompts for the user's favorite number. Use json.dumps() to store this number in a file. Write a separate program that reads in this value and prints the message "I know your favorite number! It's _____."

10-12. Favorite Number Remembered: Combine the two programs you wrote in Exercise 10-11 into one file. If the number is already stored, report the favorite number to the user. If not, prompt for the user's favorite number and store it in a file. Run the program twice to see that it works.

10-13. User Dictionary: The *remember_me.py* example only stores one piece of information, the username. Expand this example by asking for two more pieces of information about the user, then store all the information you collect in a dictionary. Write this dictionary to a file using json.dumps(), and read it back in using json.loads(). Print a summary showing exactly what your program remembers about the user.

10-14. Verify User: The final listing for *remember_me.py* assumes either that the user has already entered their username or that the program is running for the first time. We should modify it in case the current user is not the person who last used the program.

Before printing a welcome back message in greet_user(), ask the user if this is the correct username. If it's not, call get_new_username() to get the correct username.

Summary

In this chapter, you learned how to work with files. You learned to read the entire contents of a file, and then work through the contents one line at a time if you need to. You learned to write as much text as you want to a file. You also read about exceptions and how to handle the exceptions you're likely to see in your programs. Finally, you learned how to store Python data structures so you can save information your users provide, preventing them from having to start over each time they run a program.

In Chapter 11, you'll learn efficient ways to test your code. This will help you trust that the code you develop is correct, and it will help you identify bugs that are introduced as you continue to build on the programs you've written.

11

TESTING YOUR CODE

When you write a function or a class, you can also write tests for that code. Testing proves that your code works as it's supposed to in response to all the kinds of input it's designed to receive. When you write tests, you can be confident that your code will work correctly as more people begin to use your programs. You'll also be able to test new code as you add it, to make sure your changes don't break your program's existing behavior. Every programmer makes mistakes, so every programmer must test their code often, to catch problems before users encounter them.

In this chapter, you'll learn to test your code using `pytest`. The `pytest` library is a collection of tools that will help you write your first tests quickly and simply, while supporting your tests as they grow in complexity along with your projects. Python doesn't include `pytest` by default, so you'll learn to install external libraries. Knowing how to install external libraries will make a wide variety of well-designed code available to you. These libraries will expand the kinds of projects you can work on immensely.

You'll learn to build a series of tests and check that each set of inputs results in the output you want. You'll see what a passing test looks like and what a failing test looks like, and you'll learn how a failing test can help you improve your code. You'll learn to test functions and classes, and you'll start to understand how many tests to write for a project.

Installing pytest with pip

While Python includes a lot of functionality in the standard library, Python developers also depend heavily on third-party packages. A *third-party package* is a library that's developed outside the core Python language. Some popular third-party libraries are eventually adopted into the standard library, and end up being included in most Python installations from that point forward. This happens most often with libraries that are unlikely to change much once they've had their initial bugs worked out. These kinds of libraries can evolve at the same pace as the overall language.

Many packages, however, are kept out of the standard library so they can be developed on a timeline independent of the language itself. These packages tend to be updated more frequently than they would be if they were tied to Python's development schedule. This is true of `pytest` and most of the libraries we'll use in the second half of this book. You shouldn't blindly trust every third-party package, but you also shouldn't be put off by the fact that a lot of important functionality is implemented through such packages.

Updating pip

Python includes a tool called pip that's used to install third-party packages. Because pip helps install packages from external resources, it's updated often to address potential security issues. So, we'll start by updating pip.

Open a new terminal window and issue the following command:

```
$ python -m pip install --upgrade pip
❶ Requirement already satisfied: pip in /.../python3.11/site-packages (22.0.4)
--snip--
❷ Successfully installed pip-22.1.2
```

The first part of this command, **`python -m pip`**, tells Python to run the module `pip`. The second part, **`install --upgrade`**, tells pip to update a package that's already been installed. The last part, **`pip`**, specifies which third-party package should be updated. The output shows that my current version of pip, version 22.0.4 ❶, was replaced by the latest version at the time of this writing, 22.1.2 ❷.

You can use this command to update any third-party package installed on your system:

```
$ python -m pip install --upgrade package_name
```

NOTE *If you're using Linux, pip may not be included with your installation of Python. If you get an error when trying to upgrade pip, see the instructions in Appendix A.*

Installing pytest

Now that pip is up to date, we can install pytest:

```
$ python -m pip install --user pytest
Collecting pytest
  --snip--
Successfully installed attrs-21.4.0 iniconfig-1.1.1 ...pytest-7.x.x
```

We're still using the core command **pip install**, without the **--upgrade** flag this time. Instead, we're using the **--user** flag, which tells Python to install this package for the current user only. The output shows that the latest version of pytest was successfully installed, along with a number of other packages that pytest depends on.

You can use this command to install many third-party packages:

```
$ python -m pip install --user package_name
```

If you have any difficulty running this command, try running the same command without the **--user** *flag.*

Testing a Function

To learn about testing, we need code to test. Here's a simple function that takes in a first and last name, and returns a neatly formatted full name:

name_function.py

```
def get_formatted_name(first, last):
    """Generate a neatly formatted full name."""
    full_name = f"{first} {last}"
    return full_name.title()
```

The function get_formatted_name() combines the first and last name with a space in between to complete a full name, and then capitalizes and returns the full name. To check that get_formatted_name() works, let's make a program that uses this function. The program *names.py* lets users enter a first and last name, and see a neatly formatted full name:

names.py

```
from name_function import get_formatted_name

print("Enter 'q' at any time to quit.")
while True:
    first = input("\nPlease give me a first name: ")
    if first == 'q':
        break
    last = input("Please give me a last name: ")
    if last == 'q':
        break

    formatted_name = get_formatted_name(first, last)
    print(f"\tNeatly formatted name: {formatted_name}.")
```

This program imports get_formatted_name() from *name_function.py*. The user can enter a series of first and last names and see the formatted full names that are generated:

```
Enter 'q' at any time to quit.

Please give me a first name: janis
Please give me a last name: joplin
        Neatly formatted name: Janis Joplin.

Please give me a first name: bob
Please give me a last name: dylan
        Neatly formatted name: Bob Dylan.

Please give me a first name: q
```

We can see that the names generated here are correct. But say we want to modify get_formatted_name() so it can also handle middle names. As we do so, we want to make sure we don't break the way the function handles names that have only a first and last name. We could test our code by running *names.py* and entering a name like Janis Joplin every time we modify get_formatted_name(), but that would become tedious. Fortunately, pytest provides an efficient way to automate the testing of a function's output. If we automate the testing of get_formatted_name(), we can always be confident that the function will work when given the kinds of names we've written tests for.

Unit Tests and Test Cases

There is a wide variety of approaches to testing software. One of the simplest kinds of test is a unit test. A *unit test* verifies that one specific aspect of a function's behavior is correct. A *test case* is a collection of unit tests that together prove that a function behaves as it's supposed to, within the full range of situations you expect it to handle.

A good test case considers all the possible kinds of input a function could receive and includes tests to represent each of these situations. A test case with *full coverage* includes a full range of unit tests covering all the possible ways you can use a function. Achieving full coverage on a large project can be daunting. It's often good enough to write tests for your code's critical behaviors and then aim for full coverage only if the project starts to see widespread use.

A Passing Test

With pytest, writing your first unit test is pretty straightforward. We'll write a single test function. The test function will call the function we're testing, and we'll make an assertion about the value that's returned. If our assertion is correct, the test will pass; if the assertion is incorrect, the test will fail.

Here's the first test of the function `get_formatted_name()`:

test_name_function.py

```
from name_function import get_formatted_name

❶ def test_first_last_name():
      """Do names like 'Janis Joplin' work?"""
❷     formatted_name = get_formatted_name('janis', 'joplin')
❸     assert formatted_name == 'Janis Joplin'
```

Before we run the test, let's take a closer look at this function. The name of a test file is important; it must start with *test_*. When we ask `pytest` to run the tests we've written, it will look for any file that begins with *test_*, and run all of the tests it finds in that file.

In the test file, we first import the function that we want to test: `get_formatted_name()`. Then we define a test function: in this case, `test_first_last_name()` ❶. This is a longer function name than we've been using, for a good reason. First, test functions need to start with the word *test*, followed by an underscore. Any function that starts with `test_` will be *discovered* by `pytest`, and will be run as part of the testing process.

Also, test names should be longer and more descriptive than a typical function name. You'll never call the function yourself; `pytest` will find the function and run it for you. Test function names should be long enough that if you see the function name in a test report, you'll have a good sense of what behavior was being tested.

Next, we call the function we're testing ❷. Here we call `get_formatted_name()` with the arguments `'janis'` and `'joplin'`, just like we used when we ran *names.py*. We assign the return value of this function to `formatted_name`.

Finally, we make an assertion ❸. An *assertion* is a claim about a condition. Here we're claiming that the value of `formatted_name` should be `'Janis Joplin'`.

Running a Test

If you run the file *test_name_function.py* directly, you won't get any output because we never called the test function. Instead, we'll have `pytest` run the test file for us.

To do this, open a terminal window and navigate to the folder that contains the test file. If you're using VS Code, you can open the folder containing the test file and use the terminal that's embedded in the editor window. In the terminal window, enter the command **`pytest`**. Here's what you should see:

```
$ pytest
========================= test session starts =========================
❶ platform darwin -- Python 3.x.x, pytest-7.x.x, pluggy-1.x.x
❷ rootdir: /.../python_work/chapter_11
❸ collected 1 item

❹ test_name_function.py .                                        [100%]
========================== 1 passed in 0.00s ==========================
```

Let's try to make sense of this output. First of all, we see some information about the system the test is running on ❶. I'm testing this on a macOS system, so you may see some different output here. Most importantly, we can see which versions of Python, pytest, and other packages are being used to run the test.

Next, we see the directory where the test is being run from ❷: in my case, *python_work/chapter_11*. We can see that pytest found one test to run ❸, and we can see the test file that's being run ❹. The single dot after the name of the file tells us that a single test passed, and the 100% makes it clear that all of the tests have been run. A large project can have hundreds or thousands of tests, and the dots and percentage-complete indicator can be helpful in monitoring the overall progress of the test run.

The last line tells us that one test passed, and it took less than 0.01 seconds to run the test.

This output indicates that the function get_formatted_name() will always work for names that have a first and last name, unless we modify the function. When we modify get_formatted_name(), we can run this test again. If the test passes, we know the function will still work for names like Janis Joplin.

NOTE *If you're not sure how to navigate to the right location in the terminal, see "Running Python Programs from a Terminal" on page 11. Also, if you see a message that the* pytest *command was not found, use the command* python -m pytest *instead.*

A Failing Test

What does a failing test look like? Let's modify get_formatted_name() so it can handle middle names, but let's do so in a way that breaks the function for names with just a first and last name, like Janis Joplin.

Here's a new version of get_formatted_name() that requires a middle name argument:

name_function.py

```
def get_formatted_name(first, middle, last):
    """Generate a neatly formatted full name."""
    full_name = f"{first} {middle} {last}"
    return full_name.title()
```

This version should work for people with middle names, but when we test it, we see that we've broken the function for people with just a first and last name.

This time, running **pytest** gives the following output:

```
$ pytest
======================== test session starts ========================
--snip--
❶ test_name_function.py F                                        [100%]
❷ ============================ FAILURES ============================
❸ ________________________ test_first_last_name ________________________
    def test_first_last_name():
        """Do names like 'Janis Joplin' work?"""
❹ >       formatted_name = get_formatted_name('janis', 'joplin')
❺ E       TypeError: get_formatted_name() missing 1 required positional
            argument: 'last'
```

```
test_name_function.py:5: TypeError
======================= short test summary info =======================
FAILED test_name_function.py::test_first_last_name - TypeError:
    get_formatted_name() missing 1 required positional argument: 'last'
========================== 1 failed in 0.04s ==========================
```

There's a lot of information here because there's a lot you might need to know when a test fails. The first item of note in the output is a single `F` ❶, which tells us that one test failed. We then see a section that focuses on `FAILURES` ❷, because failed tests are usually the most important thing to focus on in a test run. Next, we see that `test_first_last_name()` was the test function that failed ❸. An angle bracket ❹ indicates the line of code that caused the test to fail. The `E` on the next line ❺ shows the actual error that caused the failure: a `TypeError` due to a missing required positional argument, `last`. The most important information is repeated in a shorter summary at the end, so when you're running many tests, you can get a quick sense of which tests failed and why.

Responding to a Failed Test

What do you do when a test fails? Assuming you're checking the right conditions, a passing test means the function is behaving correctly and a failing test means there's an error in the new code you wrote. So when a test fails, don't change the test. If you do, your tests might pass, but any code that calls your function like the test does will suddenly stop working. Instead, fix the code that's causing the test to fail. Examine the changes you just made to the function, and figure out how those changes broke the desired behavior.

In this case, `get_formatted_name()` used to require only two parameters: a first name and a last name. Now it requires a first name, middle name, and last name. The addition of that mandatory middle name parameter broke the original behavior of `get_formatted_name()`. The best option here is to make the middle name optional. Once we do, our test for names like `Janis Joplin` should pass again, and we should be able to accept middle names as well. Let's modify `get_formatted_name()` so middle names are optional and then run the test case again. If it passes, we'll move on to making sure the function handles middle names properly.

To make middle names optional, we move the parameter `middle` to the end of the parameter list in the function definition and give it an empty default value. We also add an `if` test that builds the full name properly, depending on whether a middle name is provided:

name_function.py

```
def get_formatted_name(first, last, middle=''):
    """Generate a neatly formatted full name."""
    if middle:
        full_name = f"{first} {middle} {last}"
    else:
        full_name = f"{first} {last}"
    return full_name.title()
```

In this new version of get_formatted_name(), the middle name is optional. If a middle name is passed to the function, the full name will contain a first, middle, and last name. Otherwise, the full name will consist of just a first and last name. Now the function should work for both kinds of names. To find out if the function still works for names like Janis Joplin, let's run the test again:

```
$ pytest
======================== test session starts ========================
--snip--
test_name_function.py .                                          [100%]
========================= 1 passed in 0.00s =========================
```

The test passes now. This is ideal; it means the function works for names like Janis Joplin again, without us having to test the function manually. Fixing our function was easier because the failed test helped us identify how the new code broke existing behavior.

Adding New Tests

Now that we know get_formatted_name() works for simple names again, let's write a second test for people who include a middle name. We do this by adding another test function to the file *test_name_function.py*:

test_name_function.py

```
from name_function import get_formatted_name

def test_first_last_name():
    --snip--

def test_first_last_middle_name():
    """Do names like 'Wolfgang Amadeus Mozart' work?"""
❶   formatted_name = get_formatted_name(
        'wolfgang', 'mozart', 'amadeus')
❷   assert formatted_name == 'Wolfgang Amadeus Mozart'
```

We name this new function test_first_last_middle_name(). The function name must start with test_ so the function runs automatically when we run **pytest**. We name the function to make it clear which behavior of get_formatted_name() we're testing. As a result, if the test fails, we'll know right away what kinds of names are affected.

To test the function, we call get_formatted_name() with a first, last, and middle name ❶, and then we make an assertion ❷ that the returned full name matches the full name (first, middle, and last) that we expect. When we run **pytest** again, both tests pass:

```
$ pytest
======================== test session starts ========================
--snip--
collected 2 items

❶ test_name_function.py ..                                       [100%]
========================= 2 passed in 0.01s =========================
```

The two dots ❶ indicate that two tests passed, which is also clear from the last line of output. This is great! We now know that the function still works for names like `Janis Joplin`, and we can be confident that it will work for names like `Wolfgang Amadeus Mozart` as well.

TRY IT YOURSELF

11-1. City, Country: Write a function that accepts two parameters: a city name and a country name. The function should return a single string of the form *`City, Country`*, such as `Santiago, Chile`. Store the function in a module called *city_functions.py*, and save this file in a new folder so `pytest` won't try to run the tests we've already written.

Create a file called *test_cities.py* that tests the function you just wrote. Write a function called `test_city_country()` to verify that calling your function with values such as `'santiago'` and `'chile'` results in the correct string. Run the test, and make sure `test_city_country()` passes.

11-2. Population: Modify your function so it requires a third parameter, `population`. It should now return a single string of the form *`City, Country - population xxx`*, such as `Santiago, Chile - population 5000000`. Run the test again, and make sure `test_city_country()` fails this time.

Modify the function so the `population` parameter is optional. Run the test, and make sure `test_city_country()` passes again.

Write a second test called `test_city_country_population()` that verifies you can call your function with the values `'santiago'`, `'chile'`, and `'population =5000000'`. Run the tests one more time, and make sure this new test passes.

Testing a Class

In the first part of this chapter, you wrote tests for a single function. Now you'll write tests for a class. You'll use classes in many of your own programs, so it's helpful to be able to prove that your classes work correctly. If you have passing tests for a class you're working on, you can be confident that improvements you make to the class won't accidentally break its current behavior.

A Variety of Assertions

So far, you've seen just one kind of assertion: a claim that a string has a specific value. When writing a test, you can make any claim that can be expressed as a conditional statement. If the condition is `True` as expected, your assumption about how that part of your program behaves will be confirmed; you can be confident that no errors exist. If the condition you assume is True is actually False, the test will fail and you'll know there's an issue to resolve. Table 11-1 shows some of the most useful kinds of assertions you can include in your initial tests.

Table 11-1: Commonly Used Assertion Statements in Tests

Assertion	Claim
`assert a == b`	Assert that two values are equal.
`assert a != b`	Assert that two values are not equal.
`assert a`	Assert that a evaluates to True.
`assert not a`	Assert that a evaluates to False.
assert *element* in *list*	Assert that an element is in a list.
assert *element* not in *list*	Assert that an element is not in a list.

These are just a few examples; anything that can be expressed as a conditional statement can be included in a test.

A Class to Test

Testing a class is similar to testing a function, because much of the work involves testing the behavior of the methods in the class. However, there are a few differences, so let's write a class to test. Consider a class that helps administer anonymous surveys:

survey.py

```
class AnonymousSurvey:
    """Collect anonymous answers to a survey question."""

❶   def __init__(self, question):
        """Store a question, and prepare to store responses."""
        self.question = question
        self.responses = []

❷   def show_question(self):
        """Show the survey question."""
        print(self.question)

❸   def store_response(self, new_response):
        """Store a single response to the survey."""
        self.responses.append(new_response)

❹   def show_results(self):
        """Show all the responses that have been given."""
        print("Survey results:")
        for response in self.responses:
            print(f"- {response}")
```

This class starts with a survey question that you provide ❶ and includes an empty list to store responses. The class has methods to print the survey question ❷, add a new response to the response list ❸, and print all the responses stored in the list ❹. To create an instance from this class, all you have to provide is a question. Once you have an instance representing a particular survey, you display the survey question with `show_question()`, store a response using `store_response()`, and show results with `show_results()`.

To show that the `AnonymousSurvey` class works, let's write a program that uses the class:

language_survey.py

```
from survey import AnonymousSurvey

# Define a question, and make a survey.
question = "What language did you first learn to speak?"
language_survey = AnonymousSurvey(question)

# Show the question, and store responses to the question.
language_survey.show_question()
print("Enter 'q' at any time to quit.\n")
while True:
    response = input("Language: ")
    if response == 'q':
        break
    language_survey.store_response(response)

# Show the survey results.
print("\nThank you to everyone who participated in the survey!")
language_survey.show_results()
```

This program defines a question (`"What language did you first learn to speak?"`) and creates an `AnonymousSurvey` object with that question. The program calls `show_question()` to display the question and then prompts for responses. Each response is stored as it is received. When all responses have been entered (the user inputs `q` to quit), `show_results()` prints the survey results:

```
What language did you first learn to speak?
Enter 'q' at any time to quit.

Language: English
Language: Spanish
Language: English
Language: Mandarin
Language: q

Thank you to everyone who participated in the survey!
Survey results:
- English
- Spanish
- English
- Mandarin
```

This class works for a simple anonymous survey, but say we want to improve `AnonymousSurvey` and the module it's in, `survey`. We could allow each user to enter more than one response, we could write a method to list only unique responses and to report how many times each response was given, or we could even write another class to manage non-anonymous surveys.

Implementing such changes would risk affecting the current behavior of the class AnonymousSurvey. For example, it's possible that while trying to allow each user to enter multiple responses, we could accidentally change how single responses are handled. To ensure we don't break existing behavior as we develop this module, we can write tests for the class.

Testing the AnonymousSurvey Class

Let's write a test that verifies one aspect of the way AnonymousSurvey behaves. We'll write a test to verify that a single response to the survey question is stored properly:

test_survey.py

```
from survey import AnonymousSurvey

❶ def test_store_single_response():
    """Test that a single response is stored properly."""
    question = "What language did you first learn to speak?"
❷   language_survey = AnonymousSurvey(question)
    language_survey.store_response('English')
❸   assert 'English' in language_survey.responses
```

We start by importing the class we want to test, AnonymousSurvey. The first test function will verify that when we store a response to the survey question, the response will end up in the survey's list of responses. A good descriptive name for this function is test_store_single_response() ❶. If this test fails, we'll know from the function name in the test summary that there was a problem storing a single response to the survey.

To test the behavior of a class, we need to make an instance of the class. We create an instance called language_survey ❷ with the question "What language did you first learn to speak?" We store a single response, English, using the store_response() method. Then we verify that the response was stored correctly by asserting that English is in the list language_survey .responses ❸.

By default, running the command **pytest** with no arguments will run all the tests that pytest discovers in the current directory. To focus on the tests in one file, pass the name of the test file you want to run. Here we'll run just the one test we wrote for AnonymousSurvey:

```
$ pytest test_survey.py
========================= test session starts =========================
--snip--
test_survey.py .                                                 [100%]
========================== 1 passed in 0.01s ==========================
```

This is a good start, but a survey is useful only if it generates more than one response. Let's verify that three responses can be stored correctly. To do this, we add another function to *test_survey.py*:

```
from survey import AnonymousSurvey

def test_store_single_response():
```

```
    --snip--

def test_store_three_responses():
    """Test that three individual responses are stored properly."""
    question = "What language did you first learn to speak?"
    language_survey = AnonymousSurvey(question)
❶   responses = ['English', 'Spanish', 'Mandarin']
    for response in responses:
        language_survey.store_response(response)

❷   for response in responses:
        assert response in language_survey.responses
```

We call the new function `test_store_three_responses()`. We create a survey object just like we did in `test_store_single_response()`. We define a list containing three different responses ❶, and then we call `store_response()` for each of these responses. Once the responses have been stored, we write another loop and assert that each response is now in `language_survey.responses` ❷.

When we run the test file again, both tests (for a single response and for three responses) pass:

```
$ pytest test_survey.py
========================= test session starts =========================
--snip--
test_survey.py ..                                                [100%]
========================== 2 passed in 0.01s ==========================
```

This works perfectly. However, these tests are a bit repetitive, so we'll use another feature of `pytest` to make them more efficient.

Using Fixtures

In *test_survey.py*, we created a new instance of `AnonymousSurvey` in each test function. This is fine in the short example we're working with, but in a real-world project with tens or hundreds of tests, this would be problematic.

In testing, a *fixture* helps set up a test environment. Often, this means creating a resource that's used by more than one test. We create a fixture in `pytest` by writing a function with the decorator `@pytest.fixture`. A *decorator* is a directive placed just before a function definition; Python applies this directive to the function before it runs, to alter how the function code behaves. Don't worry if this sounds complicated; you can start to use decorators from third-party packages before learning to write them yourself.

Let's use a fixture to create a single survey instance that can be used in both test functions in *test_survey.py*:

```
import pytest
from survey import AnonymousSurvey

❶ @pytest.fixture
❷ def language_survey():
    """A survey that will be available to all test functions."""
```

```
    question = "What language did you first learn to speak?"
    language_survey = AnonymousSurvey(question)
    return language_survey

❸ def test_store_single_response(language_survey):
    """Test that a single response is stored properly."""
❹     language_survey.store_response('English')
    assert 'English' in language_survey.responses

❺ def test_store_three_responses(language_survey):
    """Test that three individual responses are stored properly."""
    responses = ['English', 'Spanish', 'Mandarin']
    for response in responses:
❻         language_survey.store_response(response)

    for response in responses:
        assert response in language_survey.responses
```

We need to import `pytest` now, because we're using a decorator that's defined in `pytest`. We apply the `@pytest.fixture` decorator ❶ to the new function `language_survey()` ❷. This function builds an `AnonymousSurvey` object and returns the new survey.

Notice that the definitions of both test functions have changed ❸ ❺; each test function now has a parameter called `language_survey`. When a parameter in a test function matches the name of a function with the `@pytest.fixture` decorator, the fixture will be run automatically and the return value will be passed to the test function. In this example, the function `language_survey()` supplies both `test_store_single_response()` and `test_store_three_responses()` with a `language_survey` instance.

There's no new code in either of the test functions, but notice that two lines have been removed from each function ❹ ❻: the line that defined a question and the line that created an `AnonymousSurvey` object.

When we run the test file again, both tests still pass. These tests would be particularly useful when trying to expand `AnonymousSurvey` to handle multiple responses for each person. After modifying the code to accept multiple responses, you could run these tests and make sure you haven't affected the ability to store a single response or a series of individual responses.

The structure above will almost certainly look complicated; it contains some of the most abstract code you've seen so far. You don't need to use fixtures right away; it's better to write tests that have a lot of repetitive code than to write no tests at all. Just know that when you've written enough tests that the repetition is getting in the way, there's a well-established way to deal with the repetition. Also, fixtures in simple examples like this one don't really make the code any shorter or simpler to follow. But in projects with many tests, or in situations where it takes many lines to build a resource that's used in multiple tests, fixtures can drastically improve your test code.

When you want to write a fixture, write a function that generates the resource that's used by multiple test functions. Add the `@pytest.fixture` decorator to the new function, and add the name of this function as a parameter for each test function that uses this resource. Your tests will be shorter and easier to write and maintain from that point forward.

> **TRY IT YOURSELF**
>
> **11-3. Employee:** Write a class called `Employee`. The `__init__()` method should take in a first name, a last name, and an annual salary, and store each of these as attributes. Write a method called `give_raise()` that adds $5,000 to the annual salary by default but also accepts a different raise amount.
>
> Write a test file for `Employee` with two test functions, `test_give_default_raise()` and `test_give_custom_raise()`. Write your tests once without using a fixture, and make sure they both pass. Then write a fixture so you don't have to create a new employee instance in each test function. Run the tests again, and make sure both tests still pass.

Summary

In this chapter, you learned to write tests for functions and classes using tools in the `pytest` module. You learned to write test functions that verify specific behaviors your functions and classes should exhibit. You saw how fixtures can be used to efficiently create resources that can be used in multiple test functions in a test file.

Testing is an important topic that many newer programmers aren't exposed to. You don't have to write tests for all the simple projects you try as a new programmer. But as soon as you start to work on projects that involve significant development effort, you should test the critical behaviors of your functions and classes. You'll be more confident that new work on your project won't break the parts that work, and this will give you the freedom to make improvements to your code. If you accidentally break existing functionality, you'll know right away, so you can still fix the problem easily. Responding to a failed test that you ran is much easier than responding to a bug report from an unhappy user.

Other programmers will respect your projects more if you include some initial tests. They'll feel more comfortable experimenting with your code and be more willing to work with you on projects. If you want to contribute to a project that other programmers are working on, you'll be expected to show that your code passes existing tests and you'll usually be expected to write tests for any new behavior you introduce to the project.

Play around with tests to become familiar with the process of testing your code. Write tests for the most critical behaviors of your functions and classes, but don't aim for full coverage in early projects unless you have a specific reason to do so.

PART II

PROJECTS

Congratulations! You now know enough about Python to start building interactive and meaningful projects. Creating your own projects will teach you new skills and solidify your understanding of the concepts introduced in Part I.

Part II contains three kinds of projects, and you can choose to do any or all of these projects in whichever order you like. Here's a brief description of each project to help you decide which to dig into first.

Alien Invasion: Making a Game with Python

In the Alien Invasion project (**Chapters 12**, **13**, and **14**), you'll use the Pygame package to develop a 2D game. The goal of the game is to shoot down a fleet of aliens as they drop down the screen, in levels that increase in speed and difficulty. At the end of the project, you'll have learned skills that will enable you to develop your own 2D games in Pygame.

Data Visualization

The Data Visualization projects start in **Chapter 15**, where you'll learn to generate data and create a series of functional and beautiful visualizations of that data using Matplotlib and Plotly. **Chapter 16** teaches you to access data from online sources and feed it into a visualization package to create plots of weather data and a map of global earthquake activity. Finally, **Chapter 17** shows you how to write a program to automatically download and visualize data. Learning to make visualizations allows you to explore the field of data science, which is one of the highest-demand areas of programming today.

Web Applications

In the Web Application project (**Chapters 18**, **19**, and **20**), you'll use the Django package to create a simple web application that allows users to keep a journal about different topics they've been learning about. Users will create an account with a username and password, enter a topic, and then make entries about what they're learning. You'll also deploy your app to a remote server so anyone in the world can access it.

After completing this project, you'll be able to start building your own simple web applications, and you'll be ready to delve into more thorough resources on building applications with Django.

12

A SHIP THAT FIRES BULLETS

Let's build a game called *Alien Invasion*! We'll use Pygame, a collection of fun, powerful Python modules that manage graphics, animation, and even sound, making it easier for you to build sophisticated games. With Pygame handling tasks like drawing images to the screen, you can focus on the higher-level logic of game dynamics.

In this chapter, you'll set up Pygame and then create a rocket ship that moves right and left and fires bullets in response to player input. In the next two chapters, you'll create a fleet of aliens to destroy, and then continue to refine the game by setting limits on the number of ships you can use and adding a scoreboard.

While building this game, you'll also learn how to manage large projects that span multiple files. We'll refactor a lot of code and manage file contents to organize the project and make the code efficient.

Making games is an ideal way to have fun while learning a language. It's deeply satisfying to play a game you wrote, and writing a simple game will teach you a lot about how professionals develop games. As you work through this chapter, enter and run the code to identify how each code block contributes to overall gameplay. Experiment with different values and settings to better understand how to refine interactions in your games.

NOTE Alien Invasion *spans a number of different files, so make a new* alien_invasion *folder on your system. Be sure to save all files for the project to this folder so your* `import` *statements will work correctly.*

Also, if you feel comfortable using version control, you might want to use it for this project. If you haven't used version control before, see Appendix D for an overview.

Planning Your Project

When you're building a large project, it's important to prepare a plan before you begin to write code. Your plan will keep you focused and make it more likely that you'll complete the project.

Let's write a description of the general gameplay. Although the following description doesn't cover every detail of *Alien Invasion*, it provides a clear idea of how to start building the game:

> In *Alien Invasion*, the player controls a rocket ship that appears at the bottom center of the screen. The player can move the ship right and left using the arrow keys and shoot bullets using the spacebar. When the game begins, a fleet of aliens fills the sky and moves across and down the screen. The player shoots and destroys the aliens. If the player destroys all the aliens, a new fleet appears that moves faster than the previous fleet. If any alien hits the player's ship or reaches the bottom of the screen, the player loses a ship. If the player loses three ships, the game ends.

For the first development phase, we'll make a ship that can move right and left when the player presses the arrow keys and fire bullets when the player presses the spacebar. After setting up this behavior, we can create the aliens and refine the gameplay.

Installing Pygame

Before you begin coding, install Pygame. We'll do this the same way we installed pytest in Chapter 11: with pip. If you skipped Chapter 11 or need a refresher on pip, see "Installing pytest with pip" on page 210.

To install Pygame, enter the following command at a terminal prompt:

```
$ python -m pip install --user pygame
```

If you use a command other than **python** to run programs or start a terminal session, such as **python3**, make sure you use that command instead.

Starting the Game Project

We'll begin building the game by creating an empty Pygame window. Later, we'll draw the game elements, such as the ship and the aliens, on this window. We'll also make our game respond to user input, set the background color, and load a ship image.

Creating a Pygame Window and Responding to User Input

We'll make an empty Pygame window by creating a class to represent the game. In your text editor, create a new file and save it as *alien_invasion.py*; then enter the following:

alien_invasion.py

```
import sys

import pygame

class AlienInvasion:
    """Overall class to manage game assets and behavior."""

    def __init__(self):
        """Initialize the game, and create game resources."""
❶       pygame.init()

❷       self.screen = pygame.display.set_mode((1200, 800))
        pygame.display.set_caption("Alien Invasion")

    def run_game(self):
        """Start the main loop for the game."""
❸       while True:
            # Watch for keyboard and mouse events.
❹           for event in pygame.event.get():
❺               if event.type == pygame.QUIT:
                    sys.exit()

            # Make the most recently drawn screen visible.
❻           pygame.display.flip()

if __name__ == '__main__':
    # Make a game instance, and run the game.
    ai = AlienInvasion()
    ai.run_game()
```

First, we import the sys and pygame modules. The pygame module contains the functionality we need to make a game. We'll use tools in the sys module to exit the game when the player quits.

Alien Invasion starts as a class called AlienInvasion. In the __init__() method, the pygame.init() function initializes the background settings that Pygame needs to work properly ❶. Then we call pygame.display.set_mode() to create a display window ❷, on which we'll draw all the game's graphical elements. The argument (1200, 800) is a tuple that defines the dimensions of the game window, which will be 1,200 pixels wide by 800 pixels high. (You can adjust these values depending on your display size.) We assign this display window to the attribute `self.screen`, so it will be available in all methods in the class.

The object we assigned to `self.screen` is called a surface. A *surface* in Pygame is a part of the screen where a game element can be displayed. Each element in the game, like an alien or a ship, is its own surface. The surface returned by `display.set_mode()` represents the entire game window. When we activate the game's animation loop, this surface will be redrawn on every pass through the loop, so it can be updated with any changes triggered by user input.

The game is controlled by the `run_game()` method. This method contains a `while` loop ❸ that runs continually. The `while` loop contains an event loop and code that manages screen updates. An *event* is an action that the user performs while playing the game, such as pressing a key or moving the mouse. To make our program respond to events, we write an *event loop* to *listen* for events and perform appropriate tasks depending on the kinds of events that occur. The `for` loop ❹ nested inside the `while` loop is an event loop.

To access the events that Pygame detects, we'll use the `pygame.event.get()` function. This function returns a list of events that have taken place since the last time this function was called. Any keyboard or mouse event will cause this `for` loop to run. Inside the loop, we'll write a series of `if` statements to detect and respond to specific events. For example, when the player clicks the game window's close button, a `pygame.QUIT` event is detected and we call `sys.exit()` to exit the game ❺.

The call to `pygame.display.flip()` ❻ tells Pygame to make the most recently drawn screen visible. In this case, it simply draws an empty screen on each pass through the `while` loop, erasing the old screen so only the new screen is visible. When we move the game elements around, `pygame.display.flip()` continually updates the display to show the new positions of game elements and hide the old ones, creating the illusion of smooth movement.

At the end of the file, we create an instance of the game and then call `run_game()`. We place `run_game()` in an `if` block that only runs if the file is called directly. When you run this *alien_invasion.py* file, you should see an empty Pygame window.

Controlling the Frame Rate

Ideally, games should run at the same speed, or *frame rate*, on all systems. Controlling the frame rate of a game that can run on multiple systems is a complex issue, but Pygame offers a relatively simple way to accomplish this goal. We'll make a clock, and ensure the clock ticks once on each pass through the main loop. Anytime the loop processes faster than the rate we define, Pygame will calculate the correct amount of time to pause so that the game runs at a consistent rate.

We'll define the clock in the `__init__()` method:

alien_invasion.py

```
    def __init__(self):
        """Initialize the game, and create game resources."""
        pygame.init()
```

```
        self.clock = pygame.time.Clock()
        --snip--
```

After initializing pygame, we create an instance of the class Clock, from the pygame.time module. Then we'll make the clock tick at the end of the while loop in run_game():

```
    def run_game(self):
        """Start the main loop for the game."""
        while True:
            --snip--
            pygame.display.flip()
            self.clock.tick(60)
```

The tick() method takes one argument: the frame rate for the game. Here I'm using a value of 60, so Pygame will do its best to make the loop run exactly 60 times per second.

NOTE *Pygame's clock should help the game run consistently on most systems. If it makes the game run less consistently on your system, you can try different values for the frame rate. If you can't find a good frame rate on your system, you can leave the clock out entirely and adjust the game's settings so it runs well on your system.*

Setting the Background Color

Pygame creates a black screen by default, but that's boring. Let's set a different background color. We'll do this at the end of the __init__() method.

alien_invasion.py

```
    def __init__(self):
        --snip--
        pygame.display.set_caption("Alien Invasion")

        # Set the background color.
❶       self.bg_color = (230, 230, 230)

    def run_game(self):
        --snip--
            for event in pygame.event.get():
                if event.type == pygame.QUIT:
                    sys.exit()

            # Redraw the screen during each pass through the loop.
❷           self.screen.fill(self.bg_color)

            # Make the most recently drawn screen visible.
            pygame.display.flip()
            self.clock.tick(60)
```

Colors in Pygame are specified as RGB colors: a mix of red, green, and blue. Each color value can range from 0 to 255. The color value (255, 0, 0) is red, (0, 255, 0) is green, and (0, 0, 255) is blue. You can mix different RGB values to create up to 16 million colors. The color value (230, 230, 230) mixes equal amounts of red, blue, and green, which produces a light gray background color. We assign this color to `self.bg_color` ❶.

We fill the screen with the background color using the `fill()` method ❷, which acts on a surface and takes only one argument: a color.

Creating a Settings Class

Each time we introduce new functionality into the game, we'll typically create some new settings as well. Instead of adding settings throughout the code, let's write a module called `settings` that contains a class called `Settings` to store all these values in one place. This approach allows us to work with just one `settings` object anytime we need to access an individual setting. This also makes it easier to modify the game's appearance and behavior as our project grows. To modify the game, we'll change the relevant values in *settings.py*, which we'll create next, instead of searching for different settings throughout the project.

Create a new file named *settings.py* inside your *alien_invasion* folder, and add this initial `Settings` class:

settings.py

```
class Settings:
    """A class to store all settings for Alien Invasion."""

    def __init__(self):
        """Initialize the game's settings."""
        # Screen settings
        self.screen_width = 1200
        self.screen_height = 800
        self.bg_color = (230, 230, 230)
```

To make an instance of `Settings` in the project and use it to access our settings, we need to modify *alien_invasion.py* as follows:

alien_invasion.py

```
--snip--
import pygame

from settings import Settings

class AlienInvasion:
    """Overall class to manage game assets and behavior."""

    def __init__(self):
        """Initialize the game, and create game resources."""
        pygame.init()
        self.clock = pygame.time.Clock()
❶       self.settings = Settings()

❷       self.screen = pygame.display.set_mode(
            (self.settings.screen_width, self.settings.screen_height))
        pygame.display.set_caption("Alien Invasion")
```

```
    def run_game(self):
            --snip--
            # Redraw the screen during each pass through the loop.
❸           self.screen.fill(self.settings.bg_color)

            # Make the most recently drawn screen visible.
            pygame.display.flip()
            self.clock.tick(60)
--snip--
```

We import Settings into the main program file. Then we create an instance of Settings and assign it to self.settings ❶, after making the call to pygame.init(). When we create a screen ❷, we use the screen_width and screen_height attributes of self.settings, and then we use self.settings to access the background color when filling the screen ❸ as well.

When you run *alien_invasion.py* now you won't yet see any changes, because all we've done is move the settings we were already using elsewhere. Now we're ready to start adding new elements to the screen.

Adding the Ship Image

Let's add the ship to our game. To draw the player's ship on the screen, we'll load an image and then use the Pygame blit() method to draw the image.

When you're choosing artwork for your games, be sure to pay attention to licensing. The safest and cheapest way to start is to use freely licensed graphics that you can use and modify, from a website like *https://opengameart.org*.

You can use almost any type of image file in your game, but it's easiest when you use a bitmap (*.bmp*) file because Pygame loads bitmaps by default. Although you can configure Pygame to use other file types, some file types depend on certain image libraries that must be installed on your computer. Most images you'll find are in *.jpg* or *.png* formats, but you can convert them to bitmaps using tools like Photoshop, GIMP, and Paint.

Pay particular attention to the background color in your chosen image. Try to find a file with a transparent or solid background that you can replace with any background color, using an image editor. Your games will look best if the image's background color matches your game's background color. Alternatively, you can match your game's background to the image's background.

For *Alien Invasion*, you can use the file *ship.bmp* (Figure 12-1), which is available in this book's resources at *https://ehmatthes.github.io/pcc_3e*. The file's background color matches the settings we're using in this project. Make a folder called *images* inside your main *alien_invasion* project folder. Save the file *ship.bmp* in the *images* folder.

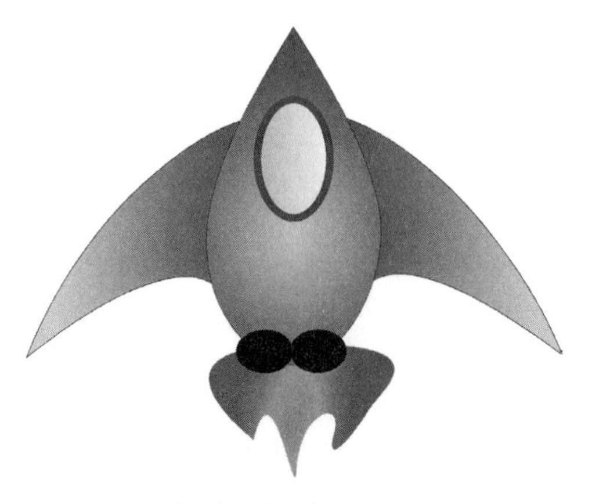

Figure 12-1: The ship for Alien Invasion

Creating the Ship Class

After choosing an image for the ship, we need to display it on the screen. To use our ship, we'll create a new `ship` module that will contain the class `Ship`. This class will manage most of the behavior of the player's ship:

ship.py

```
import pygame

class Ship:
    """A class to manage the ship."""

    def __init__(self, ai_game):
        """Initialize the ship and set its starting position."""
❶       self.screen = ai_game.screen
❷       self.screen_rect = ai_game.screen.get_rect()

        # Load the ship image and get its rect.
❸       self.image = pygame.image.load('images/ship.bmp')
        self.rect = self.image.get_rect()

        # Start each new ship at the bottom center of the screen.
❹       self.rect.midbottom = self.screen_rect.midbottom

❺   def blitme(self):
        """Draw the ship at its current location."""
        self.screen.blit(self.image, self.rect)
```

Pygame is efficient because it lets you treat all game elements like rectangles (*rects*), even if they're not exactly shaped like rectangles. Treating an element as a rectangle is efficient because rectangles are simple geometric shapes. When Pygame needs to figure out whether two game elements have collided, for example, it can do this more quickly if it treats each object as a rectangle. This approach usually works well enough that no one playing the game will notice that we're not working with the exact shape of each game element. We'll treat the ship and the screen as rectangles in this class.

We import the `pygame` module before defining the class. The `__init__()` method of `Ship` takes two parameters: the `self` reference and a reference to the current instance of the `AlienInvasion` class. This will give `Ship` access to all the game resources defined in `AlienInvasion`. We then assign the screen to an attribute of `Ship` ❶, so we can access it easily in all the methods in this class. We access the screen's `rect` attribute using the `get_rect()` method and assign it to `self.screen_rect` ❷. Doing so allows us to place the ship in the correct location on the screen.

To load the image, we call `pygame.image.load()` ❸ and give it the location of our ship image. This function returns a surface representing the ship, which we assign to `self.image`. When the image is loaded, we call `get_rect()` to access the ship surface's `rect` attribute so we can later use it to place the ship.

When you're working with a `rect` object, you can use the *x*- and *y*-coordinates of the top, bottom, left, and right edges of the rectangle, as well as the center, to place the object. You can set any of these values to establish the current position of the `rect`. When you're centering a game element, work with the `center`, `centerx`, or `centery` attributes of a `rect`. When you're working at an edge of the screen, work with the `top`, `bottom`, `left`, or `right` attributes. There are also attributes that combine these properties, such as `midbottom`, `midtop`, `midleft`, and `midright`. When you're adjusting the horizontal or vertical placement of the `rect`, you can just use the `x` and `y` attributes, which are the *x*- and *y*-coordinates of its top-left corner. These attributes spare you from having to do calculations that game developers formerly had to do manually, and you'll use them often.

NOTE *In Pygame, the origin (0, 0) is at the top-left corner of the screen, and coordinates increase as you go down and to the right. On a 1200×800 screen, the origin is at the top-left corner, and the bottom-right corner has the coordinates (1200, 800). These coordinates refer to the game window, not the physical screen.*

We'll position the ship at the bottom center of the screen. To do so, make the value of `self.rect.midbottom` match the `midbottom` attribute of the screen's `rect` ❹. Pygame uses these `rect` attributes to position the ship image so it's centered horizontally and aligned with the bottom of the screen.

Finally, we define the `blitme()` method ❺, which draws the image to the screen at the position specified by `self.rect`.

Drawing the Ship to the Screen

Now let's update *alien_invasion.py* so it creates a ship and calls the ship's `blitme()` method:

alien_invasion.py

```
--snip--
from settings import Settings
from ship import Ship
```

```
class AlienInvasion:
    """Overall class to manage game assets and behavior."""

    def __init__(self):
        --snip--
        pygame.display.set_caption("Alien Invasion")

❶       self.ship = Ship(self)

    def run_game(self):
            --snip--
            # Redraw the screen during each pass through the loop.
            self.screen.fill(self.settings.bg_color)
❷           self.ship.blitme()

            # Make the most recently drawn screen visible.
            pygame.display.flip()
            self.clock.tick(60)
--snip--
```

We import `Ship` and then make an instance of `Ship` after the screen has been created ❶. The call to `Ship()` requires one argument: an instance of `AlienInvasion`. The `self` argument here refers to the current instance of `AlienInvasion`. This is the parameter that gives `Ship` access to the game's resources, such as the `screen` object. We assign this `Ship` instance to `self.ship`.

After filling the background, we draw the ship on the screen by calling `ship.blitme()`, so the ship appears on top of the background ❷.

When you run *alien_invasion.py* now, you should see an empty game screen with the rocket ship sitting at the bottom center, as shown in Figure 12-2.

Figure 12-2: Alien Invasion *with the ship at the bottom center of the screen*

Refactoring: The _check_events() and _update_screen() Methods

In large projects, you'll often refactor code you've written before adding more code. Refactoring simplifies the structure of the code you've already written, making it easier to build on. In this section, we'll break the `run_game()` method, which is getting lengthy, into two helper methods. A *helper method* does work inside a class but isn't meant to be used by code outside the class. In Python, a single leading underscore indicates a helper method.

The _check_events() Method

We'll move the code that manages events to a separate method called `_check_events()`. This will simplify `run_game()` and isolate the event management loop. Isolating the event loop allows you to manage events separately from other aspects of the game, such as updating the screen.

Here's the `AlienInvasion` class with the new `_check_events()` method, which only affects the code in `run_game()`:

alien_invasion.py

```
    def run_game(self):
        """Start the main loop for the game."""
        while True:
❶           self._check_events()

            # Redraw the screen during each pass through the loop.
            --snip--

❷   def _check_events(self):
        """Respond to keypresses and mouse events."""
        for event in pygame.event.get():
            if event.type == pygame.QUIT:
                sys.exit()
```

We make a new `_check_events()` method ❷ and move the lines that check whether the player has clicked to close the window into this new method.

To call a method from within a class, use dot notation with the variable `self` and the name of the method ❶. We call the method from inside the `while` loop in `run_game()`.

The _update_screen() Method

To further simplify `run_game()`, we'll move the code for updating the screen to a separate method called `_update_screen()`:

alien_invasion.py

```
    def run_game(self):
        """Start the main loop for the game."""
        while True:
            self._check_events()
            self._update_screen()
            self.clock.tick(60)

    def _check_events(self):
        --snip--
```

```
    def _update_screen(self):
        """Update images on the screen, and flip to the new screen."""
        self.screen.fill(self.settings.bg_color)
        self.ship.blitme()

        pygame.display.flip()
```

We moved the code that draws the background and the ship and flips the screen to `_update_screen()`. Now the body of the main loop in `run_game()` is much simpler. It's easy to see that we're looking for new events, updating the screen, and ticking the clock on each pass through the loop.

If you've already built a number of games, you'll probably start out by breaking your code into methods like these. But if you've never tackled a project like this, you probably won't know exactly how to structure your code at first. This approach gives you an idea of a realistic development process: you start out writing your code as simply as possible, and then refactor it as your project becomes more complex.

Now that we've restructured the code to make it easier to add to, we can work on the dynamic aspects of the game!

TRY IT YOURSELF

12-1. Blue Sky: Make a Pygame window with a blue background.

12-2. Game Character: Find a bitmap image of a game character you like or convert an image to a bitmap. Make a class that draws the character at the center of the screen, then match the background color of the image to the background color of the screen or vice versa.

Piloting the Ship

Next, we'll give the player the ability to move the ship right and left. We'll write code that responds when the player presses the right or left arrow key. We'll focus first on movement to the right, and then we'll apply the same principles to control movement to the left. As we add this code, you'll learn how to control the movement of images on the screen and respond to user input.

Responding to a Keypress

Whenever the player presses a key, that keypress is registered in Pygame as an event. Each event is picked up by the `pygame.event.get()` method. We need to specify in our `_check_events()` method what kinds of events we want the game to check for. Each keypress is registered as a `KEYDOWN` event.

When Pygame detects a KEYDOWN event, we need to check whether the key that was pressed is one that triggers a certain action. For example, if the player presses the right arrow key, we want to increase the ship's rect.x value to move the ship to the right:

alien_invasion.py

```
    def _check_events(self):
        """Respond to keypresses and mouse events."""
        for event in pygame.event.get():
            if event.type == pygame.QUIT:
                sys.exit()
❶           elif event.type == pygame.KEYDOWN:
❷               if event.key == pygame.K_RIGHT:
                    # Move the ship to the right.
❸                   self.ship.rect.x += 1
```

Inside _check_events() we add an elif block to the event loop, to respond when Pygame detects a KEYDOWN event ❶. We check whether the key pressed, event.key, is the right arrow key ❷. The right arrow key is represented by pygame.K_RIGHT. If the right arrow key was pressed, we move the ship to the right by increasing the value of self.ship.rect.x by 1 ❸.

When you run *alien_invasion.py* now, the ship should move to the right one pixel every time you press the right arrow key. That's a start, but it's not an efficient way to control the ship. Let's improve this control by allowing continuous movement.

Allowing Continuous Movement

When the player holds down the right arrow key, we want the ship to continue moving right until the player releases the key. We'll have the game detect a pygame.KEYUP event so we'll know when the right arrow key is released; then we'll use the KEYDOWN and KEYUP events together with a flag called moving_right to implement continuous motion.

When the moving_right flag is False, the ship will be motionless. When the player presses the right arrow key, we'll set the flag to True, and when the player releases the key, we'll set the flag to False again.

The Ship class controls all attributes of the ship, so we'll give it an attribute called moving_right and an update() method to check the status of the moving_right flag. The update() method will change the position of the ship if the flag is set to True. We'll call this method once on each pass through the while loop to update the position of the ship.

Here are the changes to Ship:

ship.py

```
class Ship:
    """A class to manage the ship."""

    def __init__(self, ai_game):
        --snip--
        # Start each new ship at the bottom center of the screen.
        self.rect.midbottom = self.screen_rect.midbottom
```

```
        # Movement flag; start with a ship that's not moving.
❶       self.moving_right = False

❷   def update(self):
        """Update the ship's position based on the movement flag."""
        if self.moving_right:
            self.rect.x += 1

    def blitme(self):
        --snip--
```

We add a self.moving_right attribute in the __init__() method and set it to False initially ❶. Then we add update(), which moves the ship right if the flag is True ❷. The update() method will be called from outside the class, so it's not considered a helper method.

Now we need to modify _check_events() so that moving_right is set to True when the right arrow key is pressed and False when the key is released:

alien_invasion.py

```
    def _check_events(self):
        """Respond to keypresses and mouse events."""
        for event in pygame.event.get():
            --snip--
            elif event.type == pygame.KEYDOWN:
                if event.key == pygame.K_RIGHT:
❶                   self.ship.moving_right = True
❷           elif event.type == pygame.KEYUP:
                if event.key == pygame.K_RIGHT:
                    self.ship.moving_right = False
```

Here, we modify how the game responds when the player presses the right arrow key: instead of changing the ship's position directly, we merely set moving_right to True ❶. Then we add a new elif block, which responds to KEYUP events ❷. When the player releases the right arrow key (K_RIGHT), we set moving_right to False.

Next, we modify the while loop in run_game() so it calls the ship's update() method on each pass through the loop:

alien_invasion.py

```
    def run_game(self):
        """Start the main loop for the game."""
        while True:
            self._check_events()
            self.ship.update()
            self._update_screen()
            self.clock.tick(60)
```

The ship's position will be updated after we've checked for keyboard events and before we update the screen. This allows the ship's position to be updated in response to player input and ensures the updated position will be used when drawing the ship to the screen.

When you run *alien_invasion.py* and hold down the right arrow key, the ship should move continuously to the right until you release the key.

Moving Both Left and Right

Now that the ship can move continuously to the right, adding movement to the left is straightforward. Again, we'll modify the Ship class and the _check_events() method. Here are the relevant changes to __init__() and update() in Ship:

ship.py

```
    def __init__(self, ai_game):
        --snip--
        # Movement flags; start with a ship that's not moving.
        self.moving_right = False
        self.moving_left = False

    def update(self):
        """Update the ship's position based on movement flags."""
        if self.moving_right:
            self.rect.x += 1
        if self.moving_left:
            self.rect.x -= 1
```

In __init__(), we add a self.moving_left flag. In update(), we use two separate if blocks, rather than an elif, to allow the ship's rect.x value to be increased and then decreased when both arrow keys are held down. This results in the ship standing still. If we used elif for motion to the left, the right arrow key would always have priority. Using two if blocks makes the movements more accurate when the player might momentarily hold down both keys when changing directions.

We have to make two additions to _check_events():

alien_invasion.py

```
    def _check_events(self):
        """Respond to keypresses and mouse events."""
        for event in pygame.event.get():
            --snip--
            elif event.type == pygame.KEYDOWN:
                if event.key == pygame.K_RIGHT:
                    self.ship.moving_right = True
                elif event.key == pygame.K_LEFT:
                    self.ship.moving_left = True

            elif event.type == pygame.KEYUP:
                if event.key == pygame.K_RIGHT:
                    self.ship.moving_right = False
                elif event.key == pygame.K_LEFT:
                    self.ship.moving_left = False
```

If a KEYDOWN event occurs for the K_LEFT key, we set moving_left to True. If a KEYUP event occurs for the K_LEFT key, we set moving_left to False. We can use elif blocks here because each event is connected to only one key. If the player presses both keys at once, two separate events will be detected.

When you run *alien_invasion.py* now, you should be able to move the ship continuously to the right and left. If you hold down both keys, the ship should stop moving.

Next, we'll further refine the ship's movement. Let's adjust the ship's speed and limit how far the ship can move so it can't disappear off the sides of the screen.

Adjusting the Ship's Speed

Currently, the ship moves one pixel per cycle through the `while` loop, but we can take finer control of the ship's speed by adding a `ship_speed` attribute to the `Settings` class. We'll use this attribute to determine how far to move the ship on each pass through the loop. Here's the new attribute in *settings.py*:

settings.py

```
class Settings:
    """A class to store all settings for Alien Invasion."""

    def __init__(self):
        --snip--

        # Ship settings
        self.ship_speed = 1.5
```

We set the initial value of `ship_speed` to `1.5`. When the ship moves now, its position is adjusted by 1.5 pixels (rather than 1 pixel) on each pass through the loop.

We're using a float for the speed setting to give us finer control of the ship's speed when we increase the tempo of the game later on. However, rect attributes such as `x` store only integer values, so we need to make some modifications to `Ship`:

ship.py

```
class Ship:
    """A class to manage the ship."""

    def __init__(self, ai_game):
        """Initialize the ship and set its starting position."""
        self.screen = ai_game.screen
❶       self.settings = ai_game.settings
        --snip--

        # Start each new ship at the bottom center of the screen.
        self.rect.midbottom = self.screen_rect.midbottom

        # Store a float for the ship's exact horizontal position.
❷       self.x = float(self.rect.x)

        # Movement flags; start with a ship that's not moving.
        self.moving_right = False
        self.moving_left = False

    def update(self):
        """Update the ship's position based on movement flags."""
        # Update the ship's x value, not the rect.
        if self.moving_right:
❸           self.x += self.settings.ship_speed
```

```
        if self.moving_left:
            self.x -= self.settings.ship_speed

        # Update rect object from self.x.
❹       self.rect.x = self.x

    def blitme(self):
        --snip--
```

We create a settings attribute for Ship, so we can use it in update() ❶. Because we're adjusting the position of the ship by fractions of a pixel, we need to assign the position to a variable that can have a float assigned to it. You can use a float to set an attribute of a rect, but the rect will only keep the integer portion of that value. To keep track of the ship's position accurately, we define a new self.x ❷. We use the float() function to convert the value of self.rect.x to a float and assign this value to self.x.

Now when we change the ship's position in update(), the value of self.x is adjusted by the amount stored in settings.ship_speed ❸. After self.x has been updated, we use the new value to update self.rect.x, which controls the position of the ship ❹. Only the integer portion of self.x will be assigned to self.rect.x, but that's fine for displaying the ship.

Now we can change the value of ship_speed, and any value greater than 1 will make the ship move faster. This will help make the ship respond quickly enough to shoot down aliens, and it will let us change the tempo of the game as the player progresses in gameplay.

Limiting the Ship's Range

At this point, the ship will disappear off either edge of the screen if you hold down an arrow key long enough. Let's correct this so the ship stops moving when it reaches the screen's edge. We do this by modifying the update() method in Ship:

ship.py

```
    def update(self):
        """Update the ship's position based on movement flags."""
        # Update the ship's x value, not the rect.
❶       if self.moving_right and self.rect.right < self.screen_rect.right:
            self.x += self.settings.ship_speed
❷       if self.moving_left and self.rect.left > 0:
            self.x -= self.settings.ship_speed

        # Update rect object from self.x.
        self.rect.x = self.x
```

This code checks the position of the ship before changing the value of self.x. The code self.rect.right returns the *x*-coordinate of the right edge of the ship's rect. If this value is less than the value returned by self.screen_rect.right, the ship hasn't reached the right edge of the screen ❶. The same goes for the left edge: if the value of the left side of the rect is greater than 0, the ship hasn't reached the left edge of the screen ❷. This ensures the ship is within these bounds before adjusting the value of self.x.

When you run *alien_invasion.py* now, the ship should stop moving at either edge of the screen. This is pretty cool; all we've done is add a conditional test in an `if` statement, but it feels like the ship hits a wall or force field at either edge of the screen!

Refactoring _check_events()

The `_check_events()` method will increase in length as we continue to develop the game, so let's break `_check_events()` into two separate methods: one that handles `KEYDOWN` events and another that handles `KEYUP` events:

alien_invasion.py

```
    def _check_events(self):
        """Respond to keypresses and mouse events."""
        for event in pygame.event.get():
            if event.type == pygame.QUIT:
                sys.exit()
            elif event.type == pygame.KEYDOWN:
                self._check_keydown_events(event)
            elif event.type == pygame.KEYUP:
                self._check_keyup_events(event)

    def _check_keydown_events(self, event):
        """Respond to keypresses."""
        if event.key == pygame.K_RIGHT:
            self.ship.moving_right = True
        elif event.key == pygame.K_LEFT:
            self.ship.moving_left = True

    def _check_keyup_events(self, event):
        """Respond to key releases."""
        if event.key == pygame.K_RIGHT:
            self.ship.moving_right = False
        elif event.key == pygame.K_LEFT:
            self.ship.moving_left = False
```

We make two new helper methods: `_check_keydown_events()` and `_check_keyup_events()`. Each needs a `self` parameter and an `event` parameter. The bodies of these two methods are copied from `_check_events()`, and we've replaced the old code with calls to the new methods. The `_check_events()` method is simpler now with this cleaner code structure, which will make it easier to develop further responses to player input.

Pressing Q to Quit

Now that we're responding to keypresses efficiently, we can add another way to quit the game. It gets tedious to click the X at the top of the game window to end the game every time you test a new feature, so we'll add a keyboard shortcut to end the game when the player presses Q:

alien_invasion.py

```
    def _check_keydown_events(self, event):
        --snip--
        elif event.key == pygame.K_LEFT:
```

```
            self.ship.moving_left = True
        elif event.key == pygame.K_q:
            sys.exit()
```

In _check_keydown_events(), we add a new block that ends the game when the player presses Q. Now, when testing, you can press Q to close the game instead of using your cursor to close the window.

Running the Game in Fullscreen Mode

Pygame has a fullscreen mode that you might like better than running the game in a regular window. Some games look better in fullscreen mode, and on some systems, the game may perform better overall in fullscreen mode.

To run the game in fullscreen mode, make the following changes in __init__():

alien_invasion.py

```
    def __init__(self):
        """Initialize the game, and create game resources."""
        pygame.init()
        self.settings = Settings()

❶       self.screen = pygame.display.set_mode((0, 0), pygame.FULLSCREEN)
❷       self.settings.screen_width = self.screen.get_rect().width
        self.settings.screen_height = self.screen.get_rect().height
        pygame.display.set_caption("Alien Invasion")
```

When creating the screen surface, we pass a size of (0, 0) and the parameter pygame.FULLSCREEN ❶. This tells Pygame to figure out a window size that will fill the screen. Because we don't know the width and height of the screen ahead of time, we update these settings after the screen is created ❷. We use the width and height attributes of the screen's rect to update the settings object.

If you like how the game looks or behaves in fullscreen mode, keep these settings. If you liked the game better in its own window, you can revert back to the original approach where we set a specific screen size for the game.

NOTE *Make sure you can quit by pressing Q before running the game in fullscreen mode; Pygame offers no default way to quit a game while in fullscreen mode.*

A Quick Recap

In the next section, we'll add the ability to shoot bullets, which involves adding a new file called *bullet.py* and making some modifications to some of the files we're already using. Right now, we have three files containing a number of classes and methods. To be clear about how the project is organized, let's review each of these files before adding more functionality.

alien_invasion.py

The main file, *alien_invasion.py*, contains the `AlienInvasion` class. This class creates a number of important attributes used throughout the game: the settings are assigned to `settings`, the main display surface is assigned to `screen`, and a `ship` instance is created in this file as well. The main loop of the game, a `while` loop, is also stored in this module. The `while` loop calls `_check_events()`, `ship.update()`, and `_update_screen()`. It also ticks the clock on each pass through the loop.

The `_check_events()` method detects relevant events, such as keypresses and releases, and processes each of these types of events through the methods `_check_keydown_events()` and `_check_keyup_events()`. For now, these methods manage the ship's movement. The `AlienInvasion` class also contains `_update_screen()`, which redraws the screen on each pass through the main loop.

The *alien_invasion.py* file is the only file you need to run when you want to play *Alien Invasion*. The other files, *settings.py* and *ship.py*, contain code that is imported into this file.

settings.py

The *settings.py* file contains the `Settings` class. This class only has an `__init__()` method, which initializes attributes controlling the game's appearance and the ship's speed.

ship.py

The *ship.py* file contains the `Ship` class. The `Ship` class has an `__init__()` method, an `update()` method to manage the ship's position, and a `blitme()` method to draw the ship to the screen. The image of the ship is stored in *ship.bmp*, which is in the *images* folder.

TRY IT YOURSELF

12-3. Pygame Documentation: We're far enough into the game now that you might want to look at some of the Pygame documentation. The Pygame home page is at *https://pygame.org*, and the home page for the documentation is at *https://pygame.org/docs*. Just skim the documentation for now. You won't need it to complete this project, but it will help if you want to modify *Alien Invasion* or make your own game afterward.

12-4. Rocket: Make a game that begins with a rocket in the center of the screen. Allow the player to move the rocket up, down, left, or right using the four arrow keys. Make sure the rocket never moves beyond any edge of the screen.

12-5. Keys: Make a Pygame file that creates an empty screen. In the event loop, print the `event.key` attribute whenever a `pygame.KEYDOWN` event is detected. Run the program and press various keys to see how Pygame responds.

Shooting Bullets

Now let's add the ability to shoot bullets. We'll write code that fires a bullet, which is represented by a small rectangle, when the player presses the spacebar. Bullets will then travel straight up the screen until they disappear off the top of the screen.

Adding the Bullet Settings

At the end of the `__init__()` method, we'll update *settings.py* to include the values we'll need for a new `Bullet` class:

settings.py

```
    def __init__(self):
        --snip--
        # Bullet settings
        self.bullet_speed = 2.0
        self.bullet_width = 3
        self.bullet_height = 15
        self.bullet_color = (60, 60, 60)
```

These settings create dark gray bullets with a width of `3` pixels and a height of `15` pixels. The bullets will travel slightly faster than the ship.

Creating the Bullet Class

Now create a *bullet.py* file to store our `Bullet` class. Here's the first part of *bullet.py*:

bullet.py

```
import pygame
from pygame.sprite import Sprite

class Bullet(Sprite):
    """A class to manage bullets fired from the ship."""

    def __init__(self, ai_game):
        """Create a bullet object at the ship's current position."""
        super().__init__()
        self.screen = ai_game.screen
        self.settings = ai_game.settings
        self.color = self.settings.bullet_color

        # Create a bullet rect at (0, 0) and then set correct position.
❶       self.rect = pygame.Rect(0, 0, self.settings.bullet_width,
            self.settings.bullet_height)
❷       self.rect.midtop = ai_game.ship.rect.midtop

        # Store the bullet's position as a float.
❸       self.y = float(self.rect.y)
```

The `Bullet` class inherits from `Sprite`, which we import from the `pygame.sprite` module. When you use sprites, you can group related elements in your game and act on all the grouped elements at once. To create a bullet instance, `__init__()` needs the current instance of `AlienInvasion`, and we call `super()` to inherit properly from `Sprite`. We also set attributes for the screen and settings objects, and for the bullet's color.

Next we create the bullet's `rect` attribute ❶. The bullet isn't based on an image, so we have to build a rect from scratch using the `pygame.Rect()` class. This class requires the *x*- and *y*-coordinates of the top-left corner of the `rect`, and the width and height of the `rect`. We initialize the `rect` at (0, 0), but we'll move it to the correct location in the next line, because the bullet's position depends on the ship's position. We get the width and height of the bullet from the values stored in `self.settings`.

We set the bullet's `midtop` attribute to match the ship's `midtop` attribute ❷. This will make the bullet emerge from the top of the ship, making it look like the bullet is fired from the ship. We use a float for the bullet's *y*-coordinate so we can make fine adjustments to the bullet's speed ❸.

Here's the second part of *bullet.py*, `update()` and `draw_bullet()`:

bullet.py

```
    def update(self):
        """Move the bullet up the screen."""
        # Update the exact position of the bullet.
❶       self.y -= self.settings.bullet_speed
        # Update the rect position.
❷       self.rect.y = self.y

    def draw_bullet(self):
        """Draw the bullet to the screen."""
❸       pygame.draw.rect(self.screen, self.color, self.rect)
```

The `update()` method manages the bullet's position. When a bullet is fired, it moves up the screen, which corresponds to a decreasing *y*-coordinate value. To update the position, we subtract the amount stored in `settings.bullet_speed` from `self.y` ❶. We then use the value of `self.y` to set the value of `self.rect.y` ❷.

The `bullet_speed` setting allows us to increase the speed of the bullets as the game progresses or as needed to refine the game's behavior. Once a bullet is fired, we never change the value of its *x*-coordinate, so it will travel vertically in a straight line even if the ship moves.

When we want to draw a bullet, we call `draw_bullet()`. The `draw.rect()` function fills the part of the screen defined by the bullet's `rect` with the color stored in `self.color` ❸.

Storing Bullets in a Group

Now that we have a `Bullet` class and the necessary settings defined, we can write code to fire a bullet each time the player presses the spacebar. We'll create a group in `AlienInvasion` to store all the active bullets so we can manage the bullets that have already been fired. This group will be an instance of the `pygame.sprite.Group` class, which behaves like a list with some extra functionality that's helpful when building games. We'll use this group to draw bullets to the screen on each pass through the main loop and to update each bullet's position.

First, we'll import the new Bullet class:

alien_invasion.py

```
--snip--
from ship import Ship
from bullet import Bullet
```

Next we'll create the group that holds the bullets in __init__():

alien_invasion.py

```
    def __init__(self):
        --snip--
        self.ship = Ship(self)
        self.bullets = pygame.sprite.Group()
```

Then we need to update the position of the bullets on each pass through the while loop:

alien_invasion.py

```
    def run_game(self):
        """Start the main loop for the game."""
        while True:
            self._check_events()
            self.ship.update()
            self.bullets.update()
            self._update_screen()
            self.clock.tick(60)
```

When you call update() on a group, the group automatically calls update() for each sprite in the group. The line self.bullets.update() calls bullet.update() for each bullet we place in the group bullets.

Firing Bullets

In AlienInvasion, we need to modify _check_keydown_events() to fire a bullet when the player presses the spacebar. We don't need to change _check_keyup_events() because nothing happens when the spacebar is released. We also need to modify _update_screen() to make sure each bullet is drawn to the screen before we call flip().

There will be a bit of work to do when we fire a bullet, so let's write a new method, _fire_bullet(), to handle this work:

alien_invasion.py

```
    def _check_keydown_events(self, event):
        --snip--
        elif event.key == pygame.K_q:
            sys.exit()
❶       elif event.key == pygame.K_SPACE:
            self._fire_bullet()

    def _check_keyup_events(self, event):
        --snip--

    def _fire_bullet(self):
        """Create a new bullet and add it to the bullets group."""
❷       new_bullet = Bullet(self)
❸       self.bullets.add(new_bullet)
```

```
    def _update_screen(self):
        """Update images on the screen, and flip to the new screen."""
        self.screen.fill(self.settings.bg_color)
❹       for bullet in self.bullets.sprites():
            bullet.draw_bullet()
        self.ship.blitme()

        pygame.display.flip()
--snip--
```

We call `_fire_bullet()` when the spacebar is pressed ❶. In `_fire_bullet()`, we make an instance of `Bullet` and call it `new_bullet` ❷. We then add it to the group `bullets` using the `add()` method ❸. The `add()` method is similar to `append()`, but it's written specifically for Pygame groups.

The `bullets.sprites()` method returns a list of all sprites in the group `bullets`. To draw all fired bullets to the screen, we loop through the sprites in `bullets` and call `draw_bullet()` on each one ❹. We place this loop before the line that draws the ship, so the bullets don't start out on top of the ship.

When you run *alien_invasion.py* now, you should be able to move the ship right and left and fire as many bullets as you want. The bullets travel up the screen and disappear when they reach the top, as shown in Figure 12-3. You can alter the size, color, and speed of the bullets in *settings.py*.

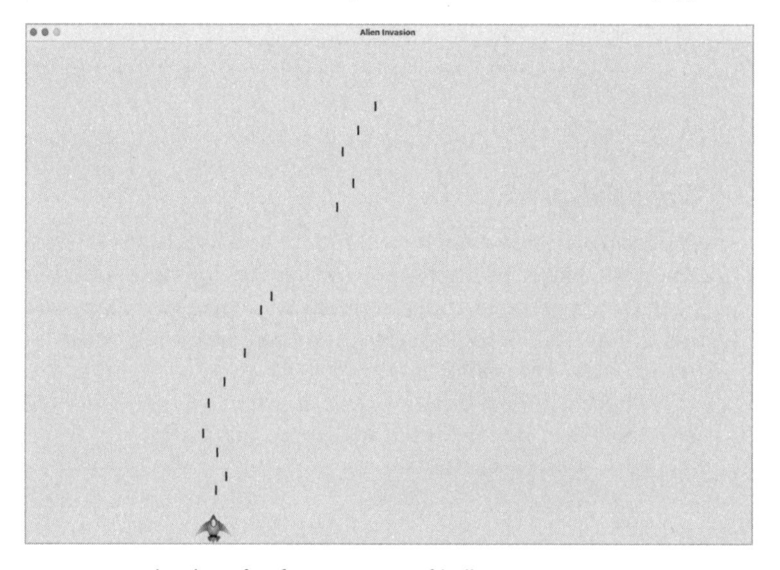

Figure 12-3: The ship after firing a series of bullets

Deleting Old Bullets

At the moment, the bullets disappear when they reach the top, but only because Pygame can't draw them above the top of the screen. The bullets actually continue to exist; their *y*-coordinate values just grow increasingly negative. This is a problem because they continue to consume memory and processing power.

We need to get rid of these old bullets, or the game will slow down from doing so much unnecessary work. To do this, we need to detect when the bottom value of a bullet's rect has a value of 0, which indicates the bullet has passed off the top of the screen:

alien_invasion.py

```
    def run_game(self):
        """Start the main loop for the game."""
        while True:
            self._check_events()
            self.ship.update()
            self.bullets.update()

            # Get rid of bullets that have disappeared.
❶           for bullet in self.bullets.copy():
❷               if bullet.rect.bottom <= 0:
❸                   self.bullets.remove(bullet)
❹           print(len(self.bullets))

            self._update_screen()
            self.clock.tick(60)
```

When you use a for loop with a list (or a group in Pygame), Python expects that the list will stay the same length as long as the loop is running. That means you can't remove items from a list or group within a for loop, so we have to loop over a copy of the group. We use the copy() method to set up the for loop ❶, which leaves us free to modify the original bullets group inside the loop. We check each bullet to see whether it has disappeared off the top of the screen ❷. If it has, we remove it from bullets ❸. We insert a print() call to show how many bullets currently exist in the game and verify they're being deleted when they reach the top of the screen ❹.

If this code works correctly, we can watch the terminal output while firing bullets and see that the number of bullets decreases to zero after each series of bullets has cleared the top of the screen. After you run the game and verify that bullets are being deleted properly, remove the print() call. If you leave it in, the game will slow down significantly because it takes more time to write output to the terminal than it does to draw graphics to the game window.

Limiting the Number of Bullets

Many shooting games limit the number of bullets a player can have on the screen at one time; doing so encourages players to shoot accurately. We'll do the same in *Alien Invasion*.

First, store the number of bullets allowed in *settings.py*:

settings.py

```
        # Bullet settings
        --snip--
        self.bullet_color = (60, 60, 60)
        self.bullets_allowed = 3
```

This limits the player to three bullets at a time. We'll use this setting in `AlienInvasion` to check how many bullets exist before creating a new bullet in `_fire_bullet()`:

alien_invasion.py

```
    def _fire_bullet(self):
        """Create a new bullet and add it to the bullets group."""
        if len(self.bullets) < self.settings.bullets_allowed:
            new_bullet = Bullet(self)
            self.bullets.add(new_bullet)
```

When the player presses the spacebar, we check the length of `bullets`. If `len(self.bullets)` is less than three, we create a new bullet. But if three bullets are already active, nothing happens when the spacebar is pressed. When you run the game now, you should only be able to fire bullets in groups of three.

Creating the _update_bullets() Method

We want to keep the `AlienInvasion` class reasonably well organized, so now that we've written and checked the bullet management code, we can move it to a separate method. We'll create a new method called `_update_bullets()` and add it just before `_update_screen()`:

alien_invasion.py

```
    def _update_bullets(self):
        """Update position of bullets and get rid of old bullets."""
        # Update bullet positions.
        self.bullets.update()

        # Get rid of bullets that have disappeared.
        for bullet in self.bullets.copy():
            if bullet.rect.bottom <= 0:
                 self.bullets.remove(bullet)
```

The code for `_update_bullets()` is cut and pasted from `run_game()`; all we've done here is clarify the comments.

The `while` loop in `run_game()` looks simple again:

alien_invasion.py

```
        while True:
            self._check_events()
            self.ship.update()
            self._update_bullets()
            self._update_screen()
            self.clock.tick(60)
```

Now our main loop contains only minimal code, so we can quickly read the method names and understand what's happening in the game. The main loop checks for player input, and then updates the position of the ship and any bullets that have been fired. We then use the updated positions to draw a new screen and tick the clock at the end of each pass through the loop.

Run *alien_invasion.py* one more time, and make sure you can still fire bullets without errors.

> **TRY IT YOURSELF**
>
> **12-6. Sideways Shooter:** Write a game that places a ship on the left side of the screen and allows the player to move the ship up and down. Make the ship fire a bullet that travels right across the screen when the player presses the spacebar. Make sure bullets are deleted once they disappear off the screen.

Summary

In this chapter, you learned to make a plan for a game and learned the basic structure of a game written in Pygame. You learned to set a background color and store settings in a separate class where you can adjust them more easily. You saw how to draw an image to the screen and give the player control over the movement of game elements. You created elements that move on their own, like bullets flying up a screen, and you deleted objects that are no longer needed. You also learned to refactor code in a project on a regular basis to facilitate ongoing development.

In Chapter 13, we'll add aliens to *Alien Invasion*. By the end of the chapter, you'll be able to shoot down aliens, hopefully before they reach your ship!

13

ALIENS!

In this chapter, we'll add aliens to *Alien Invasion*. We'll add one alien near the top of the screen and then generate a whole fleet of aliens. We'll make the fleet advance sideways and down, and we'll get rid of any aliens hit by a bullet. Finally, we'll limit the number of ships a player has and end the game when the player runs out of ships.

As you work through this chapter, you'll learn more about Pygame and about managing a large project. You'll also learn to detect collisions between game objects, like bullets and aliens. Detecting collisions helps you define interactions between elements in your games. For example, you can confine a character inside the walls of a maze or pass a ball between two characters. We'll continue to work from a plan that we revisit occasionally to maintain the focus of our code-writing sessions.

Before we start writing new code to add a fleet of aliens to the screen, let's look at the project and update our plan.

Reviewing the Project

When you're beginning a new phase of development on a large project, it's always a good idea to revisit your plan and clarify what you want to accomplish with the code you're about to write. In this chapter, we'll do the following:

- Add a single alien to the top-left corner of the screen, with appropriate spacing around it.
- Fill the upper portion of the screen with as many aliens as we can fit horizontally. We'll then create additional rows of aliens until we have a full fleet.
- Make the fleet move sideways and down until the entire fleet is shot down, an alien hits the ship, or an alien reaches the ground. If the entire fleet is shot down, we'll create a new fleet. If an alien hits the ship or the ground, we'll destroy the ship and create a new fleet.
- Limit the number of ships the player can use, and end the game when the player has used up the allotted number of ships.

We'll refine this plan as we implement features, but this is specific enough to start writing code.

You should also review your existing code when you begin working on a new series of features in a project. Because each new phase typically makes a project more complex, it's best to clean up any cluttered or inefficient code. We've been refactoring as we go, so there isn't any code that we need to refactor at this point.

Creating the First Alien

Placing one alien on the screen is like placing a ship on the screen. Each alien's behavior is controlled by a class called `Alien`, which we'll structure like the `Ship` class. We'll continue using bitmap images for simplicity. You can find your own image for an alien or use the one shown in Figure 13-1, which is available in the book's resources at *https://ehmatthes.github.io/pcc_3e*. This image has a gray background, which matches the screen's background color. Make sure you save the image file you choose in the *images* folder.

Figure 13-1: The alien we'll use to build the fleet

Creating the Alien Class

Now we'll write the Alien class and save it as *alien.py*:

alien.py

```
import pygame
from pygame.sprite import Sprite

class Alien(Sprite):
    """A class to represent a single alien in the fleet."""

    def __init__(self, ai_game):
        """Initialize the alien and set its starting position."""
        super().__init__()
        self.screen = ai_game.screen

        # Load the alien image and set its rect attribute.
        self.image = pygame.image.load('images/alien.bmp')
        self.rect = self.image.get_rect()

        # Start each new alien near the top left of the screen.
❶       self.rect.x = self.rect.width
        self.rect.y = self.rect.height

        # Store the alien's exact horizontal position.
❷       self.x = float(self.rect.x)
```

Most of this class is like the Ship class, except for the alien's placement on the screen. We initially place each alien near the top-left corner of the screen; we add a space to the left of it that's equal to the alien's width and a space above it equal to its height ❶, so it's easy to see. We're mainly concerned with the aliens' horizontal speed, so we'll track the horizontal position of each alien precisely ❷.

This Alien class doesn't need a method for drawing it to the screen; instead, we'll use a Pygame group method that automatically draws all the elements of a group to the screen.

Creating an Instance of the Alien

We want to create an instance of Alien so we can see the first alien on the screen. Because it's part of our setup work, we'll add the code for this instance at the end of the __init__() method in AlienInvasion. Eventually, we'll create an entire fleet of aliens, which will be quite a bit of work, so we'll make a new helper method called _create_fleet().

The order of methods in a class doesn't matter, as long as there's some consistency to how they're placed. I'll place _create_fleet() just before the _update_screen() method, but anywhere in AlienInvasion will work. First, we'll import the Alien class.

Here are the updated import statements for *alien_invasion.py*:

alien_invasion.py

```
--snip--
from bullet import Bullet
from alien import Alien
```

And here's the updated `__init__()` method:

alien_invasion.py

```
    def __init__(self):
        --snip--
        self.ship = Ship(self)
        self.bullets = pygame.sprite.Group()
        self.aliens = pygame.sprite.Group()

        self._create_fleet()
```

We create a group to hold the fleet of aliens, and we call `_create_fleet()`, which we're about to write.

Here's the new `_create_fleet()` method:

alien_invasion.py

```
    def _create_fleet(self):
        """Create the fleet of aliens."""
        # Make an alien.
        alien = Alien(self)
        self.aliens.add(alien)
```

In this method, we're creating one instance of `Alien` and then adding it to the group that will hold the fleet. The alien will be placed in the default upper-left area of the screen.

To make the alien appear, we need to call the group's `draw()` method in `_update_screen()`:

alien_invasion.py

```
    def _update_screen(self):
        --snip--
        self.ship.blitme()
        self.aliens.draw(self.screen)

        pygame.display.flip()
```

When you call `draw()` on a group, Pygame draws each element in the group at the position defined by its `rect` attribute. The `draw()` method requires one argument: a surface on which to draw the elements from the group. Figure 13-2 shows the first alien on the screen.

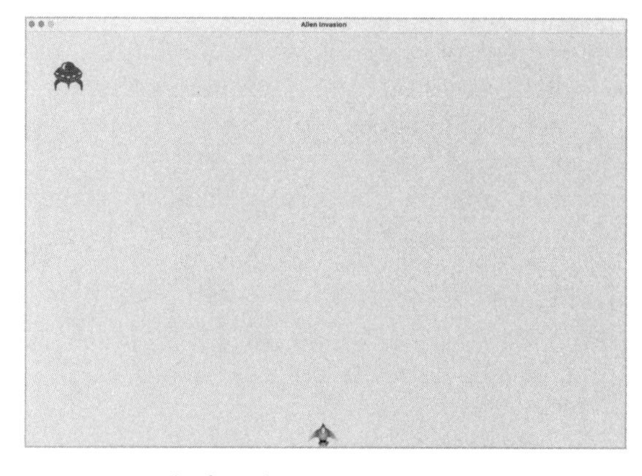

Figure 13-2: The first alien appears.

Now that the first alien appears correctly, we'll write the code to draw an entire fleet.

Building the Alien Fleet

To draw a fleet, we need to figure out how to fill the upper portion of the screen with aliens, without overcrowding the game window. There are a number of ways to accomplish this goal. We'll approach it by adding aliens across the top of the screen, until there's no space left for a new alien. Then we'll repeat this process, as long as we have enough vertical space to add a new row.

Creating a Row of Aliens

Now we're ready to generate a full row of aliens. To make a full row, we'll first make a single alien so we have access to the alien's width. We'll place an alien on the left side of the screen and then keep adding aliens until we run out of space:

alien_invasion.py

```
    def _create_fleet(self):
        """Create the fleet of aliens."""
        # Create an alien and keep adding aliens until there's no room left.
        # Spacing between aliens is one alien width.
        alien = Alien(self)
        alien_width = alien.rect.width

❶       current_x = alien_width
❷       while current_x < (self.settings.screen_width - 2 * alien_width):
❸           new_alien = Alien(self)
❹           new_alien.x = current_x
            new_alien.rect.x = current_x
            self.aliens.add(new_alien)
❺           current_x += 2 * alien_width
```

We get the alien's width from the first alien we created, and then define a variable called `current_x` ❶. This refers to the horizontal position of the next alien we intend to place on the screen. We initially set this to one alien width, to offset the first alien in the fleet from the left edge of the screen.

Next, we begin the `while` loop ❷; we're going to keep adding aliens *while* there's enough room to place one. To determine whether there's room to place another alien, we'll compare `current_x` to some maximum value. A first attempt at defining this loop might look like this:

```
while current_x < self.settings.screen_width:
```

This seems like it might work, but it would place the final alien in the row at the far-right edge of the screen. So we add a little margin on the right side of the screen. As long as there's at least two alien widths' worth of space at the right edge of the screen, we enter the loop and add another alien to the fleet.

Whenever there's enough horizontal space to continue the loop, we want to do two things: create an alien at the correct position, and define the horizontal position of the next alien in the row. We create an alien and assign it to `new_alien` ❸. Then we set the precise horizontal position to the current value of `current_x` ❹. We also position the alien's `rect` at this same *x*-value, and add the new alien to the group `self.aliens`.

Finally, we increment the value of `current_x` ❺. We add two alien widths to the horizontal position, to move past the alien we just added and to leave some space between the aliens as well. Python will re-evaluate the condition at the start of the `while` loop and decide if there's room for another alien. When there's no room left, the loop will end, and we should have a full row of aliens.

When you run *Alien Invasion* now, you should see the first row of aliens appear, as in Figure 13-3.

Figure 13-3: The first row of aliens

NOTE *It's not always obvious exactly how to construct a loop like the one shown in this section. One nice thing about programming is that your initial ideas for how to approach a problem like this don't have to be correct. It's perfectly reasonable to start a loop like this with the aliens positioned too far to the right, and then modify the loop until you have an appropriate amount of space on the screen.*

Refactoring _create_fleet()

If the code we've written so far was all we needed to create a fleet, we'd probably leave `_create_fleet()` as is. But we have more work to do, so let's clean up the method a bit. We'll add a new helper method, `_create_alien()`, and call it from `_create_fleet()`:

alien_invasion.py

```
    def _create_fleet(self):
        --snip--
        while current_x < (self.settings.screen_width - 2 * alien_width):
            self._create_alien(current_x)
            current_x += 2 * alien_width
```

```
❶     def _create_alien(self, x_position):
          """Create an alien and place it in the row."""
          new_alien = Alien(self)
          new_alien.x = x_position
          new_alien.rect.x = x_position
          self.aliens.add(new_alien)
```

The method _create_alien() requires one parameter in addition to self: the *x*-value that specifies where the alien should be placed ❶. The code in the body of _create_alien() is the same code that was in _create_fleet(), except we use the parameter name x_position in place of current_x. This refactoring will make it easier to add new rows and create an entire fleet.

Adding Rows

To finish the fleet, we'll keep adding more rows until we run out of room. We'll use a nested loop—we'll wrap another while loop around the current one. The inner loop will place aliens horizontally in a row by focusing on the aliens' *x*-values. The outer loop will place aliens vertically by focusing on the *y*-values. We'll stop adding rows when we get near the bottom of the screen, leaving enough space for the ship and some room to start firing at the aliens.

Here's how to nest the two while loops in _create_fleet():

```
      def _create_fleet(self):
          """Create the fleet of aliens."""
          # Create an alien and keep adding aliens until there's no room left.
          # Spacing between aliens is one alien width and one alien height.
          alien = Alien(self)
❶         alien_width, alien_height = alien.rect.size

❷         current_x, current_y = alien_width, alien_height
❸         while current_y < (self.settings.screen_height - 3 * alien_height):
              while current_x < (self.settings.screen_width - 2 * alien_width):
❹                 self._create_alien(current_x, current_y)
                  current_x += 2 * alien_width

❺             # Finished a row; reset x value, and increment y value.
              current_x = alien_width
              current_y += 2 * alien_height
```

We'll need to know the alien's height in order to place rows, so we grab the alien's width and height using the size attribute of an alien rect ❶. A rect's size attribute is a tuple containing its width and height.

Next, we set the initial *x*- and *y*-values for the placement of the first alien in the fleet ❷. We place it one alien width in from the left and one alien height down from the top. Then we define the while loop that controls how many rows are placed onto the screen ❸. As long as the *y*-value for the next row is less than the screen height, minus three alien heights, we'll keep adding rows. (If this doesn't leave the right amount of space, we can adjust it later.)

We call `_create_alien()`, and pass it the *y*-value as well as its *x*-position ❹. We'll modify `_create_alien()` in a moment.

Notice the indentation of the last two lines of code ❺. They're inside the outer `while` loop, but outside the inner `while` loop. This block runs after the inner loop is finished; it runs once after each row is created. After each row has been added, we reset the value of `current_x` so the first alien in the next row will be placed at the same position as the first alien in the previous rows. Then we add two alien heights to the current value of `current_y`, so the next row will be placed further down the screen. Indentation is really important here; if you don't see the correct fleet when you run *alien_invasion.py* at the end of this section, check the indentation of all the lines in these nested loops.

We need to modify `_create_alien()` to set the vertical position of the alien correctly:

```
def _create_alien(self, x_position, y_position):
    """Create an alien and place it in the fleet."""
    new_alien = Alien(self)
    new_alien.x = x_position
    new_alien.rect.x = x_position
    new_alien.rect.y = y_position
    self.aliens.add(new_alien)
```

We modify the definition of the method to accept the *y*-value for the new alien, and we set the vertical position of the rect in the body of the method.

When you run the game now, you should see a full fleet of aliens, as shown in Figure 13-4.

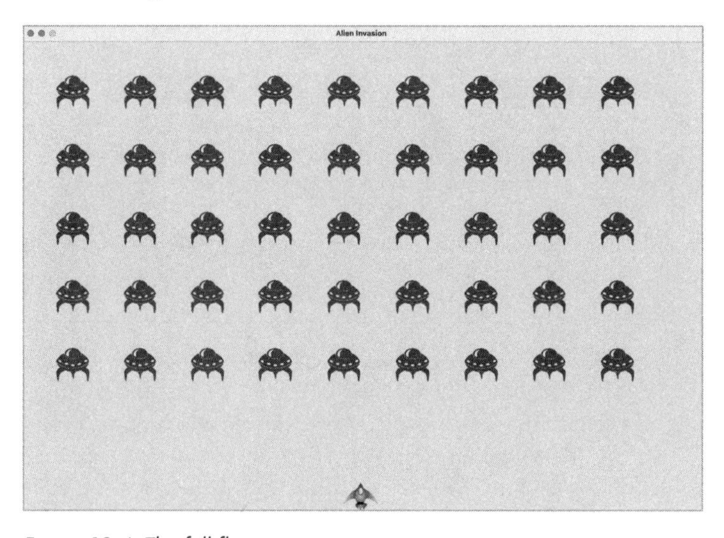

Figure 13-4: The full fleet appears.

In the next section, we'll make the fleet move!

TRY IT YOURSELF

13-1. Stars: Find an image of a star. Make a grid of stars appear on the screen.

13-2. Better Stars: You can make a more realistic star pattern by introducing randomness when you place each star. Recall from Chapter 9 that you can get a random number like this:

```
from random import randint
random_number = randint(-10, 10)
```

This code returns a random integer between –10 and 10. Using your code in Exercise 13-1, adjust each star's position by a random amount.

Making the Fleet Move

Now let's make the fleet of aliens move to the right across the screen until it hits the edge, and then make it drop a set amount and move in the other direction. We'll continue this movement until all aliens have been shot down, one collides with the ship, or one reaches the bottom of the screen. Let's begin by making the fleet move to the right.

Moving the Aliens Right

To move the aliens, we'll use an `update()` method in *alien.py*, which we'll call for each alien in the group of aliens. First, add a setting to control the speed of each alien:

settings.py

```
    def __init__(self):
        --snip--
        # Alien settings
        self.alien_speed = 1.0
```

Then use this setting to implement `update()` in *alien.py*:

alien.py

```
    def __init__(self, ai_game):
        """Initialize the alien and set its starting position."""
        super().__init__()
        self.screen = ai_game.screen
        self.settings = ai_game.settings
        --snip--

    def update(self):
        """Move the alien to the right."""
❶       self.x += self.settings.alien_speed
❷       self.rect.x = self.x
```

We create a `settings` parameter in `__init__()` so we can access the alien's speed in `update()`. Each time we update an alien's position, we move it to the right by the amount stored in alien_speed. We track the alien's exact position with the self.x attribute, which can hold float values ❶. We then use the value of self.x to update the position of the alien's rect ❷.

In the main while loop, we already have calls to update the ship and bullet positions. Now we'll add a call to update the position of each alien as well:

alien_invasion.py

```
        while True:
            self._check_events()
            self.ship.update()
            self._update_bullets()
            self._update_aliens()
            self._update_screen()
            self.clock.tick(60)
```

We're about to write some code to manage the movement of the fleet, so we create a new method called _update_aliens(). We update the aliens' positions after the bullets have been updated, because we'll soon be checking to see whether any bullets hit any aliens.

Where you place this method in the module is not critical. But to keep the code organized, I'll place it just after _update_bullets() to match the order of method calls in the while loop. Here's the first version of _update_aliens():

alien_invasion.py

```
    def _update_aliens(self):
        """Update the positions of all aliens in the fleet."""
        self.aliens.update()
```

We use the update() method on the aliens group, which calls each alien's update() method. When you run *Alien Invasion* now, you should see the fleet move right and disappear off the side of the screen.

Creating Settings for Fleet Direction

Now we'll create the settings that will make the fleet move down the screen and to the left when it hits the right edge of the screen. Here's how to implement this behavior:

settings.py

```
        # Alien settings
        self.alien_speed = 1.0
        self.fleet_drop_speed = 10
        # fleet_direction of 1 represents right; -1 represents left.
        self.fleet_direction = 1
```

The setting fleet_drop_speed controls how quickly the fleet drops down the screen each time an alien reaches either edge. It's helpful to separate this speed from the aliens' horizontal speed so you can adjust the two speeds independently.

To implement the setting fleet_direction, we could use a text value such as 'left' or 'right', but we'd end up with if-elif statements testing for the fleet direction. Instead, because we only have two directions to deal with, let's use the values 1 and –1 and switch between them each time the fleet changes direction. (Using numbers also makes sense because moving right involves adding to each alien's *x*-coordinate value, and moving left involves subtracting from each alien's *x*-coordinate value.)

Checking Whether an Alien Has Hit the Edge

We need a method to check whether an alien is at either edge, and we need to modify `update()` to allow each alien to move in the appropriate direction. This code is part of the `Alien` class:

alien.py

```
    def check_edges(self):
        """Return True if alien is at edge of screen."""
        screen_rect = self.screen.get_rect()
❶       return (self.rect.right >= screen_rect.right) or (self.rect.left <= 0)

    def update(self):
        """Move the alien right or left."""
❷       self.x += self.settings.alien_speed * self.settings.fleet_direction
        self.rect.x = self.x
```

We can call the new method `check_edges()` on any alien to see whether it's at the left or right edge. The alien is at the right edge if the `right` attribute of its `rect` is greater than or equal to the `right` attribute of the screen's `rect`. It's at the left edge if its `left` value is less than or equal to 0 ❶. Rather than put this conditional test in an `if` block, we put the test directly in the `return` statement. This method will return `True` if the alien is at the right or left edge, and `False` if it is not at either edge.

We modify the method `update()` to allow motion to the left or right by multiplying the alien's speed by the value of `fleet_direction` ❷. If `fleet_direction` is 1, the value of `alien_speed` will be added to the alien's current position, moving the alien to the right; if `fleet_direction` is –1, the value will be subtracted from the alien's position, moving the alien to the left.

Dropping the Fleet and Changing Direction

When an alien reaches the edge, the entire fleet needs to drop down and change direction. Therefore, we need to add some code to `AlienInvasion` because that's where we'll check whether any aliens are at the left or right edge. We'll make this happen by writing the methods `_check_fleet_edges()` and `_change_fleet_direction()`, and then modifying `_update_aliens()`. I'll put these new methods after `_create_alien()`, but again, the placement of these methods in the class isn't critical.

alien_invasion.py

```
    def _check_fleet_edges(self):
        """Respond appropriately if any aliens have reached an edge."""
❶       for alien in self.aliens.sprites():
            if alien.check_edges():
❷               self._change_fleet_direction()
                break

    def _change_fleet_direction(self):
```

```
        """Drop the entire fleet and change the fleet's direction."""
        for alien in self.aliens.sprites():
❸           alien.rect.y += self.settings.fleet_drop_speed
        self.settings.fleet_direction *= -1
```

In _check_fleet_edges(), we loop through the fleet and call check_edges() on each alien ❶. If check_edges() returns True, we know an alien is at an edge and the whole fleet needs to change direction; so we call _change_fleet_direction() and break out of the loop ❷. In _change_fleet_direction(), we loop through all the aliens and drop each one using the setting fleet_drop_speed ❸; then we change the value of fleet_direction by multiplying its current value by –1. The line that changes the fleet's direction isn't part of the for loop. We want to change each alien's vertical position, but we only want to change the direction of the fleet once.

Here are the changes to _update_aliens():

alien_invasion.py

```
    def _update_aliens(self):
        """Check if the fleet is at an edge, then update positions."""
        self._check_fleet_edges()
        self.aliens.update()
```

We've modified the method by calling _check_fleet_edges() before updating each alien's position.

When you run the game now, the fleet should move back and forth between the edges of the screen and drop down every time it hits an edge. Now we can start shooting down aliens and watch for any aliens that hit the ship or reach the bottom of the screen.

TRY IT YOURSELF

13-3. Raindrops: Find an image of a raindrop and create a grid of raindrops. Make the raindrops fall toward the bottom of the screen until they disappear.

13-4. Steady Rain: Modify your code in Exercise 13-3 so when a row of raindrops disappears off the bottom of the screen, a new row appears at the top of the screen and begins to fall.

Shooting Aliens

We've built our ship and a fleet of aliens, but when the bullets reach the aliens, they simply pass through because we aren't checking for collisions. In game programming, *collisions* happen when game elements overlap. To make the bullets shoot down aliens, we'll use the function sprite.groupcollide() to look for collisions between members of two groups.

Detecting Bullet Collisions

We want to know right away when a bullet hits an alien so we can make an alien disappear as soon as it's hit. To do this, we'll look for collisions immediately after updating the position of all the bullets.

The `sprite.groupcollide()` function compares the rects of each element in one group with the rects of each element in another group. In this case, it compares each bullet's rect with each alien's rect and returns a dictionary containing the bullets and aliens that have collided. Each key in the dictionary will be a bullet, and the corresponding value will be the alien that was hit. (We'll also use this dictionary when we implement a scoring system in Chapter 14.)

Add the following code to the end of `_update_bullets()` to check for collisions between bullets and aliens:

alien_invasion.py

```
    def _update_bullets(self):
        """Update position of bullets and get rid of old bullets."""
        --snip--

        # Check for any bullets that have hit aliens.
        #   If so, get rid of the bullet and the alien.
        collisions = pygame.sprite.groupcollide(
                self.bullets, self.aliens, True, True)
```

The new code we added compares the positions of all the bullets in `self.bullets` and all the aliens in `self.aliens`, and identifies any that overlap. Whenever the rects of a bullet and alien overlap, `groupcollide()` adds a key-value pair to the dictionary it returns. The two `True` arguments tell Pygame to delete the bullets and aliens that have collided. (To make a high-powered bullet that can travel to the top of the screen, destroying every alien in its path, you could set the first Boolean argument to `False` and keep the second Boolean argument set to `True`. The aliens hit would disappear, but all bullets would stay active until they disappeared off the top of the screen.)

When you run *Alien Invasion* now, aliens you hit should disappear. Figure 13-5 shows a fleet that has been partially shot down.

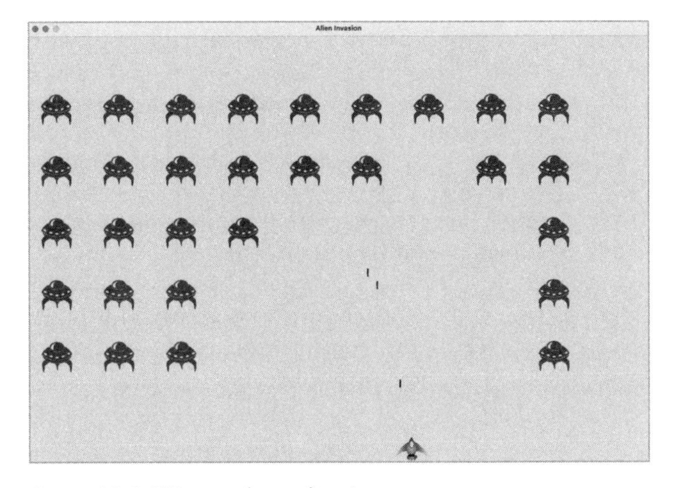

Figure 13-5: We can shoot aliens!

Making Larger Bullets for Testing

You can test many features of *Alien Invasion* simply by running the game, but some features are tedious to test in the normal version of the game. For example, it's a lot of work to shoot down every alien on the screen multiple times to test whether your code responds to an empty fleet correctly.

To test particular features, you can change certain game settings to focus on a particular area. For example, you might shrink the screen so there are fewer aliens to shoot down or increase the bullet speed and give yourself lots of bullets at once.

My favorite change for testing *Alien Invasion* is to use really wide bullets that remain active even after they've hit an alien (see Figure 13-6). Try setting `bullet_width` to 300, or even 3,000, to see how quickly you can shoot down the fleet!

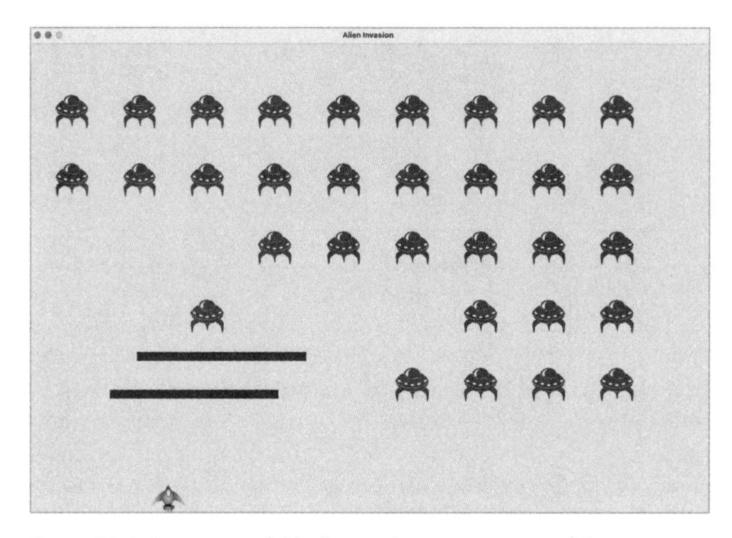

Figure 13-6: Extra-powerful bullets make some aspects of the game easier to test.

Changes like these will help you test the game more efficiently and possibly spark ideas for giving players bonus powers. Just remember to restore the settings to normal when you're finished testing a feature.

Repopulating the Fleet

One key feature of *Alien Invasion* is that the aliens are relentless: every time the fleet is destroyed, a new fleet should appear.

To make a new fleet of aliens appear after a fleet has been destroyed, we first check whether the `aliens` group is empty. If it is, we make a call to `_create_fleet()`. We'll perform this check at the end of `_update_bullets()`, because that's where individual aliens are destroyed.

alien_invasion.py

```
    def _update_bullets(self):
        --snip--
❶       if not self.aliens:
```

```
        # Destroy existing bullets and create new fleet.
❷       self.bullets.empty()
        self._create_fleet()
```

We check whether the aliens group is empty ❶. An empty group evaluates to False, so this is a simple way to check whether the group is empty. If it is, we get rid of any existing bullets by using the empty() method, which removes all the remaining sprites from a group ❷. We also call _create_fleet(), which fills the screen with aliens again.

Now a new fleet appears as soon as you destroy the current fleet.

Speeding Up the Bullets

If you've tried firing at the aliens in the game's current state, you might find that the bullets aren't traveling at the best speed for gameplay. They might be a little too slow or a little too fast. At this point, you can modify the settings to make the gameplay more interesting. Keep in mind that the game is going to get progressively faster, so don't make the game too fast at the beginning.

We modify the speed of the bullets by adjusting the value of bullet_speed in *settings.py*. On my system, I'll adjust the value of bullet_speed to 2.5, so the bullets travel a little faster:

settings.py

```
        # Bullet settings
        self.bullet_speed = 2.5
        self.bullet_width = 3
        --snip--
```

The best value for this setting depends on your experience of the game, so find a value that works for you. You can adjust other settings as well.

Refactoring _update_bullets()

Let's refactor _update_bullets() so it's not doing so many different tasks. We'll move the code for dealing with bullet-alien collisions to a separate method:

alien_invasion.py

```
    def _update_bullets(self):
        --snip--
        # Get rid of bullets that have disappeared.
        for bullet in self.bullets.copy():
            if bullet.rect.bottom <= 0:
                 self.bullets.remove(bullet)

        self._check_bullet_alien_collisions()

    def _check_bullet_alien_collisions(self):
        """Respond to bullet-alien collisions."""
        # Remove any bullets and aliens that have collided.
        collisions = pygame.sprite.groupcollide(
                self.bullets, self.aliens, True, True)
```

```
        if not self.aliens:
            # Destroy existing bullets and create new fleet.
            self.bullets.empty()
            self._create_fleet()
```

We've created a new method, `_check_bullet_alien_collisions()`, to look for collisions between bullets and aliens and to respond appropriately if the entire fleet has been destroyed. Doing so keeps `_update_bullets()` from growing too long and simplifies further development.

> **TRY IT YOURSELF**
>
> **13-5. Sideways Shooter Part 2:** We've come a long way since Exercise 12-6, *Sideways Shooter*. For this exercise, try to develop *Sideways Shooter* to the same point we've brought *Alien Invasion* to. Add a fleet of aliens, and make them move sideways toward the ship. Or, write code that places aliens at random positions along the right side of the screen and then sends them toward the ship. Also, write code that makes the aliens disappear when they're hit.

Ending the Game

What's the fun and challenge in playing a game you can't lose? If the player doesn't shoot down the fleet quickly enough, we'll have the aliens destroy the ship when they make contact. At the same time, we'll limit the number of ships a player can use, and we'll destroy the ship when an alien reaches the bottom of the screen. The game will end when the player has used up all their ships.

Detecting Alien-Ship Collisions

We'll start by checking for collisions between aliens and the ship so we can respond appropriately when an alien hits it. We'll check for alien-ship collisions immediately after updating the position of each alien in `AlienInvasion`:

alien_invasion.py

```
    def _update_aliens(self):
        --snip--
        self.aliens.update()

        # Look for alien-ship collisions.
❶       if pygame.sprite.spritecollideany(self.ship, self.aliens):
❷           print("Ship hit!!!")
```

The `spritecollideany()` function takes two arguments: a sprite and a group. The function looks for any member of the group that has collided with the sprite and stops looping through the group as soon as it finds one member that has collided with the sprite. Here, it loops through the group `aliens` and returns the first alien it finds that has collided with `ship`.

If no collisions occur, spritecollideany() returns None and the if block won't execute ❶. If it finds an alien that has collided with the ship, it returns that alien and the if block executes: it prints Ship hit!!! ❷. When an alien hits the ship, we'll need to do a number of tasks: delete all remaining aliens and bullets, recenter the ship, and create a new fleet. Before we write code to do all this, we want to know that our approach to detecting alien-ship collisions works correctly. Writing a print() call is a simple way to ensure we're detecting these collisions properly.

Now when you run *Alien Invasion,* the message *Ship hit!!!* should appear in the terminal whenever an alien runs into the ship. When you're testing this feature, set fleet_drop_speed to a higher value, such as 50 or 100, so the aliens reach your ship faster.

Responding to Alien-Ship Collisions

Now we need to figure out exactly what will happen when an alien collides with the ship. Instead of destroying the ship instance and creating a new one, we'll count how many times the ship has been hit by tracking statistics for the game. Tracking statistics will also be useful for scoring.

Let's write a new class, GameStats, to track game statistics, and let's save it as *game_stats.py*:

game_stats.py

```
class GameStats:
    """Track statistics for Alien Invasion."""

    def __init__(self, ai_game):
        """Initialize statistics."""
        self.settings = ai_game.settings
❶       self.reset_stats()

    def reset_stats(self):
        """Initialize statistics that can change during the game."""
        self.ships_left = self.settings.ship_limit
```

We'll make one GameStats instance for the entire time *Alien Invasion* is running, but we'll need to reset some statistics each time the player starts a new game. To do this, we'll initialize most of the statistics in the reset_stats() method, instead of directly in __init__(). We'll call this method from __init__() so the statistics are set properly when the GameStats instance is first created ❶. But we'll also be able to call reset_stats() anytime the player starts a new game. Right now we have only one statistic, ships_left, the value of which will change throughout the game.

The number of ships the player starts with should be stored in *settings.py* as ship_limit:

settings.py

```
        # Ship settings
        self.ship_speed = 1.5
        self.ship_limit = 3
```

We also need to make a few changes in *alien_invasion.py* to create an instance of GameStats. First, we'll update the import statements at the top of the file:

alien_invasion.py

```
import sys
from time import sleep

import pygame

from settings import Settings
from game_stats import GameStats
from ship import Ship
--snip--
```

We import the sleep() function from the time module in the Python standard library, so we can pause the game for a moment when the ship is hit. We also import GameStats.

We'll create an instance of GameStats in __init__():

alien_invasion.py

```
    def __init__(self):
        --snip--
        self.screen = pygame.display.set_mode(
            (self.settings.screen_width, self.settings.screen_height))
        pygame.display.set_caption("Alien Invasion")

        # Create an instance to store game statistics.
        self.stats = GameStats(self)

        self.ship = Ship(self)
        --snip--
```

We make the instance after creating the game window but before defining other game elements, such as the ship.

When an alien hits the ship, we'll subtract 1 from the number of ships left, destroy all existing aliens and bullets, create a new fleet, and reposition the ship in the middle of the screen. We'll also pause the game for a moment so the player can notice the collision and regroup before a new fleet appears.

Let's put most of this code in a new method called _ship_hit(). We'll call this method from _update_aliens() when an alien hits the ship:

alien_invasion.py

```
    def _ship_hit(self):
        """Respond to the ship being hit by an alien."""
        # Decrement ships_left.
❶       self.stats.ships_left -= 1

        # Get rid of any remaining bullets and aliens.
❷       self.bullets.empty()
        self.aliens.empty()

        # Create a new fleet and center the ship.
❸       self._create_fleet()
        self.ship.center_ship()
```

```
        # Pause.
❹       sleep(0.5)
```

The new method _ship_hit() coordinates the response when an alien hits a ship. Inside _ship_hit(), the number of ships left is reduced by 1 ❶, after which we empty the groups bullets and aliens ❷.

Next, we create a new fleet and center the ship ❸. (We'll add the method center_ship() to Ship in a moment.) Then we add a pause after the updates have been made to all the game elements but before any changes have been drawn to the screen, so the player can see that their ship has been hit ❹. The sleep() call pauses program execution for half a second, long enough for the player to see that the alien has hit the ship. When the sleep() function ends, code execution moves on to the _update_screen() method, which draws the new fleet to the screen.

In _update_aliens(), we replace the print() call with a call to _ship_hit() when an alien hits the ship:

alien_invasion.py

```
def _update_aliens(self):
    --snip--
    if pygame.sprite.spritecollideany(self.ship, self.aliens):
        self._ship_hit()
```

Here's the new method center_ship(), which belongs in *ship.py*:

ship.py

```
def center_ship(self):
    """Center the ship on the screen."""
    self.rect.midbottom = self.screen_rect.midbottom
    self.x = float(self.rect.x)
```

We center the ship the same way we did in __init__(). After centering it, we reset the self.x attribute, which allows us to track the ship's exact position.

NOTE *Notice that we never make more than one ship; we make only one ship instance for the whole game and recenter it whenever the ship has been hit. The statistic ships_left will tell us when the player has run out of ships.*

Run the game, shoot a few aliens, and let an alien hit the ship. The game should pause, and a new fleet should appear with the ship centered at the bottom of the screen again.

Aliens That Reach the Bottom of the Screen

If an alien reaches the bottom of the screen, we'll have the game respond the same way it does when an alien hits the ship. To check when this happens, add a new method in *alien_invasion.py*:

alien_invasion.py

```
def _check_aliens_bottom(self):
    """Check if any aliens have reached the bottom of the screen."""
    for alien in self.aliens.sprites():
```

```
❶           if alien.rect.bottom >= self.settings.screen_height:
                # Treat this the same as if the ship got hit.
                self._ship_hit()
                break
```

The method _check_aliens_bottom() checks whether any aliens have reached the bottom of the screen. An alien reaches the bottom when its rect.bottom value is greater than or equal to the screen's height ❶. If an alien reaches the bottom, we call _ship_hit(). If one alien hits the bottom, there's no need to check the rest, so we break out of the loop after calling _ship_hit().

We'll call this method from _update_aliens():

alien_invasion.py

```
    def _update_aliens(self):
        --snip--
        # Look for alien-ship collisions.
        if pygame.sprite.spritecollideany(self.ship, self.aliens):
            self._ship_hit()

        # Look for aliens hitting the bottom of the screen.
        self._check_aliens_bottom()
```

We call _check_aliens_bottom() after updating the positions of all the aliens and after looking for alien-ship collisions. Now a new fleet will appear every time the ship is hit by an alien or an alien reaches the bottom of the screen.

Game Over!

Alien Invasion feels more complete now, but the game never ends. The value of ships_left just grows increasingly negative. Let's add a game_active flag, so we can end the game when the player runs out of ships. We'll set this flag at the end of the __init__() method in AlienInvasion:

alien_invasion.py

```
    def __init__(self):
        --snip--
        # Start Alien Invasion in an active state.
        self.game_active = True
```

Now we add code to _ship_hit() that sets game_active to False when the player has used up all their ships:

alien_invasion.py

```
    def _ship_hit(self):
        """Respond to ship being hit by alien."""
        if self.stats.ships_left > 0:
            # Decrement ships_left.
            self.stats.ships_left -= 1
            --snip--
            # Pause.
            sleep(0.5)
        else:
            self.game_active = False
```

Most of _ship_hit() is unchanged. We've moved all the existing code into an if block, which tests to make sure the player has at least one ship remaining. If so, we create a new fleet, pause, and move on. If the player has no ships left, we set game_active to False.

Identifying When Parts of the Game Should Run

We need to identify the parts of the game that should always run and the parts that should run only when the game is active:

alien_invasion.py

```
def run_game(self):
    """Start the main loop for the game."""
    while True:
        self._check_events()

        if self.game_active:
            self.ship.update()
            self._update_bullets()
            self._update_aliens()

        self._update_screen()
        self.clock.tick(60)
```

In the main loop, we always need to call _check_events(), even if the game is inactive. For example, we still need to know if the user presses Q to quit the game or clicks the button to close the window. We also continue updating the screen so we can make changes to the screen while waiting to see whether the player chooses to start a new game. The rest of the function calls need to happen only when the game is active, because when the game is inactive, we don't need to update the positions of game elements.

Now when you play *Alien Invasion*, the game should freeze when you've used up all your ships.

> **TRY IT YOURSELF**
>
> **13-6. Game Over:** In *Sideways Shooter*, keep track of the number of times the ship is hit and the number of times an alien is hit by the ship. Decide on an appropriate condition for ending the game, and stop the game when this situation occurs.

Summary

In this chapter, you learned how to add a large number of identical elements to a game by creating a fleet of aliens. You used nested loops to create a grid of elements, and you made a large set of game elements move by calling each element's update() method. You learned to control the direction of objects on the screen and to respond to specific situations, such as when the fleet reaches the edge of the screen. You detected and responded to collisions when bullets hit aliens and aliens hit the ship. You also learned how to track statistics in a game and use a `game_active` flag to determine when the game is over.

In the next and final chapter of this project, we'll add a Play button so the player can choose when to start their first game and whether to play again when the game ends. We'll speed up the game each time the player shoots down the entire fleet, and we'll add a scoring system. The final result will be a fully playable game!

14

SCORING

In this chapter, we'll finish building *Alien Invasion*. We'll add a Play button to start the game on demand and to restart the game once it ends. We'll also change the game so it speeds up when the player moves up a level, and we'll implement a scoring system. By the end of the chapter, you'll know enough to start writing games that increase in difficulty as a player progresses and that feature complete scoring systems.

Adding the Play Button

In this section, we'll add a Play button that appears before a game begins and reappears when the game ends so the player can play again.

Right now, the game begins as soon as you run *alien_invasion.py*. Let's start the game in an inactive state and then prompt the player to click a Play button to begin. To do this, modify the `__init__()` method of `AlienInvasion`:

alien_invasion.py

```
    def __init__(self):
        """Initialize the game, and create game resources."""
        pygame.init()
        --snip--

        # Start Alien Invasion in an inactive state.
        self.game_active = False
```

Now the game should start in an inactive state, with no way for the player to start it until we make a Play button.

Creating a Button Class

Because Pygame doesn't have a built-in method for making buttons, we'll write a `Button` class to create a filled rectangle with a label. You can use this code to make any button in a game. Here's the first part of the `Button` class; save it as *button.py*:

button.py

```
import pygame.font

class Button:
    """A class to build buttons for the game."""

❶   def __init__(self, ai_game, msg):
        """Initialize button attributes."""
        self.screen = ai_game.screen
        self.screen_rect = self.screen.get_rect()

        # Set the dimensions and properties of the button.
❷       self.width, self.height = 200, 50
        self.button_color = (0, 135, 0)
        self.text_color = (255, 255, 255)
❸       self.font = pygame.font.SysFont(None, 48)

        # Build the button's rect object and center it.
❹       self.rect = pygame.Rect(0, 0, self.width, self.height)
        self.rect.center = self.screen_rect.center

        # The button message needs to be prepped only once.
❺       self._prep_msg(msg)
```

First, we import the `pygame.font` module, which lets Pygame render text to the screen. The `__init__()` method takes the parameters `self`, the `ai_game` object, and `msg`, which contains the button's text ❶. We set the button dimensions ❷, set `button_color` to color the button's rect object dark green, and set `text_color` to render the text in white.

Next, we prepare a font attribute for rendering text ❸. The None argument tells Pygame to use the default font, and 48 specifies the size of the text. To center the button on the screen, we create a rect for the button ❹ and set its center attribute to match that of the screen.

Pygame works with text by rendering the string you want to display as an image. Finally, we call _prep_msg() to handle this rendering ❺.

Here's the code for _prep_msg():

button.py

```
    def _prep_msg(self, msg):
        """Turn msg into a rendered image and center text on the button."""
❶       self.msg_image = self.font.render(msg, True, self.text_color,
                self.button_color)
❷       self.msg_image_rect = self.msg_image.get_rect()
        self.msg_image_rect.center = self.rect.center
```

The _prep_msg() method needs a self parameter and the text to be rendered as an image (msg). The call to font.render() turns the text stored in msg into an image, which we then store in self.msg_image ❶. The font.render() method also takes a Boolean value to turn antialiasing on or off (*antialiasing* makes the edges of the text smoother). The remaining arguments are the specified font color and background color. We set antialiasing to True and set the text background to the same color as the button. (If you don't include a background color, Pygame will try to render the font with a transparent background.)

We center the text image on the button by creating a rect from the image and setting its center attribute to match that of the button ❷.

Finally, we create a draw_button() method that we can call to display the button onscreen:

button.py

```
    def draw_button(self):
        """Draw blank button and then draw message."""
        self.screen.fill(self.button_color, self.rect)
        self.screen.blit(self.msg_image, self.msg_image_rect)
```

We call screen.fill() to draw the rectangular portion of the button. Then we call screen.blit() to draw the text image to the screen, passing it an image and the rect object associated with the image. This completes the Button class.

Drawing the Button to the Screen

We'll use the Button class to create a Play button in AlienInvasion. First, we'll update the import statements:

alien_invasion.py

```
--snip--
from game_stats import GameStats
from button import Button
```

Because we need only one Play button, we'll create the button in the `__init__()` method of `AlienInvasion`. We can place this code at the very end of `__init__()`:

alien_invasion.py

```
    def __init__(self):
        --snip--
        self.game_active = False

        # Make the Play button.
        self.play_button = Button(self, "Play")
```

This code creates an instance of `Button` with the label `Play`, but it doesn't draw the button to the screen. To do this, we'll call the button's `draw_button()` method in `_update_screen()`:

alien_invasion.py

```
    def _update_screen(self):
        --snip--
        self.aliens.draw(self.screen)

        # Draw the play button if the game is inactive.
        if not self.game_active:
            self.play_button.draw_button()

        pygame.display.flip()
```

To make the Play button visible above all other elements on the screen, we draw it after all the other elements have been drawn but before flipping to a new screen. We include it in an `if` block, so the button only appears when the game is inactive.

Now when you run *Alien Invasion*, you should see a Play button in the center of the screen, as shown in Figure 14-1.

Figure 14-1: A Play button appears when the game is inactive.

Starting the Game

To start a new game when the player clicks Play, add the following `elif` block to the end of `_check_events()` to monitor mouse events over the button:

alien_invasion.py

```
    def _check_events(self):
        """Respond to keypresses and mouse events."""
        for event in pygame.event.get():
            if event.type == pygame.QUIT:
                --snip--
❶           elif event.type == pygame.MOUSEBUTTONDOWN:
❷               mouse_pos = pygame.mouse.get_pos()
❸               self._check_play_button(mouse_pos)
```

Pygame detects a `MOUSEBUTTONDOWN` event when the player clicks anywhere on the screen ❶, but we want to restrict our game to respond to mouse clicks only on the Play button. To accomplish this, we use `pygame.mouse.get_pos()`, which returns a tuple containing the mouse cursor's *x*- and *y*-coordinates when the mouse button is clicked ❷. We send these values to the new method `_check_play_button()` ❸.

Here's `_check_play_button()`, which I chose to place after `_check_events()`:

alien_invasion.py

```
    def _check_play_button(self, mouse_pos):
        """Start a new game when the player clicks Play."""
❶       if self.play_button.rect.collidepoint(mouse_pos):
            self.game_active = True
```

We use the `rect` method `collidepoint()` to check whether the point of the mouse click overlaps the region defined by the Play button's `rect` ❶. If so, we set `game_active` to `True`, and the game begins!

At this point, you should be able to start and play a full game. When the game ends, the value of `game_active` should become `False` and the Play button should reappear.

Resetting the Game

The Play button code we just wrote works the first time the player clicks Play. But it doesn't work after the first game ends, because the conditions that caused the game to end haven't been reset.

To reset the game each time the player clicks Play, we need to reset the game statistics, clear out the old aliens and bullets, build a new fleet, and center the ship, as shown here:

alien_invasion.py

```
    def _check_play_button(self, mouse_pos):
        """Start a new game when the player clicks Play."""
        if self.play_button.rect.collidepoint(mouse_pos):
            # Reset the game statistics.
❶           self.stats.reset_stats()
            self.game_active = True
```

```
            # Get rid of any remaining bullets and aliens.
❷           self.bullets.empty()
            self.aliens.empty()

            # Create a new fleet and center the ship.
❸           self._create_fleet()
            self.ship.center_ship()
```

We reset the game statistics ❶, which gives the player three new ships. Then we set `game_active` to `True` so the game will begin as soon as the code in this function finishes running. We empty the `aliens` and `bullets` groups ❷, and then we create a new fleet and center the ship ❸.

Now the game will reset properly each time you click Play, allowing you to play it as many times as you want!

Deactivating the Play Button

One issue with our Play button is that the button region on the screen will continue to respond to clicks even when the Play button isn't visible. If you click the Play button area by accident after a game begins, the game will restart!

To fix this, set the game to start only when `game_active` is `False`:

alien_invasion.py

```
    def _check_play_button(self, mouse_pos):
        """Start a new game when the player clicks Play."""
❶       button_clicked = self.play_button.rect.collidepoint(mouse_pos)
❷       if button_clicked and not self.game_active:
            # Reset the game statistics.
            self.stats.reset_stats()
            --snip--
```

The flag `button_clicked` stores a `True` or `False` value ❶, and the game will restart only if Play is clicked *and* the game is not currently active ❷. To test this behavior, start a new game and repeatedly click where the Play button should be. If everything works as expected, clicking the Play button area should have no effect on the gameplay.

Hiding the Mouse Cursor

We want the mouse cursor to be visible when the game is inactive, but once play begins, it just gets in the way. To fix this, we'll make it invisible when the game becomes active. We can do this at the end of the `if` block in `_check_play_button()`:

alien_invasion.py

```
    def _check_play_button(self, mouse_pos):
        """Start a new game when the player clicks Play."""
        button_clicked = self.play_button.rect.collidepoint(mouse_pos)
        if button_clicked and not self.game_active:
            --snip--
            # Hide the mouse cursor.
            pygame.mouse.set_visible(False)
```

Passing False to set_visible() tells Pygame to hide the cursor when the mouse is over the game window.

We'll make the cursor reappear once the game ends so the player can click Play again to begin a new game. Here's the code to do that:

alien_invasion.py

```
    def _ship_hit(self):
        """Respond to ship being hit by alien."""
        if self.stats.ships_left > 0:
            --snip--
        else:
            self.game_active = False
            pygame.mouse.set_visible(True)
```

We make the cursor visible again as soon as the game becomes inactive, which happens in _ship_hit(). Attention to details like this makes your game more professional looking and allows the player to focus on playing, rather than figuring out the user interface.

TRY IT YOURSELF

14-1. Press P to Play: Because *Alien Invasion* uses keyboard input to control the ship, it would be useful to start the game with a keypress. Add code that lets the player press P to start. It might help to move some code from _check_play_button() to a _start_game() method that can be called from _check_play_button() *and* _check_keydown_events().

14-2. Target Practice: Create a rectangle at the right edge of the screen that moves up and down at a steady rate. Then on the left side of the screen, create a ship that the player can move up and down while firing bullets at the rectangular target. Add a Play button that starts the game, and when the player misses the target three times, end the game and make the Play button reappear. Let the player restart the game with this Play button.

Leveling Up

In our current game, once a player shoots down the entire alien fleet, the player reaches a new level, but the game difficulty doesn't change. Let's liven things up a bit and make the game more challenging by increasing the game's speed each time a player clears the screen.

Modifying the Speed Settings

We'll first reorganize the Settings class to group the game settings into static and dynamic ones. We'll also make sure any settings that change during the game reset when we start a new game. Here's the `__init__()` method for *settings.py*:

settings.py

```
    def __init__(self):
        """Initialize the game's static settings."""
        # Screen settings
        self.screen_width = 1200
        self.screen_height = 800
        self.bg_color = (230, 230, 230)

        # Ship settings
        self.ship_limit = 3

        # Bullet settings
        self.bullet_width = 3
        self.bullet_height = 15
        self.bullet_color = 60, 60, 60
        self.bullets_allowed = 3

        # Alien settings
        self.fleet_drop_speed = 10

        # How quickly the game speeds up
❶       self.speedup_scale = 1.1

❷       self.initialize_dynamic_settings()
```

We continue to initialize settings that stay constant in the `__init__()` method. We add a `speedup_scale` setting ❶ to control how quickly the game speeds up: a value of 2 will double the game speed every time the player reaches a new level; a value of 1 will keep the speed constant. A value like `1.1` should increase the speed enough to make the game challenging but not impossible. Finally, we call the `initialize_dynamic_settings()` method to initialize the values for attributes that need to change throughout the game ❷.

Here's the code for `initialize_dynamic_settings()`:

settings.py

```
    def initialize_dynamic_settings(self):
        """Initialize settings that change throughout the game."""
        self.ship_speed = 1.5
        self.bullet_speed = 2.5
        self.alien_speed = 1.0

        # fleet_direction of 1 represents right; -1 represents left.
        self.fleet_direction = 1
```

This method sets the initial values for the ship, bullet, and alien speeds. We'll increase these speeds as the player progresses in the game and reset them each time the player starts a new game. We include `fleet_direction` in this method so the aliens always move right at the beginning of a new game. We don't need to increase the value of `fleet_drop_speed`, because when the aliens move faster across the screen, they'll also come down the screen faster.

To increase the speeds of the ship, bullets, and aliens each time the player reaches a new level, we'll write a new method called `increase_speed()`:

settings.py

```
    def increase_speed(self):
        """Increase speed settings."""
        self.ship_speed *= self.speedup_scale
        self.bullet_speed *= self.speedup_scale
        self.alien_speed *= self.speedup_scale
```

To increase the speed of these game elements, we multiply each speed setting by the value of `speedup_scale`.

We increase the game's tempo by calling `increase_speed()` in `_check_bullet_alien_collisions()` when the last alien in a fleet has been shot down:

alien_invasion.py

```
    def _check_bullet_alien_collisions(self):
        --snip--
        if not self.aliens:
            # Destroy existing bullets and create new fleet.
            self.bullets.empty()
            self._create_fleet()
            self.settings.increase_speed()
```

Changing the values of the speed settings `ship_speed`, `alien_speed`, and `bullet_speed` is enough to speed up the entire game!

Resetting the Speed

Now we need to return any changed settings to their initial values each time the player starts a new game; otherwise, each new game would start with the increased speed settings of the previous game:

alien_invasion.py

```
    def _check_play_button(self, mouse_pos):
        """Start a new game when the player clicks Play."""
        button_clicked = self.play_button.rect.collidepoint(mouse_pos)
        if button_clicked and not self.game_active:
            # Reset the game settings.
            self.settings.initialize_dynamic_settings()
            --snip--
```

Playing *Alien Invasion* should be more fun and challenging now. Each time you clear the screen, the game should speed up and become slightly more difficult. If the game becomes too difficult too quickly, decrease the value of `settings.speedup_scale`. Or if the game isn't challenging enough, increase the value slightly. Find a sweet spot by ramping up the difficulty in a reasonable amount of time. The first couple of screens should be easy, the next few should be challenging but doable, and subsequent screens should be almost impossibly difficult.

TRY IT YOURSELF

14-3. Challenging Target Practice: Start with your work from Exercise 14-2 (page 283). Make the target move faster as the game progresses, and restart the target at the original speed when the player clicks Play.

14-4. Difficulty Levels: Make a set of buttons for *Alien Invasion* that allows the player to select an appropriate starting difficulty level for the game. Each button should assign the appropriate values for the attributes in `Settings` needed to create different difficulty levels.

Scoring

Let's implement a scoring system to track the game's score in real time and display the high score, level, and number of ships remaining.

The score is a game statistic, so we'll add a `score` attribute to `GameStats`:

game_stats.py

```
class GameStats:
    --snip--
    def reset_stats(self):
        """Initialize statistics that can change during the game."""
        self.ships_left = self.ai_settings.ship_limit
        self.score = 0
```

To reset the score each time a new game starts, we initialize `score` in `reset_stats()` rather than `__init__()`.

Displaying the Score

To display the score on the screen, we first create a new class, `Scoreboard`. For now, this class will just display the current score. Eventually, we'll use it to report the high score, level, and number of ships remaining as well. Here's the first part of the class; save it as *scoreboard.py*:

scoreboard.py

```
import pygame.font

class Scoreboard:
    """A class to report scoring information."""

❶   def __init__(self, ai_game):
        """Initialize scorekeeping attributes."""
        self.screen = ai_game.screen
        self.screen_rect = self.screen.get_rect()
        self.settings = ai_game.settings
        self.stats = ai_game.stats

        # Font settings for scoring information.
❷       self.text_color = (30, 30, 30)
❸       self.font = pygame.font.SysFont(None, 48)
```

```
❹         # Prepare the initial score image.
          self.prep_score()
```

Because Scoreboard writes text to the screen, we begin by importing the pygame.font module. Next, we give __init__() the ai_game parameter so it can access the settings, screen, and stats objects, which it will need to report the values we're tracking ❶. Then we set a text color ❷ and instantiate a font object ❸.

To turn the text to be displayed into an image, we call prep_score() ❹, which we define here:

scoreboard.py

```
    def prep_score(self):
        """Turn the score into a rendered image."""
❶       score_str = str(self.stats.score)
❷       self.score_image = self.font.render(score_str, True,
                self.text_color, self.settings.bg_color)

        # Display the score at the top right of the screen.
❸       self.score_rect = self.score_image.get_rect()
❹       self.score_rect.right = self.screen_rect.right - 20
❺       self.score_rect.top = 20
```

In prep_score(), we turn the numerical value stats.score into a string ❶ and then pass this string to render(), which creates the image ❷. To display the score clearly onscreen, we pass the screen's background color and the text color to render().

We'll position the score in the upper-right corner of the screen and have it expand to the left as the score increases and the width of the number grows. To make sure the score always lines up with the right side of the screen, we create a rect called score_rect ❸ and set its right edge 20 pixels from the right edge of the screen ❹. We then place the top edge 20 pixels down from the top of the screen ❺.

Then we create a show_score() method to display the rendered score image:

scoreboard.py

```
    def show_score(self):
        """Draw score to the screen."""
        self.screen.blit(self.score_image, self.score_rect)
```

This method draws the score image onscreen at the location score_rect specifies.

Making a Scoreboard

To display the score, we'll create a Scoreboard instance in AlienInvasion. First, let's update the import statements:

alien_invasion.py

```
--snip--
from game_stats import GameStats
from scoreboard import Scoreboard
--snip--
```

Next, we make an instance of Scoreboard in `__init__()`:

alien_invasion.py

```
    def __init__(self):
        --snip--
        pygame.display.set_caption("Alien Invasion")

        # Create an instance to store game statistics,
        #   and create a scoreboard.
        self.stats = GameStats(self)
        self.sb = Scoreboard(self)
        --snip--
```

Then we draw the scoreboard onscreen in `_update_screen()`:

alien_invasion.py

```
    def _update_screen(self):
        --snip--
        self.aliens.draw(self.screen)

        # Draw the score information.
        self.sb.show_score()

        # Draw the play button if the game is inactive.
        --snip--
```

We call `show_score()` just before we draw the Play button.

When you run *Alien Invasion* now, a 0 should appear at the top right of the screen. (At this point, we just want to make sure the score appears in the right place before developing the scoring system further.) Figure 14-2 shows the score as it appears before the game starts.

Next, we'll assign point values to each alien!

Figure 14-2: The score appears at the top-right corner of the screen.

Updating the Score as Aliens Are Shot Down

To write a live score onscreen, we update the value of stats.score whenever an alien is hit, and then call prep_score() to update the score image. But first, let's determine how many points a player gets each time they shoot down an alien:

settings.py

```
    def initialize_dynamic_settings(self):
        --snip--

        # Scoring settings
        self.alien_points = 50
```

We'll increase each alien's point value as the game progresses. To make sure this point value is reset each time a new game starts, we set the value in initialize_dynamic_settings().

Let's update the score in _check_bullet_alien_collisions() each time an alien is shot down:

alien_invasion.py

```
    def _check_bullet_alien_collisions(self):
        """Respond to bullet-alien collisions."""
        # Remove any bullets and aliens that have collided.
        collisions = pygame.sprite.groupcollide(
                self.bullets, self.aliens, True, True)

        if collisions:
            self.stats.score += self.settings.alien_points
            self.sb.prep_score()
        --snip--
```

When a bullet hits an alien, Pygame returns a collisions dictionary. We check whether the dictionary exists, and if it does, the alien's value is added to the score. We then call prep_score() to create a new image for the updated score.

Now when you play *Alien Invasion*, you should be able to rack up points!

Resetting the Score

Right now, we're only prepping a new score *after* an alien has been hit, which works for most of the game. But when we start a new game, we'll still see our score from the old game until the first alien is hit.

We can fix this by prepping the score when starting a new game:

alien_invasion.py

```
    def _check_play_button(self, mouse_pos):
        --snip--
        if button_clicked and not self.game_active:
            --snip--
            # Reset the game statistics.
            self.stats.reset_stats()
            self.sb.prep_score()
            --snip--
```

We call `prep_score()` after resetting the game stats when starting a new game. This preps the scoreboard with a score of 0.

Making Sure to Score All Hits

As currently written, our code could miss scoring for some aliens. For example, if two bullets collide with aliens during the same pass through the loop or if we make an extra-wide bullet to hit multiple aliens, the player will only receive points for hitting one of the aliens. To fix this, let's refine the way that bullet-alien collisions are detected.

In `_check_bullet_alien_collisions()`, any bullet that collides with an alien becomes a key in the `collisions` dictionary. The value associated with each bullet is a list of aliens it has collided with. We loop through the values in the `collisions` dictionary to make sure we award points for each alien hit:

alien_invasion.py

```
    def _check_bullet_alien_collisions(self):
        --snip--
        if collisions:
            for aliens in collisions.values():
                self.stats.score += self.settings.alien_points * len(aliens)
            self.sb.prep_score()
        --snip--
```

If the `collisions` dictionary has been defined, we loop through all values in the dictionary. Remember that each value is a list of aliens hit by a single bullet. We multiply the value of each alien by the number of aliens in each list and add this amount to the current score. To test this, change the width of a bullet to 300 pixels and verify that you receive points for each alien you hit with your extra-wide bullets; then return the bullet width to its normal value.

Increasing Point Values

Because the game gets more difficult each time a player reaches a new level, aliens in later levels should be worth more points. To implement this functionality, we'll add code to increase the point value when the game's speed increases:

settings.py

```
class Settings:
    """A class to store all settings for Alien Invasion."""

    def __init__(self):
        --snip--
        # How quickly the game speeds up
        self.speedup_scale = 1.1
        # How quickly the alien point values increase
❶       self.score_scale = 1.5

        self.initialize_dynamic_settings()

    def initialize_dynamic_settings(self):
        --snip--
```

```
    def increase_speed(self):
        """Increase speed settings and alien point values."""
        self.ship_speed *= self.speedup_scale
        self.bullet_speed *= self.speedup_scale
        self.alien_speed *= self.speedup_scale

❷       self.alien_points = int(self.alien_points * self.score_scale)
```

We define a rate at which points increase, which we call score_scale ❶. A small increase in speed (1.1) makes the game more challenging quickly. But to see a more notable difference in scoring, we need to change the alien point value by a larger amount (1.5). Now when we increase the game's speed, we also increase the point value of each hit ❷. We use the int() function to increase the point value by whole integers.

To see the value of each alien, add a print() call to the increase_speed() method in Settings:

settings.py

```
    def increase_speed(self):
        --snip--
        self.alien_points = int(self.alien_points * self.score_scale)
        print(self.alien_points)
```

The new point value should appear in the terminal every time you reach a new level.

NOTE *Be sure to remove the* print() *call after verifying that the point value is increasing, or it might affect your game's performance and distract the player.*

Rounding the Score

Most arcade-style shooting games report scores as multiples of 10, so let's follow that lead with our scores. Also, let's format the score to include comma separators in large numbers. We'll make this change in Scoreboard:

scoreboard.py

```
    def prep_score(self):
        """Turn the score into a rendered image."""
        rounded_score = round(self.stats.score, -1)
        score_str = f"{rounded_score:,}"
        self.score_image = self.font.render(score_str, True,
                self.text_color, self.settings.bg_color)
        --snip--
```

The round() function normally rounds a float to a set number of decimal places given as the second argument. However, when you pass a negative number as the second argument, round() will round the value to the nearest 10, 100, 1,000, and so on. This code tells Python to round the value of stats.score to the nearest 10 and assign it to rounded_score.

We then use a format specifier in the f-string for the score. A *format specifier* is a special sequence of characters that modifies the way a variable's value is presented. In this case the sequence :, tells Python to insert commas at appropriate places in the numerical value that's provided. This results in strings like 1,000,000 instead of 1000000.

Now when you run the game, you should see a neatly formatted, rounded score even when you rack up lots of points, as shown in Figure 14-3.

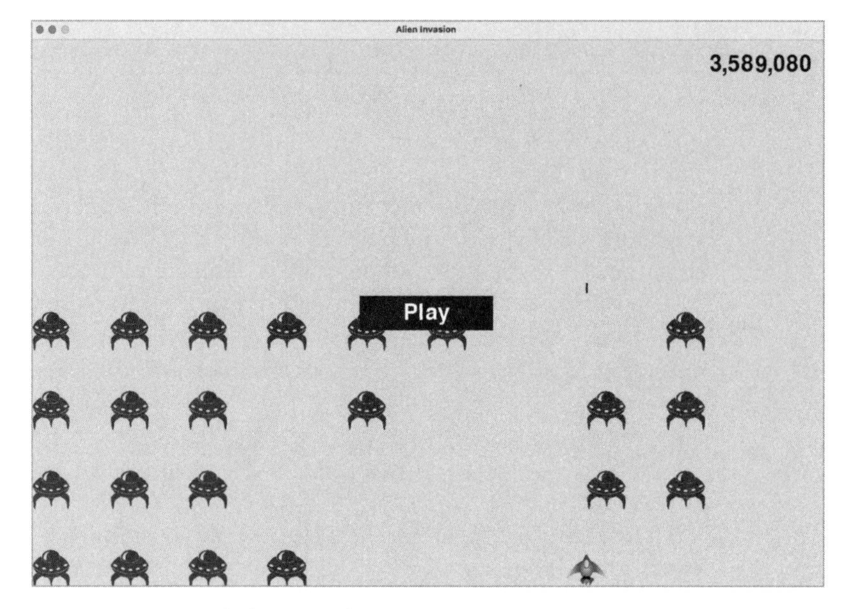

Figure 14-3: A rounded score with comma separators

High Scores

Every player wants to beat a game's high score, so let's track and report high scores to give players something to work toward. We'll store high scores in GameStats:

game_stats.py

```
    def __init__(self, ai_game):
        --snip--
        # High score should never be reset.
        self.high_score = 0
```

Because the high score should never be reset, we initialize high_score in __init__() rather than in reset_stats().

Next, we'll modify Scoreboard to display the high score. Let's start with the __init__() method:

scoreboard.py

```
    def __init__(self, ai_game):
        --snip--
        # Prepare the initial score images.
        self.prep_score()
❶       self.prep_high_score()
```

The high score will be displayed separately from the score, so we need a new method, prep_high_score(), to prepare the high-score image ❶.

Here's the prep_high_score() method:

scoreboard.py

```
    def prep_high_score(self):
        """Turn the high score into a rendered image."""
❶       high_score = round(self.stats.high_score, -1)
        high_score_str = f"{high_score:,}"
❷       self.high_score_image = self.font.render(high_score_str, True,
                self.text_color, self.settings.bg_color)

        # Center the high score at the top of the screen.
        self.high_score_rect = self.high_score_image.get_rect()
❸       self.high_score_rect.centerx = self.screen_rect.centerx
❹       self.high_score_rect.top = self.score_rect.top
```

We round the high score to the nearest 10 and format it with commas ❶. We then generate an image from the high score ❷, center the high score rect horizontally ❸, and set its top attribute to match the top of the score image ❹.

The show_score() method now draws the current score at the top right and the high score at the top center of the screen:

scoreboard.py

```
    def show_score(self):
        """Draw score to the screen."""
        self.screen.blit(self.score_image, self.score_rect)
        self.screen.blit(self.high_score_image, self.high_score_rect)
```

To check for high scores, we'll write a new method, check_high_score(), in Scoreboard:

scoreboard.py

```
    def check_high_score(self):
        """Check to see if there's a new high score."""
        if self.stats.score > self.stats.high_score:
            self.stats.high_score = self.stats.score
            self.prep_high_score()
```

The method check_high_score() checks the current score against the high score. If the current score is greater, we update the value of high_score and call prep_high_score() to update the high score's image.

We need to call check_high_score() each time an alien is hit after updating the score in _check_bullet_alien_collisions():

alien_invasion.py

```
    def _check_bullet_alien_collisions(self):
        --snip--
        if collisions:
            for aliens in collisions.values():
                self.stats.score += self.settings.alien_points * len(aliens)
            self.sb.prep_score()
            self.sb.check_high_score()
        --snip--
```

We call check_high_score() when the collisions dictionary is present, and we do so after updating the score for all the aliens that have been hit.

The first time you play *Alien Invasion*, your score will be the high score, so it will be displayed as the current score and the high score. But when you start a second game, your high score should appear in the middle and your current score should appear at the right, as shown in Figure 14-4.

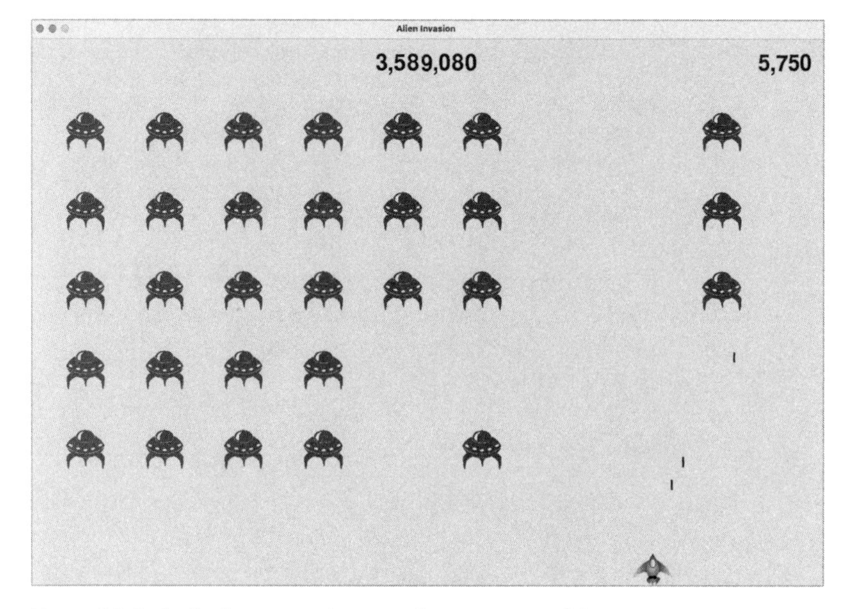

Figure 14-4: The high score is shown at the top center of the screen.

Displaying the Level

To display the player's level in the game, we first need an attribute in `GameStats` representing the current level. To reset the level at the start of each new game, initialize it in `reset_stats()`:

game_stats.py

```
def reset_stats(self):
    """Initialize statistics that can change during the game."""
    self.ships_left = self.settings.ship_limit
    self.score = 0
    self.level = 1
```

To have `Scoreboard` display the current level, we call a new method, `prep_level()`, from `__init__()`:

scoreboard.py

```
def __init__(self, ai_game):
    --snip--
    self.prep_high_score()
    self.prep_level()
```

Here's `prep_level()`:

scoreboard.py

```
def prep_level(self):
    """Turn the level into a rendered image."""
    level_str = str(self.stats.level)
```

```
❶         self.level_image = self.font.render(level_str, True,
                  self.text_color, self.settings.bg_color)

          # Position the level below the score.
          self.level_rect = self.level_image.get_rect()
❷         self.level_rect.right = self.score_rect.right
❸         self.level_rect.top = self.score_rect.bottom + 10
```

The prep_level() method creates an image from the value stored in stats.level ❶ and sets the image's right attribute to match the score's right attribute ❷. It then sets the top attribute 10 pixels beneath the bottom of the score image to leave space between the score and the level ❸.

We also need to update show_score():

scoreboard.py

```
    def show_score(self):
        """Draw scores and level to the screen."""
        self.screen.blit(self.score_image, self.score_rect)
        self.screen.blit(self.high_score_image, self.high_score_rect)
        self.screen.blit(self.level_image, self.level_rect)
```

This new line draws the level image to the screen.

We'll increment stats.level and update the level image in _check_bullet_alien_collisions():

alien_invasion.py

```
    def _check_bullet_alien_collisions(self):
        --snip--
        if not self.aliens:
            # Destroy existing bullets and create new fleet.
            self.bullets.empty()
            self._create_fleet()
            self.settings.increase_speed()

            # Increase level.
            self.stats.level += 1
            self.sb.prep_level()
```

If a fleet is destroyed, we increment the value of stats.level and call prep_level() to make sure the new level displays correctly.

To ensure the level image updates properly at the start of a new game, we also call prep_level() when the player clicks the Play button:

alien_invasion.py

```
    def _check_play_button(self, mouse_pos):
        --snip--
        if button_clicked and not self.game_active:
            --snip--
            self.sb.prep_score()
            self.sb.prep_level()
            --snip--
```

We call prep_level() right after calling prep_score().

Now you'll see how many levels you've completed, as shown in Figure 14-5.

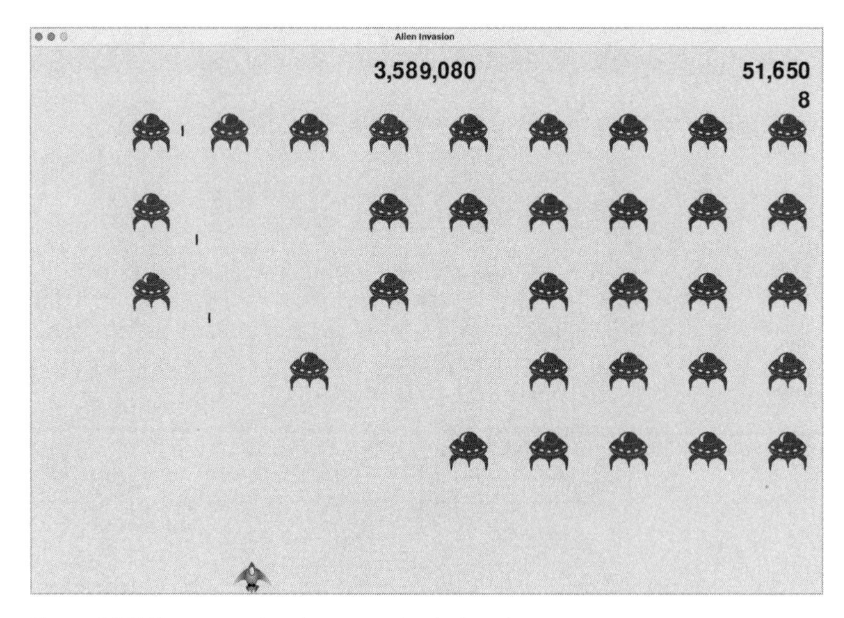

Figure 14-5: The current level appears just below the current score.

NOTE *In some classic games, the scores have labels, such as Score, High Score, and Level. We've omitted these labels because the meaning of each number becomes clear once you've played the game. To include these labels, add them to the score strings just before the calls to* `font.render()` *in* `Scoreboard`*.*

Displaying the Number of Ships

Finally, let's display the number of ships the player has left, but this time, let's use a graphic. To do so, we'll draw ships in the upper-left corner of the screen to represent how many ships are left, just as many classic arcade games do.

First, we need to make `Ship` inherit from `Sprite` so we can create a group of ships:

ship.py

```
import pygame
from pygame.sprite import Sprite

❶ class Ship(Sprite):
    """A class to manage the ship."""

    def __init__(self, ai_game):
        """Initialize the ship and set its starting position."""
❷       super().__init__()
        --snip--
```

Here we import `Sprite`, make sure `Ship` inherits from `Sprite` ❶, and call `super()` at the beginning of `__init__()` ❷.

Next, we need to modify Scoreboard to create a group of ships we can display. Here are the import statements for Scoreboard:

scoreboard.py

```
import pygame.font
from pygame.sprite import Group

from ship import Ship
```

Because we're making a group of ships, we import the Group and Ship classes.

Here's __init__():

scoreboard.py

```
    def __init__(self, ai_game):
        """Initialize scorekeeping attributes."""
        self.ai_game = ai_game
        self.screen = ai_game.screen
        --snip--
        self.prep_level()
        self.prep_ships()
```

We assign the game instance to an attribute, because we'll need it to create some ships. We call prep_ships() after the call to prep_level().

Here's prep_ships():

scoreboard.py

```
    def prep_ships(self):
        """Show how many ships are left."""
❶       self.ships = Group()
❷       for ship_number in range(self.stats.ships_left):
            ship = Ship(self.ai_game)
❸           ship.rect.x = 10 + ship_number * ship.rect.width
❹           ship.rect.y = 10
❺           self.ships.add(ship)
```

The prep_ships() method creates an empty group, self.ships, to hold the ship instances ❶. To fill this group, a loop runs once for every ship the player has left ❷. Inside the loop, we create a new ship and set each ship's *x*-coordinate value so the ships appear next to each other with a 10-pixel margin on the left side of the group of ships ❸. We set the *y*-coordinate value 10 pixels down from the top of the screen so the ships appear in the upper-left corner of the screen ❹. Then we add each new ship to the group ships ❺.

Now we need to draw the ships to the screen:

scoreboard.py

```
    def show_score(self):
        """Draw scores, level, and ships to the screen."""
        self.screen.blit(self.score_image, self.score_rect)
        self.screen.blit(self.high_score_image, self.high_score_rect)
        self.screen.blit(self.level_image, self.level_rect)
        self.ships.draw(self.screen)
```

To display the ships on the screen, we call draw() on the group, and Pygame draws each ship.

To show the player how many ships they have to start with, we call prep_ships() when a new game starts. We do this in _check_play_button() in AlienInvasion:

alien_invasion.py

```
    def _check_play_button(self, mouse_pos):
        --snip--
        if button_clicked and not self.game_active:
            --snip--
            self.sb.prep_level()
            self.sb.prep_ships()
            --snip--
```

We also call prep_ships() when a ship is hit, to update the display of ship images when the player loses a ship:

alien_invasion.py

```
    def _ship_hit(self):
        """Respond to ship being hit by alien."""
        if self.stats.ships_left > 0:
            # Decrement ships_left, and update scoreboard.
            self.stats.ships_left -= 1
            self.sb.prep_ships()
            --snip--
```

We call prep_ships() after decreasing the value of ships_left, so the correct number of remaining ships displays each time a ship is destroyed.

Figure 14-6 shows the complete scoring system, with the remaining ships displayed at the top left of the screen.

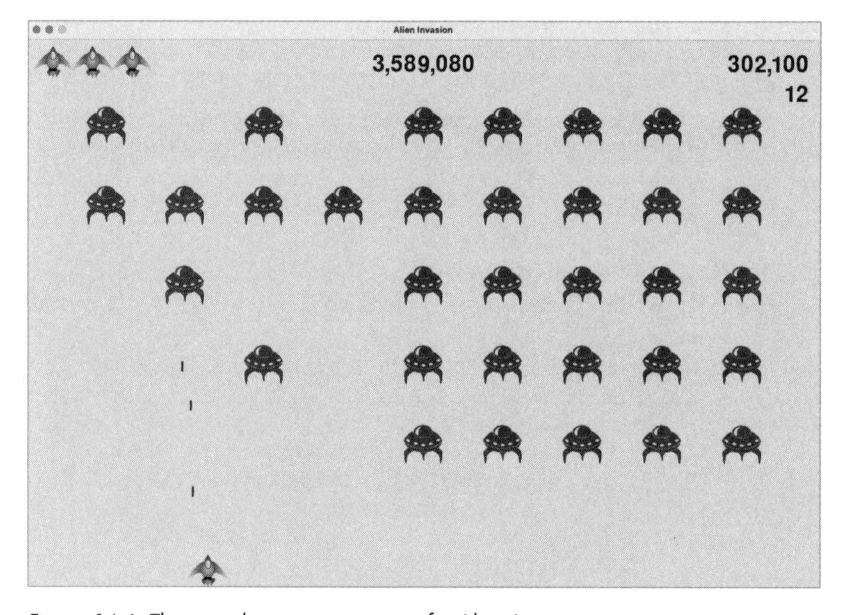

Figure 14-6: The complete scoring system for Alien Invasion

TRY IT YOURSELF

14-5. All-Time High Score: The high score is reset every time a player closes and restarts *Alien Invasion*. Fix this by writing the high score to a file before calling `sys.exit()` and reading in the high score when initializing its value in `GameStats`.

14-6. Refactoring: Look for methods that are doing more than one task, and refactor them to organize your code and make it efficient. For example, move some of the code in `_check_bullet_alien_collisions()`, which starts a new level when the fleet of aliens has been destroyed, to a function called `start_new_level()`. Also, move the four separate method calls in the `__init__()` method in `Scoreboard` to a method called `prep_images()` to shorten `__init__()`. The `prep_images()` method could also help simplify `_check_play_button()` or `start_game()` if you've already refactored `_check_play_button()`.

NOTE *Before attempting to refactor the project, see Appendix D to learn how to restore the project to a working state if you introduce bugs while refactoring.*

14-7. Expanding the Game: Think of a way to expand *Alien Invasion*. For example, you could program the aliens to shoot bullets down at your ship. You can also add shields for your ship to hide behind, which can be destroyed by bullets from either side. Or you can use something like the `pygame.mixer` module to add sound effects, such as explosions and shooting sounds.

14-8. Sideways Shooter, Final Version: Continue developing *Sideways Shooter*, using everything we've done in this project. Add a Play button, make the game speed up at appropriate points, and develop a scoring system. Be sure to refactor your code as you work, and look for opportunities to customize the game beyond what has been shown in this chapter.

Summary

In this chapter, you learned how to implement a Play button to start a new game. You also learned how to detect mouse events and hide the cursor in active games. You can use what you've learned to create other buttons, like a Help button to display instructions on how to play your games. You also learned how to modify the speed of a game as it progresses, implement a progressive scoring system, and display information in textual and nontextual ways.

15

GENERATING DATA

Data visualization is the use of visual representations to explore and present patterns in datasets. It's closely associated with *data analysis*, which uses code to explore the patterns and connections in a dataset. A dataset can be a small list of numbers that fits in a single line of code, or it can be terabytes of data that include many different kinds of information.

Creating effective data visualizations is about more than just making information look nice. When a representation of a dataset is simple and visually appealing, its meaning becomes clear to viewers. People will see patterns and significance in your datasets that they never knew existed.

Fortunately, you don't need a supercomputer to visualize complex data. Python is so efficient that with just a laptop, you can quickly explore datasets containing millions of individual data points. These data points don't have to be numbers; with the basics you learned in the first part of this book, you can analyze non-numerical data as well.

People use Python for data-intensive work in genetics, climate research, political and economic analysis, and much more. Data scientists have written an impressive array of visualization and analysis tools in Python, many of which are available to you as well. One of the most popular tools is Matplotlib, a mathematical plotting library. In this chapter, we'll use Matplotlib to make simple plots, such as line graphs and scatter plots. Then we'll create a more interesting dataset based on the concept of a random walk—a visualization generated from a series of random decisions.

We'll also use a package called Plotly, which creates visualizations that work well on digital devices, to analyze the results of rolling dice. Plotly generates visualizations that automatically resize to fit a variety of display devices. These visualizations can also include a number of interactive features, such as emphasizing particular aspects of the dataset when users hover over different parts of the visualization. Learning to use Matplotlib and Plotly will help you get started visualizing the kinds of data you're most interested in.

Installing Matplotlib

To use Matplotlib for your initial set of visualizations, you'll need to install it using pip, just like we did with pytest in Chapter 11 (see "Installing pytest with pip" on page 210).

To install Matplotlib, enter the following command at a terminal prompt:

```
$ python -m pip install --user matplotlib
```

If you use a command other than **python** to run programs or start a terminal session, such as **python3**, your command will look like this:

```
$ python3 -m pip install --user matplotlib
```

To see the kinds of visualizations you can make with Matplotlib, visit the Matplotlib home page at *https://matplotlib.org* and click **Plot types**. When you click a visualization in the gallery, you'll see the code used to generate the plot.

Plotting a Simple Line Graph

Let's plot a simple line graph using Matplotlib and then customize it to create a more informative data visualization. We'll use the square number sequence 1, 4, 9, 16, and 25 as the data for the graph.

To make a simple line graph, specify the numbers you want to work with and let Matplotlib do the rest:

mpl_squares.py

```
import matplotlib.pyplot as plt

squares = [1, 4, 9, 16, 25]
```

```
❶ fig, ax = plt.subplots()
  ax.plot(squares)

  plt.show()
```

We first import the pyplot module using the alias plt so we don't have to type pyplot repeatedly. (You'll see this convention often in online examples, so we'll use it here.) The pyplot module contains a number of functions that help generate charts and plots.

We create a list called squares to hold the data that we'll plot. Then we follow another common Matplotlib convention by calling the subplots() function ❶. This function can generate one or more plots in the same figure. The variable fig represents the entire *figure*, which is the collection of plots that are generated. The variable ax represents a single plot in the figure; this is the variable we'll use most of the time when defining and customizing a single plot.

We then use the plot() method, which tries to plot the data it's given in a meaningful way. The function plt.show() opens Matplotlib's viewer and displays the plot, as shown in Figure 15-1. The viewer allows you to zoom and navigate the plot, and you can save any plot images you like by clicking the disk icon.

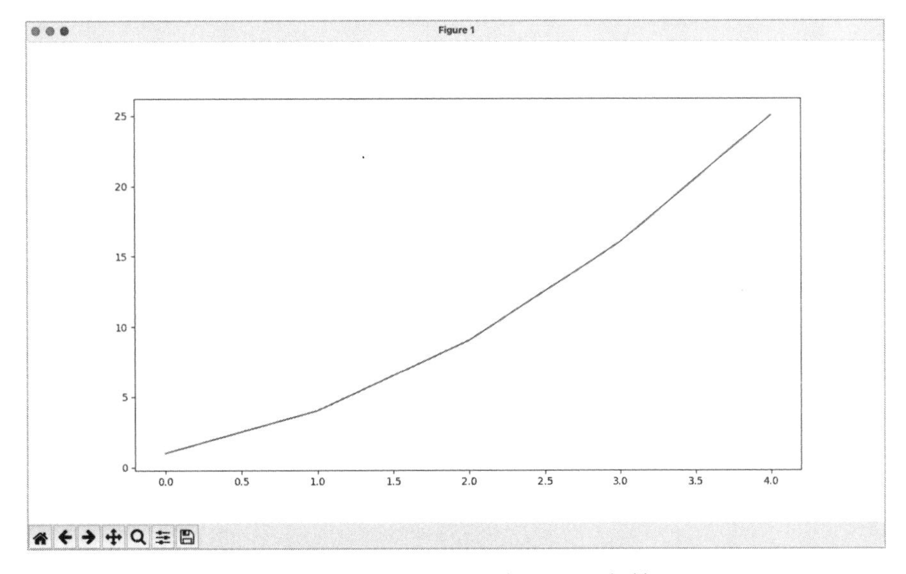

Figure 15-1: One of the simplest plots you can make in Matplotlib

Changing the Label Type and Line Thickness

Although the plot in Figure 15-1 shows that the numbers are increasing, the label type is too small and the line is a little thin to read easily. Fortunately, Matplotlib allows you to adjust every feature of a visualization.

We'll use a few of the available customizations to improve this plot's readability. Let's start by adding a title and labeling the axes:

mpl_squares.py

```
import matplotlib.pyplot as plt

squares = [1, 4, 9, 16, 25]

fig, ax = plt.subplots()
❶ ax.plot(squares, linewidth=3)

# Set chart title and label axes.
❷ ax.set_title("Square Numbers", fontsize=24)
❸ ax.set_xlabel("Value", fontsize=14)
ax.set_ylabel("Square of Value", fontsize=14)

# Set size of tick labels.
❹ ax.tick_params(labelsize=14)

plt.show()
```

The `linewidth` parameter controls the thickness of the line that `plot()` generates ❶. Once a plot has been generated, there are many methods available to modify the plot before it's presented. The `set_title()` method sets an overall title for the chart ❷. The `fontsize` parameters, which appear repeatedly throughout the code, control the size of the text in various elements on the chart.

The `set_xlabel()` and `set_ylabel()` methods allow you to set a title for each of the axes ❸, and the method `tick_params()` styles the tick marks ❹. Here `tick_params()` sets the font size of the tick mark labels to 14 on both axes.

As you can see in Figure 15-2, the resulting chart is much easier to read. The label type is bigger, and the line graph is thicker. It's often worth experimenting with these values to see what works best in the resulting graph.

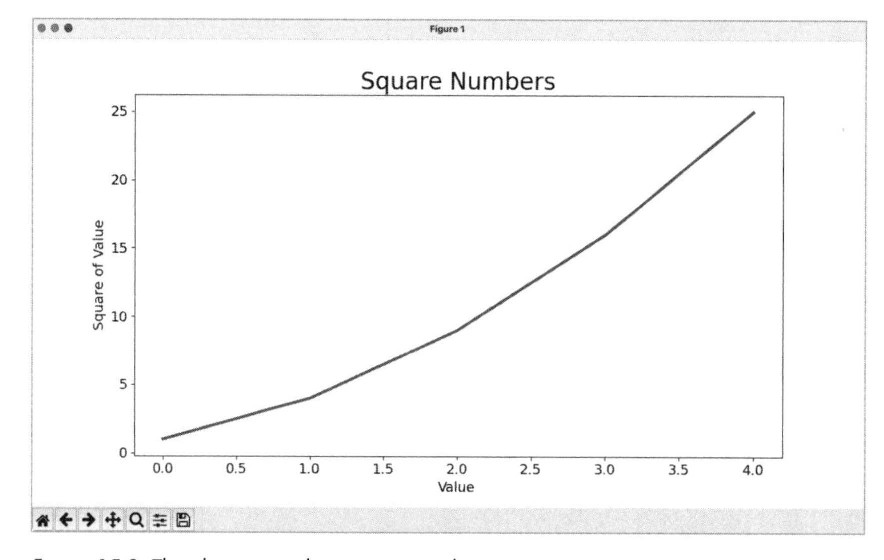

Figure 15-2: The chart is much easier to read now.

Correcting the Plot

Now that we can read the chart better, we can see that the data is not plotted correctly. Notice at the end of the graph that the square of 4.0 is shown as 25! Let's fix that.

When you give `plot()` a single sequence of numbers, it assumes the first data point corresponds to an *x*-value of 0, but our first point corresponds to an *x*-value of 1. We can override the default behavior by giving `plot()` both the input and output values used to calculate the squares:

mpl_squares.py

```
import matplotlib.pyplot as plt

input_values = [1, 2, 3, 4, 5]
squares = [1, 4, 9, 16, 25]

fig, ax = plt.subplots()
ax.plot(input_values, squares, linewidth=3)

# Set chart title and label axes.
--snip--
```

Now `plot()` doesn't have to make any assumptions about how the output numbers were generated. The resulting plot, shown in Figure 15-3, is correct.

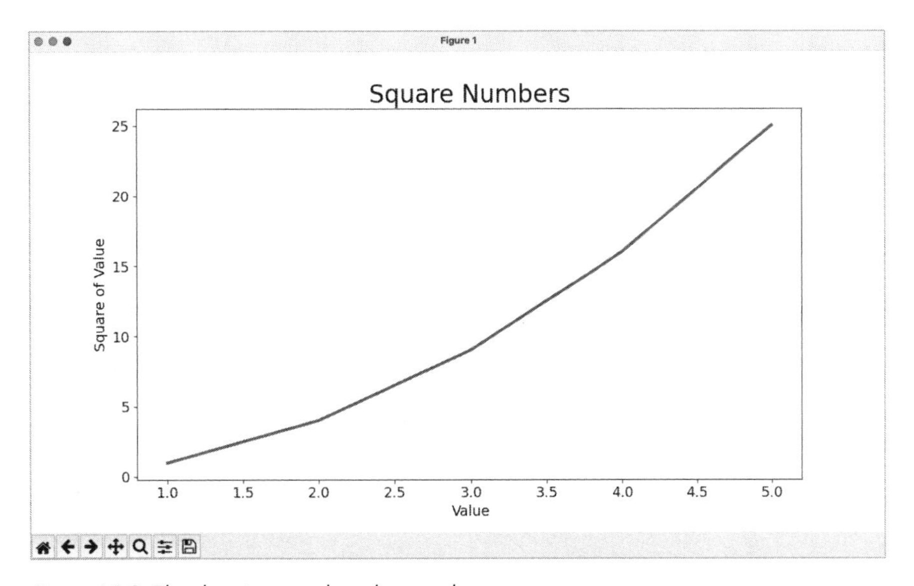

Figure 15-3: The data is now plotted correctly.

You can specify a number of arguments when calling `plot()` and use a number of methods to customize your plots after generating them. We'll continue to explore these approaches to customization as we work with more interesting datasets throughout this chapter.

Using Built-in Styles

Matplotlib has a number of predefined styles available. These styles contain a variety of default settings for background colors, gridlines, line widths, fonts, font sizes, and more. They can make your visualizations appealing without requiring much customization. To see the full list of available styles, run the following lines in a terminal session:

```
>>> import matplotlib.pyplot as plt
>>> plt.style.available
['Solarize_Light2', '_classic_test_patch', '_mpl-gallery',
--snip--
```

To use any of these styles, add one line of code before calling `subplots()`:

mpl_squares.py

```
import matplotlib.pyplot as plt

input_values = [1, 2, 3, 4, 5]
squares = [1, 4, 9, 16, 25]

plt.style.use('seaborn-v0_8')
fig, ax = plt.subplots()
--snip--
```

This code generates the plot shown in Figure 15-4. A wide variety of styles is available; play around with these styles to find some that you like.

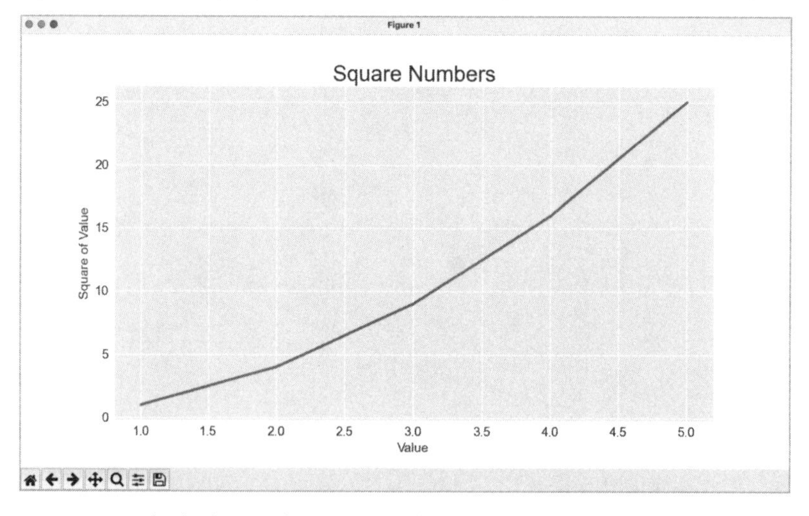

Figure 15-4: The built-in seaborn-v0_8 style

Plotting and Styling Individual Points with scatter()

Sometimes, it's useful to plot and style individual points based on certain characteristics. For example, you might plot small values in one color and larger values in a different color. You could also plot a large dataset with one set of styling options and then emphasize individual points by replotting them with different options.

To plot a single point, pass the single *x*- and *y*-values of the point to scatter():

scatter_squares.py

```
import matplotlib.pyplot as plt

plt.style.use('seaborn-v0_8')
fig, ax = plt.subplots()
ax.scatter(2, 4)

plt.show()
```

Let's style the output to make it more interesting. We'll add a title, label the axes, and make sure all the text is large enough to read:

```
import matplotlib.pyplot as plt

plt.style.use('seaborn-v0_8')
fig, ax = plt.subplots()
❶ ax.scatter(2, 4, s=200)

# Set chart title and label axes.
ax.set_title("Square Numbers", fontsize=24)
ax.set_xlabel("Value", fontsize=14)
ax.set_ylabel("Square of Value", fontsize=14)

# Set size of tick labels.
ax.tick_params(labelsize=14)

plt.show()
```

We call scatter() and use the s argument to set the size of the dots used to draw the graph ❶. When you run *scatter_squares.py* now, you should see a single point in the middle of the chart, as shown in Figure 15-5.

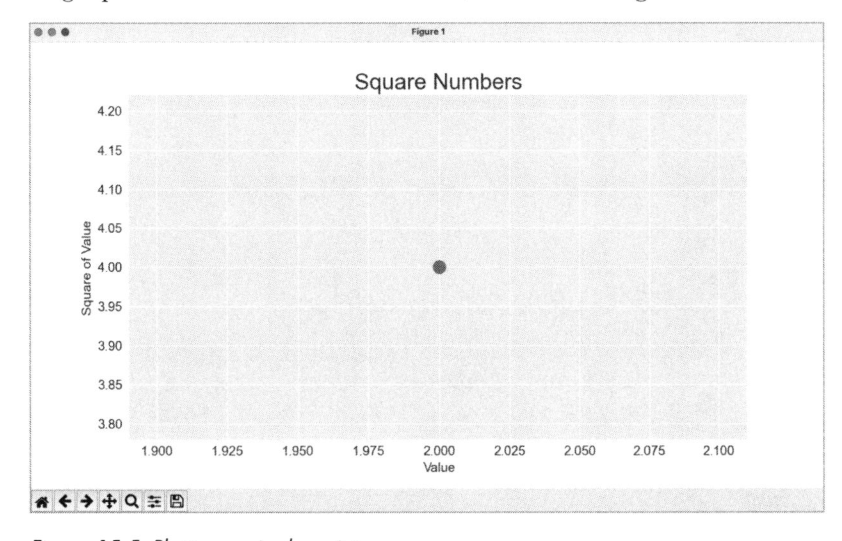

Figure 15-5: Plotting a single point

Plotting a Series of Points with scatter()

To plot a series of points, we can pass scatter() separate lists of *x*- and *y*-values, like this:

scatter_squares.py

```
import matplotlib.pyplot as plt

x_values = [1, 2, 3, 4, 5]
y_values = [1, 4, 9, 16, 25]

plt.style.use('seaborn-v0_8')
fig, ax = plt.subplots()
ax.scatter(x_values, y_values, s=100)

# Set chart title and label axes.
--snip--
```

The x_values list contains the numbers to be squared, and y_values contains the square of each number. When these lists are passed to scatter(), Matplotlib reads one value from each list as it plots each point. The points to be plotted are (1, 1), (2, 4), (3, 9), (4, 16), and (5, 25); Figure 15-6 shows the result.

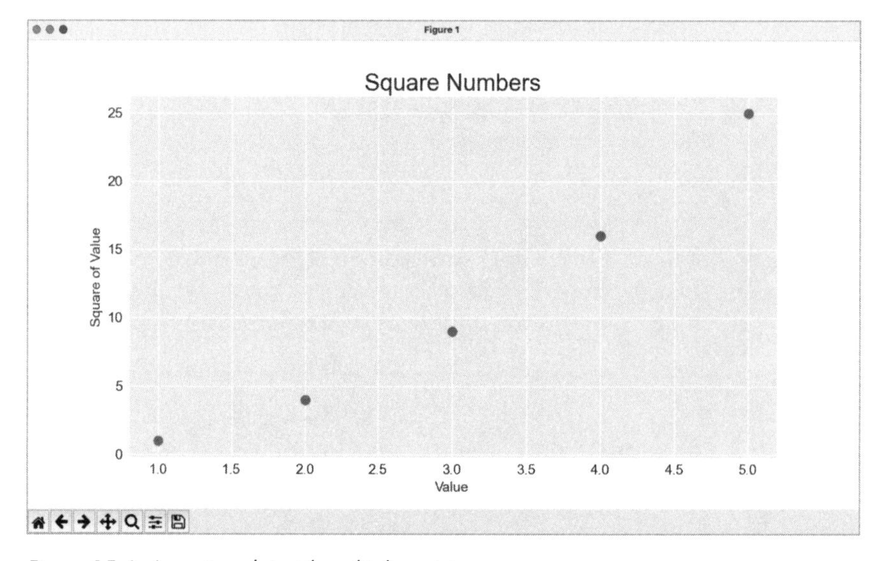

Figure 15-6: A scatter plot with multiple points

Calculating Data Automatically

Writing lists by hand can be inefficient, especially when we have many points. Rather than writing out each value, let's use a loop to do the calculations for us.

Here's how this would look with 1,000 points:

scatter_squares.py

```
import matplotlib.pyplot as plt

❶ x_values = range(1, 1001)
y_values = [x**2 for x in x_values]
```

```
    plt.style.use('seaborn-v0_8')
    fig, ax = plt.subplots()
❷ ax.scatter(x_values, y_values, s=10)

    # Set chart title and label axes.
    --snip--

    # Set the range for each axis.
❸ ax.axis([0, 1100, 0, 1_100_000])

    plt.show()
```

We start with a range of *x*-values containing the numbers 1 through 1,000 ❶. Next, a list comprehension generates the *y*-values by looping through the *x*-values (`for x in x_values`), squaring each number (`x**2`), and assigning the results to `y_values`. We then pass the input and output lists to `scatter()` ❷. Because this is a large dataset, we use a smaller point size.

Before showing the plot, we use the `axis()` method to specify the range of each axis ❸. The `axis()` method requires four values: the minimum and maximum values for the *x*-axis and the *y*-axis. Here, we run the *x*-axis from 0 to 1,100 and the *y*-axis from 0 to 1,100,000. Figure 15-7 shows the result.

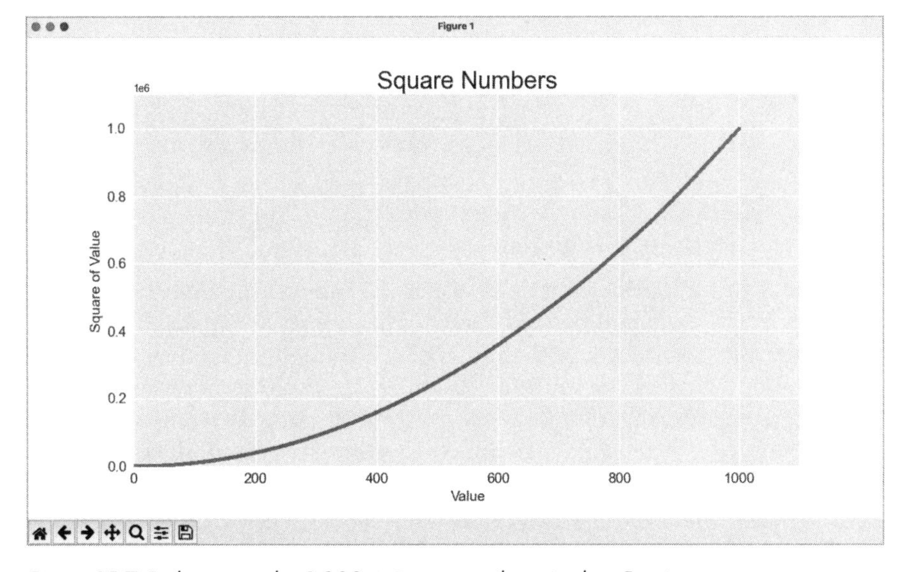

Figure 15-7: Python can plot 1,000 points as easily as it plots 5 points.

Customizing Tick Labels

When the numbers on an axis get large enough, Matplotlib defaults to scientific notation for tick labels. This is usually a good thing, because larger numbers in plain notation take up a lot of unnecessary space on a visualization.

Almost every element of a chart is customizable, so you can tell Matplotlib to keep using plain notation if you prefer:

```
--snip--
# Set the range for each axis.
ax.axis([0, 1100, 0, 1_100_000])
ax.ticklabel_format(style='plain')

plt.show()
```

The `ticklabel_format()` method allows you to override the default tick label style for any plot.

Defining Custom Colors

To change the color of the points, pass the argument `color` to `scatter()` with the name of a color to use in quotation marks, as shown here:

```
ax.scatter(x_values, y_values, color='red', s=10)
```

You can also define custom colors using the RGB color model. To define a color, pass the `color` argument a tuple with three float values (one each for red, green, and blue, in that order), using values between 0 and 1. For example, the following line creates a plot with light-green dots:

```
ax.scatter(x_values, y_values, color=(0, 0.8, 0), s=10)
```

Values closer to 0 produce darker colors, and values closer to 1 produce lighter colors.

Using a Colormap

A *colormap* is a sequence of colors in a gradient that moves from a starting to an ending color. In visualizations, colormaps are used to emphasize patterns in data. For example, you might make low values a light color and high values a darker color. Using a colormap ensures that all points in the visualization vary smoothly and accurately along a well-designed color scale.

The `pyplot` module includes a set of built-in colormaps. To use one of these colormaps, you need to specify how `pyplot` should assign a color to each point in the dataset. Here's how to assign a color to each point, based on its *y*-value:

scatter_squares.py

```
--snip--
plt.style.use('seaborn-v0_8')
fig, ax = plt.subplots()
ax.scatter(x_values, y_values, c=y_values, cmap=plt.cm.Blues, s=10)

# Set chart title and label axes.
--snip--
```

The `c` argument is similar to `color`, but it's used to associate a sequence of values with a color mapping. We pass the list of *y*-values to `c`, and then tell pyplot which colormap to use with the cmap argument. This code colors the points with lower *y*-values light blue and the points with higher *y*-values dark blue. Figure 15-8 shows the resulting plot.

NOTE *You can see all the colormaps available in* pyplot *at* https://matplotlib.org. *Go to Tutorials, scroll down to Colors, and click* ***Choosing Colormaps in Matplotlib****.*

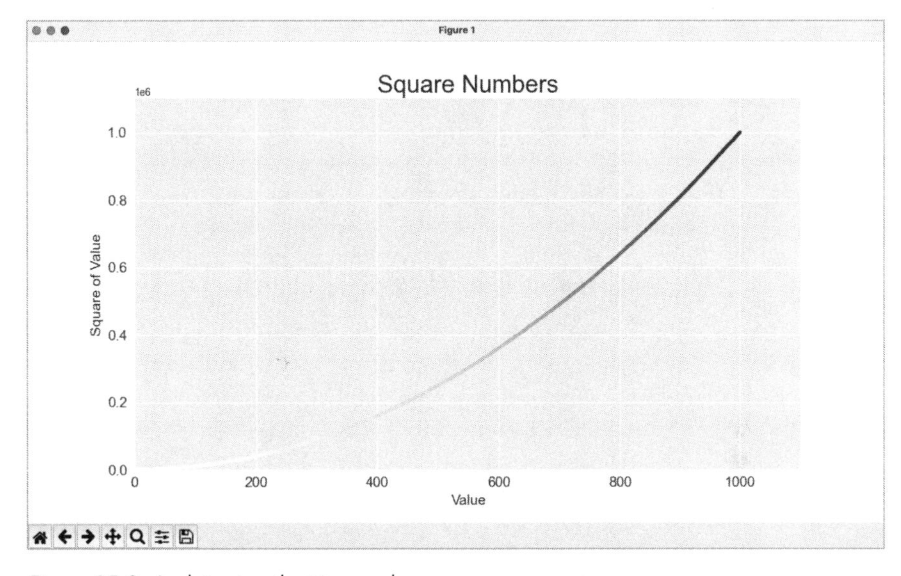

Figure 15-8: A plot using the Blues *colormap*

Saving Your Plots Automatically

If you want to save the plot to a file instead of showing it in the Matplotlib viewer, you can use plt.savefig() instead of plt.show():

```
plt.savefig('squares_plot.png', bbox_inches='tight')
```

The first argument is a filename for the plot image, which will be saved in the same directory as *scatter_squares.py*. The second argument trims extra whitespace from the plot. If you want the extra whitespace around the plot, you can omit this argument. You can also call savefig() with a Path object, and write the output file anywhere you want on your system.

TRY IT YOURSELF

15-1. Cubes: A number raised to the third power is a *cube*. Plot the first five cubic numbers, and then plot the first 5,000 cubic numbers.

15-2. Colored Cubes: Apply a colormap to your cubes plot.

Random Walks

In this section, we'll use Python to generate data for a random walk and then use Matplotlib to create a visually appealing representation of that data. A *random walk* is a path that's determined by a series of simple decisions, each of which is left entirely to chance. You might imagine a random walk as the path a confused ant would take if it took every step in a random direction.

Random walks have practical applications in nature, physics, biology, chemistry, and economics. For example, a pollen grain floating on a drop of water moves across the surface of the water because it's constantly pushed around by water molecules. Molecular motion in a water drop is random, so the path a pollen grain traces on the surface is a random walk. The code we'll write next models many real-world situations.

Creating the RandomWalk Class

To create a random walk, we'll create a `RandomWalk` class, which will make random decisions about which direction the walk should take. The class needs three attributes: one variable to track the number of points in the walk, and two lists to store the *x*- and *y*-coordinates of each point in the walk.

We'll only need two methods for the `RandomWalk` class: the `__init__()` method and `fill_walk()`, which will calculate the points in the walk. Let's start with the `__init__()` method:

random_walk.py

```
❶ from random import choice

  class RandomWalk:
      """A class to generate random walks."""

❷     def __init__(self, num_points=5000):
          """Initialize attributes of a walk."""
          self.num_points = num_points

          # All walks start at (0, 0).
❸         self.x_values = [0]
          self.y_values = [0]
```

To make random decisions, we'll store possible moves in a list and use the `choice()` function (from the `random` module) to decide which move to make each time a step is taken ❶. We set the default number of points in a walk to `5000`, which is large enough to generate some interesting patterns but small enough to generate walks quickly ❷. Then we make two lists to hold the *x*- and *y*-values, and we start each walk at the point (0, 0) ❸.

Choosing Directions

We'll use the `fill_walk()` method to determine the full sequence of points in the walk. Add this method to *random_walk.py*:

random_walk.py

```
    def fill_walk(self):
        """Calculate all the points in the walk."""
```

```
        # Keep taking steps until the walk reaches the desired length.
❶       while len(self.x_values) < self.num_points:

            # Decide which direction to go, and how far to go.
❷           x_direction = choice([1, -1])
            x_distance = choice([0, 1, 2, 3, 4])
❸           x_step = x_direction * x_distance

            y_direction = choice([1, -1])
            y_distance = choice([0, 1, 2, 3, 4])
❹           y_step = y_direction * y_distance

            # Reject moves that go nowhere.
❺           if x_step == 0 and y_step == 0:
                continue

            # Calculate the new position.
❻           x = self.x_values[-1] + x_step
            y = self.y_values[-1] + y_step

            self.x_values.append(x)
            self.y_values.append(y)
```

We first set up a loop that runs until the walk is filled with the correct number of points ❶. The main part of `fill_walk()` tells Python how to simulate four random decisions: Will the walk go right or left? How far will it go in that direction? Will it go up or down? How far will it go in that direction?

We use `choice([1, -1])` to choose a value for `x_direction`, which returns either 1 for movement to the right or –1 for movement to the left ❷. Next, `choice([0, 1, 2, 3, 4])` randomly selects a distance to move in that direction. We assign this value to `x_distance`. The inclusion of a 0 allows for the possibility of steps that have movement along only one axis.

We determine the length of each step in the *x*- and *y*-directions by multiplying the direction of movement by the distance chosen ❸ ❹. A positive result for `x_step` means move to the right, a negative result means move to the left, and 0 means move vertically. A positive result for `y_step` means move up, negative means move down, and 0 means move horizontally. If the values of both `x_step` and `y_step` are 0, the walk doesn't go anywhere; when this happens, we continue the loop ❺.

To get the next *x*-value for the walk, we add the value in `x_step` to the last value stored in `x_values` ❻ and do the same for the *y*-values. When we have the new point's coordinates, we append them to `x_values` and `y_values`.

Plotting the Random Walk

Here's the code to plot all the points in the walk:

rw_visual.py

```
import matplotlib.pyplot as plt

from random_walk import RandomWalk

# Make a random walk.
```

```
❶ rw = RandomWalk()
  rw.fill_walk()

  # Plot the points in the walk.
  plt.style.use('classic')
  fig, ax = plt.subplots()
❷ ax.scatter(rw.x_values, rw.y_values, s=15)
❸ ax.set_aspect('equal')
  plt.show()
```

We begin by importing `pyplot` and `RandomWalk`. We then create a random walk and assign it to `rw` ❶, making sure to call `fill_walk()`. To visualize the walk, we feed the walk's *x*- and *y*-values to `scatter()` and choose an appropriate dot size ❷. By default, Matplotlib scales each axis independently. But that approach would stretch most walks out horizontally or vertically. Here we use the `set_aspect()` method to specify that both axes should have equal spacing between tick marks ❸.

Figure 15-9 shows the resulting plot with 5,000 points. The images in this section omit Matplotlib's viewer, but you'll continue to see it when you run *rw_visual.py*.

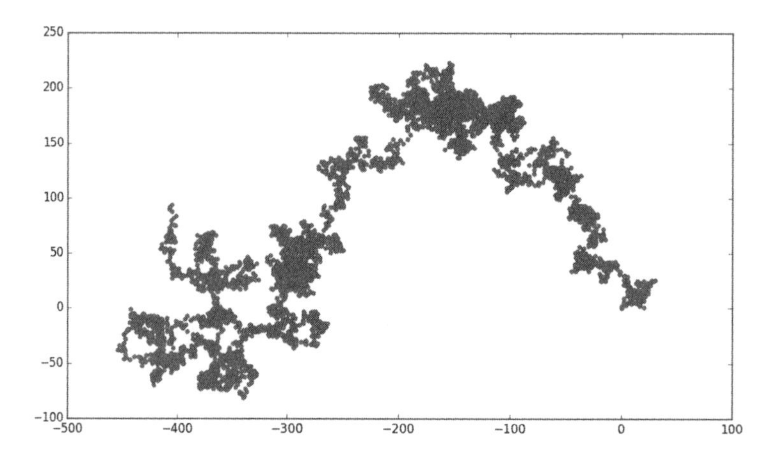

Figure 15-9: A random walk with 5,000 points

Generating Multiple Random Walks

Every random walk is different, and it's fun to explore the various patterns that can be generated. One way to use the preceding code to make multiple walks without having to run the program several times is to wrap it in a `while` loop, like this:

rw_visual.py

```
import matplotlib.pyplot as plt

from random_walk import RandomWalk

# Keep making new walks, as long as the program is active.
```

```
while True:
    # Make a random walk.
    --snip--
    plt.show()

    keep_running = input("Make another walk? (y/n): ")
    if keep_running == 'n':
        break
```

This code generates a random walk, displays it in Matplotlib's viewer, and pauses with the viewer open. When you close the viewer, you'll be asked whether you want to generate another walk. If you generate a few walks, you should see some that stay near the starting point, some that wander off mostly in one direction, some that have thin sections connecting larger groups of points, and many other kinds of walks. When you want to end the program, press N.

Styling the Walk

In this section, we'll customize our plots to emphasize the important characteristics of each walk and deemphasize distracting elements. To do so, we identify the characteristics we want to emphasize, such as where the walk began, where it ended, and the path taken. Next, we identify the characteristics to deemphasize, such as tick marks and labels. The result should be a simple visual representation that clearly communicates the path taken in each random walk.

Coloring the Points

We'll use a colormap to show the order of the points in the walk, and remove the black outline from each dot so the color of the dots will be clearer. To color the points according to their position in the walk, we pass the c argument a list containing the position of each point. Because the points are plotted in order, this list just contains the numbers from 0 to 4,999:

rw_visual.py

```
--snip--
while True:
    # Make a random walk.
    rw = RandomWalk()
    rw.fill_walk()

    # Plot the points in the walk.
    plt.style.use('classic')
    fig, ax = plt.subplots()
❶   point_numbers = range(rw.num_points)
    ax.scatter(rw.x_values, rw.y_values, c=point_numbers, cmap=plt.cm.Blues,
        edgecolors='none', s=15)
    ax.set_aspect('equal')
    plt.show()
    --snip--
```

We use range() to generate a list of numbers equal to the number of points in the walk ❶. We assign this list to point_numbers, which we'll use to set the color of each point in the walk. We pass point_numbers to the c argument, use the Blues colormap, and then pass edgecolors='none' to get rid of the black outline around each point. The result is a plot that varies from light to dark blue, showing exactly how the walk moves from its starting point to its ending point. This is shown in Figure 15-10.

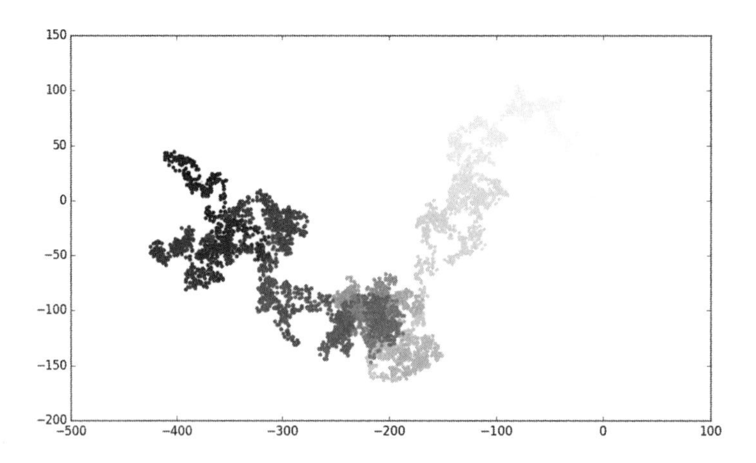

Figure 15-10: A random walk colored with the Blues colormap

Plotting the Starting and Ending Points

In addition to coloring points to show their position along the walk, it would be useful to see exactly where each walk begins and ends. To do so, we can plot the first and last points individually after the main series has been plotted. We'll make the end points larger and color them differently to make them stand out:

rw_visual.py

```
--snip--
while True:
    --snip--
    ax.scatter(rw.x_values, rw.y_values, c=point_numbers, cmap=plt.cm.Blues,
        edgecolors='none', s=15)
    ax.set_aspect('equal')

    # Emphasize the first and last points.
    ax.scatter(0, 0, c='green', edgecolors='none', s=100)
    ax.scatter(rw.x_values[-1], rw.y_values[-1], c='red', edgecolors='none',
        s=100)

    plt.show()
    --snip--
```

To show the starting point, we plot the point (0, 0) in green and in a larger size (s=100) than the rest of the points. To mark the end point, we plot the last *x*- and *y*-values in red with a size of 100 as well. Make sure you insert this code just before the call to `plt.show()` so the starting and ending points are drawn on top of all the other points.

When you run this code, you should be able to spot exactly where each walk begins and ends. If these end points don't stand out clearly enough, adjust their color and size until they do.

Cleaning Up the Axes

Let's remove the axes in this plot so they don't distract from the path of each walk. Here's how to hide the axes:

rw_visual.py

```
--snip--
while True:
    --snip--
    ax.scatter(rw.x_values[-1], rw.y_values[-1], c='red', edgecolors='none',
        s=100)

    # Remove the axes.
    ax.get_xaxis().set_visible(False)
    ax.get_yaxis().set_visible(False)

    plt.show()
    --snip--
```

To modify the axes, we use the `ax.get_xaxis()` and `ax.get_yaxis()` methods to get each axis, and then chain the `set_visible()` method to make each axis invisible. As you continue to work with visualizations, you'll frequently see this chaining of methods to customize different aspects of a visualization.

Run *rw_visual.py* now; you should see a series of plots with no axes.

Adding Plot Points

Let's increase the number of points, to give us more data to work with. To do so, we increase the value of `num_points` when we make a `RandomWalk` instance and adjust the size of each dot when drawing the plot:

rw_visual.py

```
--snip--
while True:
    # Make a random walk.
    rw = RandomWalk(50_000)
    rw.fill_walk()

    # Plot the points in the walk.
    plt.style.use('classic')
    fig, ax = plt.subplots()
    point_numbers = range(rw.num_points)
    ax.scatter(rw.x_values, rw.y_values, c=point_numbers, cmap=plt.cm.Blues,
        edgecolors='none', s=1)
    --snip--
```

This example creates a random walk with 50,000 points and plots each point at size `s=1`. The resulting walk is wispy and cloudlike, as shown in Figure 15-11. We've created a piece of art from a simple scatter plot!

Experiment with this code to see how much you can increase the number of points in a walk before your system starts to slow down significantly or the plot loses its visual appeal.

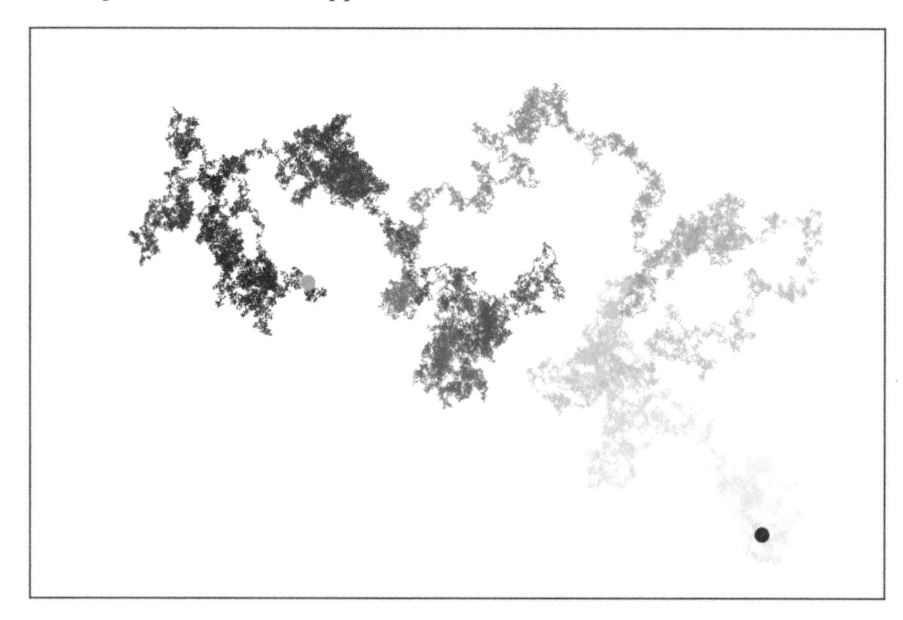

Figure 15-11: A walk with 50,000 points

Altering the Size to Fill the Screen

A visualization is much more effective at communicating patterns in data if it fits nicely on the screen. To make the plotting window better fit your screen, you can adjust the size of Matplotlib's output. This is done in the `subplots()` call:

```
fig, ax = plt.subplots(figsize=(15, 9))
```

When creating a plot, you can pass `subplots()` a `figsize` argument, which sets the size of the figure. The `figsize` parameter takes a tuple that tells Matplotlib the dimensions of the plotting window in inches.

Matplotlib assumes your screen resolution is 100 pixels per inch; if this code doesn't give you an accurate plot size, adjust the numbers as necessary. Or, if you know your system's resolution, you can pass `subplots()` the resolution using the `dpi` parameter:

```
fig, ax = plt.subplots(figsize=(10, 6), dpi=128)
```

This should help make the most efficient use of the space available on your screen.

TRY IT YOURSELF

15-3. Molecular Motion: Modify *rw_visual.py* by replacing `ax.scatter()` with `ax.plot()`. To simulate the path of a pollen grain on the surface of a drop of water, pass in the `rw.x_values` and `rw.y_values`, and include a `linewidth` argument. Use 5,000 instead of 50,000 points to keep the plot from being too busy.

15-4. Modified Random Walks: In the `RandomWalk` class, `x_step` and `y_step` are generated from the same set of conditions. The direction is chosen randomly from the list `[1, -1]` and the distance from the list `[0, 1, 2, 3, 4]`. Modify the values in these lists to see what happens to the overall shape of your walks. Try a longer list of choices for the distance, such as 0 through 8, or remove the −1 from the *x*- or *y*-direction list.

15-5. Refactoring: The `fill_walk()` method is lengthy. Create a new method called `get_step()` to determine the direction and distance for each step, and then calculate the step. You should end up with two calls to `get_step()` in `fill_walk()`:

```
x_step = self.get_step()
y_step = self.get_step()
```

This refactoring should reduce the size of `fill_walk()` and make the method easier to read and understand.

Rolling Dice with Plotly

In this section, we'll use Plotly to produce interactive visualizations. Plotly is particularly useful when you're creating visualizations that will be displayed in a browser, because the visualizations will scale automatically to fit the viewer's screen. These visualizations are also interactive; when the user hovers over certain elements on the screen, information about those elements is highlighted. We'll build our initial visualization in just a couple lines of code using *Plotly Express*, a subset of Plotly that focuses on generating plots with as little code as possible. Once we know our plot is correct, we'll customize the output just as we did with Matplotlib.

In this project, we'll analyze the results of rolling dice. When you roll one regular, six-sided die, you have an equal chance of rolling any of the numbers from 1 through 6. However, when you use two dice, you're more likely to roll certain numbers than others. We'll try to determine which numbers are most likely to occur by generating a dataset that represents rolling dice. Then we'll plot the results of a large number of rolls to determine which results are more likely than others.

This work helps model games involving dice, but the core ideas also apply to games that involve chance of any kind, such as card games. It also relates to many real-world situations where randomness plays a significant factor.

Installing Plotly

Install Plotly using pip, just as you did for Matplotlib:

```
$ python -m pip install --user plotly
$ python -m pip install --user pandas
```

Plotly Express depends on *pandas*, which is a library for working efficiently with data, so we need to install that as well. If you used **python3** or something else when installing Matplotlib, make sure you use the same command here.

To see what kind of visualizations are possible with Plotly, visit the gallery of chart types at *https://plotly.com/python*. Each example includes source code, so you can see how Plotly generates the visualizations.

Creating the Die Class

We'll create the following `Die` class to simulate the roll of one die:

die.py

```
from random import randint

class Die:
    """A class representing a single die."""

❶   def __init__(self, num_sides=6):
        """Assume a six-sided die."""
        self.num_sides = num_sides

    def roll(self):
        """"Return a random value between 1 and number of sides."""
❷       return randint(1, self.num_sides)
```

The `__init__()` method takes one optional argument ❶. With the `Die` class, when an instance of our die is created, the number of sides will be six if no argument is included. If an argument *is* included, that value will set the number of sides on the die. (Dice are named for their number of sides: a six-sided die is a D6, an eight-sided die is a D8, and so on.)

The `roll()` method uses the `randint()` function to return a random number between 1 and the number of sides ❷. This function can return the starting value (1), the ending value (`num_sides`), or any integer between the two.

Rolling the Die

Before creating a visualization based on the `Die` class, let's roll a D6, print the results, and check that the results look reasonable:

die_visual.py

```
from die import Die

# Create a D6.
❶ die = Die()
```

```
# Make some rolls, and store results in a list.
results = []
❷ for roll_num in range(100):
    result = die.roll()
    results.append(result)

print(results)
```

We create an instance of `Die` with the default six sides ❶. Then we roll the die `100` times ❷ and store the result of each roll in the list `results`. Here's a sample set of results:

```
[4, 6, 5, 6, 1, 5, 6, 3, 5, 3, 5, 3, 2, 2, 1, 3, 1, 5, 3, 6, 3, 6, 5, 4,
 1, 1, 4, 2, 3, 6, 4, 2, 6, 4, 1, 3, 2, 5, 6, 3, 6, 2, 1, 1, 3, 4, 1, 4,
 3, 5, 1, 4, 5, 5, 2, 3, 3, 1, 2, 3, 5, 6, 2, 5, 6, 1, 3, 2, 1, 1, 1, 6,
 5, 5, 2, 2, 6, 4, 1, 4, 5, 1, 1, 1, 4, 5, 3, 3, 1, 3, 5, 4, 5, 6, 5, 4,
 1, 5, 1, 2]
```

A quick scan of these results shows that the `Die` class seems to be working. We see the values 1 and 6, so we know the smallest and largest possible values are being returned, and because we don't see 0 or 7, we know all the results are in the appropriate range. We also see each number from 1 through 6, which indicates that all possible outcomes are represented. Let's determine exactly how many times each number appears.

Analyzing the Results

We'll analyze the results of rolling one D6 by counting how many times we roll each number:

die_visual.py

```
--snip--
# Make some rolls, and store results in a list.
results = []
❶ for roll_num in range(1000):
    result = die.roll()
    results.append(result)

# Analyze the results.
frequencies = []
❷ poss_results = range(1, die.num_sides+1)
for value in poss_results:
❸     frequency = results.count(value)
❹     frequencies.append(frequency)

print(frequencies)
```

Because we're no longer printing the results, we can increase the number of simulated rolls to `1000` ❶. To analyze the rolls, we create the empty list `frequencies` to store the number of times each value is rolled. We then generate all the possible results we could get; in this example, that's all the numbers from `1` to however many sides `die` has ❷. We loop through the possible values, count how many times each number appears in `results` ❸, and then append this value to frequencies ❹. We print this list before making a visualization:

```
[155, 167, 168, 170, 159, 181]
```

These results look reasonable: we see six frequencies, one for each possible number when you roll a D6. We also see that no frequency is significantly higher than any other. Now let's visualize these results.

Making a Histogram

Now that we have the data we want, we can generate a visualization in just a couple lines of code using Plotly Express:

die_visual.py

```
import plotly.express as px

from die import Die
--snip--

for value in poss_results:
    frequency = results.count(value)
    frequencies.append(frequency)

# Visualize the results.
fig = px.bar(x=poss_results, y=frequencies)
fig.show()
```

We first import the plotly.express module, using the conventional alias px. We then use the px.bar() function to create a bar graph. In the simplest use of this function, we only need to pass a set of *x*-values and a set of *y*-values. Here the *x*-values are the possible results from rolling a single die, and the *y*-values are the frequencies for each possible result.

The final line calls fig.show(), which tells Plotly to render the resulting chart as an HTML file and open that file in a new browser tab. The result is shown in Figure 15-12.

This is a really simple chart, and it's certainly not complete. But this is exactly how Plotly Express is meant to be used; you write a couple lines of code, look at the plot, and make sure it represents the data the way you want it to. If you like what you see, you can move on to customizing elements of the chart such as labels and styles. But if you want to explore other possible chart types, you can do so now, without having spent extra time on customization work. Feel free to try this now by changing px.bar() to something like px.scatter() or px.line(). You can find a full list of available chart types at *https://plotly.com/python/plotly-express.*

This chart is dynamic and interactive. If you change the size of your browser window, the chart will resize to match the available space. If you hover over any of the bars, you'll see a pop-up highlighting the specific data related to that bar.

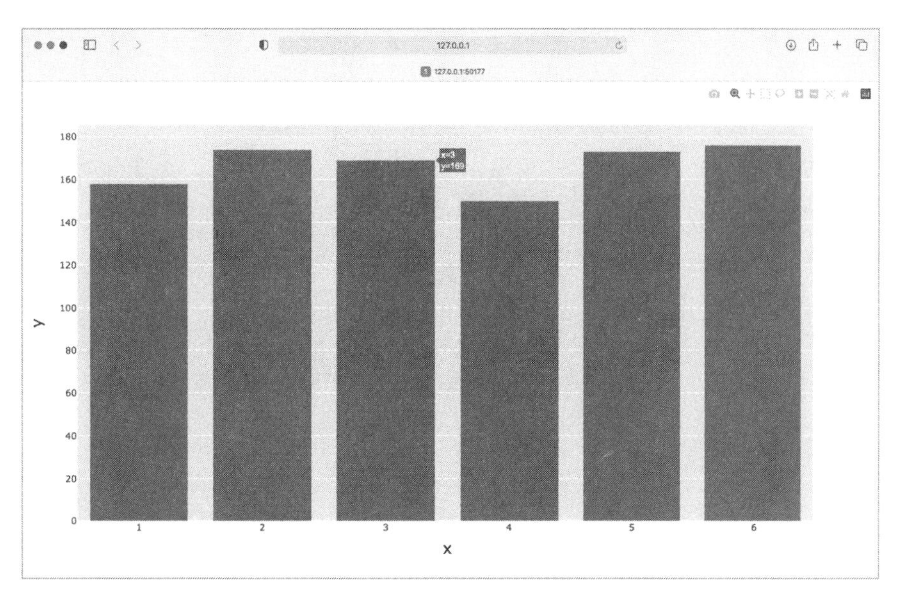

Figure 15-12: The initial plot produced by Plotly Express

Customizing the Plot

Now that we know we have the correct kind of plot and our data is being represented accurately, we can focus on adding the appropriate labels and styles for the chart.

The first way to customize a plot with Plotly is to use some optional parameters in the initial call that generates the plot, in this case, `px.bar()`. Here's how to add an overall title and a label for each axis:

die_visual.py

```
--snip--
# Visualize the results.
❶ title = "Results of Rolling One D6 1,000 Times"
❷ labels = {'x': 'Result', 'y': 'Frequency of Result'}
fig = px.bar(x=poss_results, y=frequencies, title=title, labels=labels)
fig.show()
```

We first define the title that we want, here assigned to `title` ❶. To define axis labels, we write a dictionary ❷. The keys in the dictionary refer to the labels we want to customize, and the values are the custom labels we want to use. Here we give the *x*-axis the label `Result` and the *y*-axis the label `Frequency of Result`. The call to `px.bar()` now includes the optional arguments `title` and `labels`.

Now when the plot is generated it includes an appropriate title and a label for each axis, as shown in Figure 15-13.

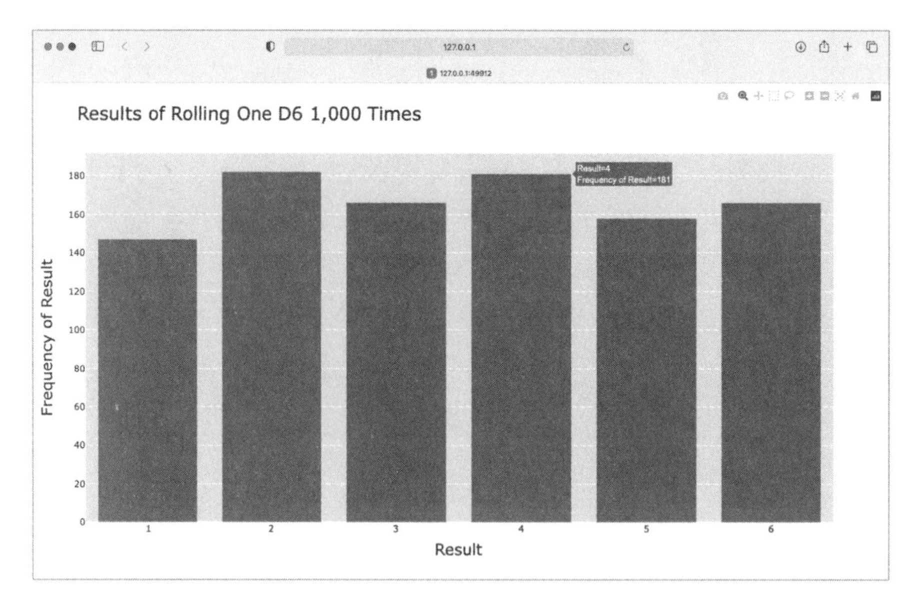

Figure 15-13: A simple bar chart created with Plotly

Rolling Two Dice

Rolling two dice results in larger numbers and a different distribution of results. Let's modify our code to create two D6 dice to simulate the way we roll a pair of dice. Each time we roll the pair, we'll add the two numbers (one from each die) and store the sum in `results`. Save a copy of *die_visual.py* as *dice_visual.py* and make the following changes:

dice_visual.py

```
import plotly.express as px

from die import Die

# Create two D6 dice.
die_1 = Die()
die_2 = Die()

# Make some rolls, and store results in a list.
results = []
for roll_num in range(1000):
❶     result = die_1.roll() + die_2.roll()
    results.append(result)

# Analyze the results.
frequencies = []
❷ max_result = die_1.num_sides + die_2.num_sides
❸ poss_results = range(2, max_result+1)
for value in poss_results:
    frequency = results.count(value)
    frequencies.append(frequency)
```

```
# Visualize the results.
title = "Results of Rolling Two D6 Dice 1,000 Times"
labels = {'x': 'Result', 'y': 'Frequency of Result'}
fig = px.bar(x=poss_results, y=frequencies, title=title, labels=labels)
fig.show()
```

After creating two instances of Die, we roll the dice and calculate the sum of the two dice for each roll ❶. The smallest possible result (2) is the sum of the smallest number on each die. The largest possible result (12) is the sum of the largest number on each die, which we assign to max_result ❷. The variable max_result makes the code for generating poss_results much easier to read ❸. We could have written range(2, 13), but this would work only for two D6 dice. When modeling real-world situations, it's best to write code that can easily model a variety of situations. This code allows us to simulate rolling a pair of dice with any number of sides.

After running this code, you should see a chart that looks like Figure 15-14.

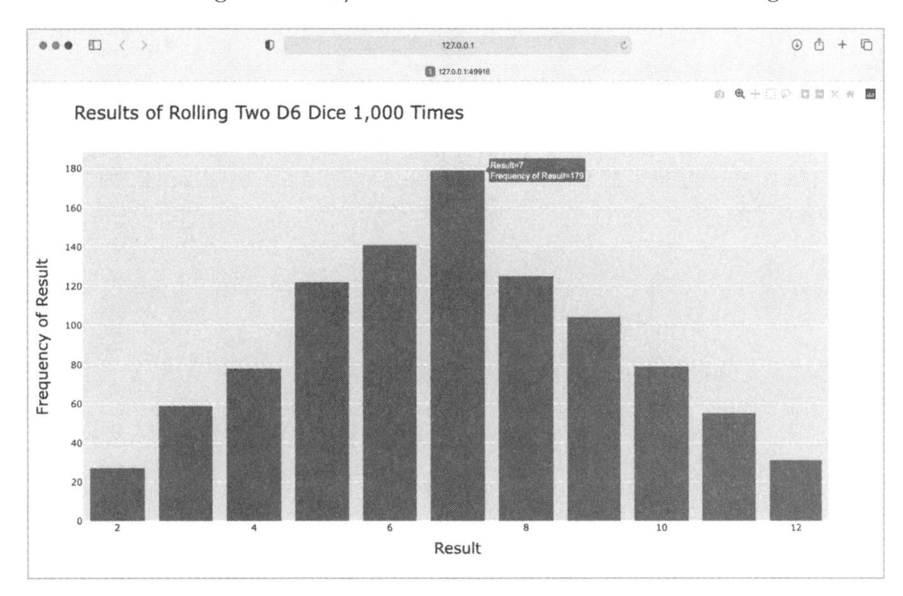

Figure 15-14: Simulated results of rolling two six-sided dice 1,000 times

This graph shows the approximate distribution of results you're likely to get when you roll a pair of D6 dice. As you can see, you're least likely to roll a 2 or a 12 and most likely to roll a 7. This happens because there are six ways to roll a 7: 1 and 6, 2 and 5, 3 and 4, 4 and 3, 5 and 2, and 6 and 1.

Further Customizations

There's one issue that we should address with the plot we just generated. Now that there are 11 bars, the default layout settings for the *x*-axis leave some of the bars unlabeled. While the default settings work well for most visualizations, this chart would look better with all of the bars labeled.

Plotly has an `update_layout()` method that can be used to make a wide variety of updates to a figure after it's been created. Here's how to tell Plotly to give each bar its own label:

dice_visual.py

```
--snip--
fig = px.bar(x=poss_results, y=frequencies, title=title, labels=labels)

# Further customize chart.
fig.update_layout(xaxis_dtick=1)

fig.show()
```

The `update_layout()` method acts on the `fig` object, which represents the overall chart. Here we use the `xaxis_dtick` argument, which specifies the distance between tick marks on the *x*-axis. We set that spacing to `1`, so that every bar is labeled. When you run *dice_visual.py* again, you should see a label on each bar.

Rolling Dice of Different Sizes

Let's create a six-sided die and a ten-sided die, and see what happens when we roll them 50,000 times:

dice_visual_d6d10.py

```
import plotly.express as px

from die import Die

# Create a D6 and a D10.
die_1 = Die()
❶ die_2 = Die(10)

# Make some rolls, and store results in a list.
results = []
for roll_num in range(50_000):
    result = die_1.roll() + die_2.roll()
    results.append(result)

# Analyze the results.
--snip--

# Visualize the results.
❷ title = "Results of Rolling a D6 and a D10 50,000 Times"
labels = {'x': 'Result', 'y': 'Frequency of Result'}
--snip--
```

To make a D10, we pass the argument `10` when creating the second `Die` instance ❶ and change the first loop to simulate 50,000 rolls instead of 1,000. We change the title of the graph as well ❷.

Figure 15-15 shows the resulting chart. Instead of one most likely result, there are five such results. This happens because there's still only one way to roll the smallest value (1 and 1) and the largest value (6 and 10), but the smaller die limits the number of ways you can generate the middle numbers. There are six ways to roll a 7, 8, 9, 10, or 11, these are the most common results, and you're equally likely to roll any one of them.

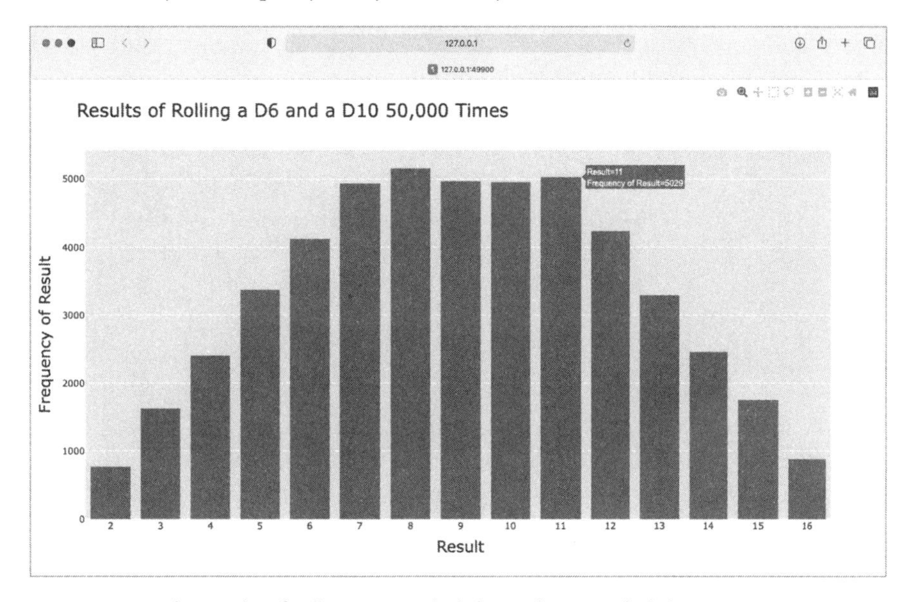

Figure 15-15: The results of rolling a six-sided die and a ten-sided die 50,000 times

Our ability to use Plotly to model the rolling of dice gives us considerable freedom in exploring this phenomenon. In just minutes, you can simulate a tremendous number of rolls using a large variety of dice.

Saving Figures

When you have a figure you like, you can always save the chart as an HTML file through your browser. But you can also do so programmatically. To save your chart as an HTML file, replace the call to `fig.show()` with a call to `fig.write_html()`:

```
fig.write_html('dice_visual_d6d10.html')
```

The `write_html()` method requires one argument: the name of the file to write to. If you only provide a filename, the file will be saved in the same directory as the *.py* file. You can also call `write_html()` with a `Path` object, and write the output file anywhere you want on your system.

TRY IT YOURSELF

15-6. Two D8s: Create a simulation showing what happens when you roll two eight-sided dice 1,000 times. Try to picture what you think the visualization will look like before you run the simulation, then see if your intuition was correct. Gradually increase the number of rolls until you start to see the limits of your system's capabilities.

15-7. Three Dice: When you roll three D6 dice, the smallest number you can roll is 3 and the largest number is 18. Create a visualization that shows what happens when you roll three D6 dice.

15-8. Multiplication: When you roll two dice, you usually add the two numbers together to get the result. Create a visualization that shows what happens if you multiply these numbers by each other instead.

15-9. Die Comprehensions: For clarity, the listings in this section use the long form of `for` loops. If you're comfortable using list comprehensions, try writing a comprehension for one or both of the loops in each of these programs.

15-10. Practicing with Both Libraries: Try using Matplotlib to make a die-rolling visualization, and use Plotly to make the visualization for a random walk. (You'll need to consult the documentation for each library to complete this exercise.)

Summary

In this chapter, you learned to generate datasets and create visualizations of that data. You created simple plots with Matplotlib and used a scatter plot to explore random walks. You also created a histogram with Plotly, and used it to explore the results of rolling dice of different sizes.

Generating your own datasets with code is an interesting and powerful way to model and explore a wide variety of real-world situations. As you continue to work through the data visualization projects that follow, keep an eye out for situations you might be able to model with code. Look at the visualizations you see in news media, and see if you can identify those that were generated using methods similar to the ones you're learning in these projects.

In Chapter 16, you'll download data from online sources and continue to use Matplotlib and Plotly to explore that data.

16

DOWNLOADING DATA

In this chapter, you'll download datasets from online sources and create working visualizations of that data. You can find an incredible variety of data online, much of which hasn't been examined thoroughly. The ability to analyze this data allows you to discover patterns and connections that no one else has found.

We'll access and visualize data stored in two common data formats: CSV and JSON. We'll use Python's `csv` module to process weather data stored in the CSV format and analyze high and low temperatures over time in two different locations. We'll then use Matplotlib to generate a chart based on our downloaded data to display variations in temperature in two dissimilar environments: Sitka, Alaska, and Death Valley, California. Later in the chapter, we'll use the `json` module to access earthquake data stored in the GeoJSON format and use Plotly to draw a world map showing the locations and magnitudes of recent earthquakes.

By the end of this chapter, you'll be prepared to work with various types of datasets in different formats, and you'll have a deeper understanding of how to build complex visualizations. Being able to access and visualize online data is essential to working with a wide variety of real-world datasets.

The CSV File Format

One simple way to store data in a text file is to write the data as a series of values separated by commas, called *comma-separated values*. The resulting files are *CSV* files. For example, here's a chunk of weather data in CSV format:

```
"USW00025333","SITKA AIRPORT, AK US","2021-01-01",,"44","40"
```

This is an excerpt of weather data from January 1, 2021, in Sitka, Alaska. It includes the day's high and low temperatures, as well as a number of other measurements from that day. CSV files can be tedious for humans to read, but programs can process and extract information from them quickly and accurately.

We'll begin with a small set of CSV-formatted weather data recorded in Sitka; it is available in this book's resources at *https://ehmatthes.github.io/pcc_3e*. Make a folder called *weather_data* inside the folder where you're saving this chapter's programs. Copy the file *sitka_weather_07-2021_simple.csv* into this new folder. (After you download this book's resources, you'll have all the files you need for this project.)

NOTE *The weather data in this project was originally downloaded from* https://ncdc.noaa.gov/cdo-web.

Parsing the CSV File Headers

Python's `csv` module in the standard library parses the lines in a CSV file and allows us to quickly extract the values we're interested in. Let's start by examining the first line of the file, which contains a series of headers for the data. These headers tell us what kind of information the data holds:

sitka_highs.py

```
from pathlib import Path
import csv

❶ path = Path('weather_data/sitka_weather_07-2021_simple.csv')
lines = path.read_text(encoding='utf-8').splitlines()

❷ reader = csv.reader(lines)
❸ header_row = next(reader)
print(header_row)
```

We first import `Path` and the `csv` module. We then build a `Path` object that looks in the *weather_data* folder, and points to the specific weather data file we want to work with ❶. We read the file and chain the `splitlines()` method to get a list of all lines in the file, which we assign to `lines`.

Next, we build a reader object ❷. This is an object that can be used to parse each line in the file. To make a reader object, call the function csv.reader() and pass it the list of lines from the CSV file.

When given a reader object, the next() function returns the next line in the file, starting from the beginning of the file. Here we call next() only once, so we get the first line of the file, which contains the file headers ❸. We assign the data that's returned to header_row. As you can see, header_row contains meaningful, weather-related headers that tell us what information each line of data holds:

```
['STATION', 'NAME', 'DATE', 'TAVG', 'TMAX', 'TMIN']
```

The reader object processes the first line of comma-separated values in the file and stores each value as an item in a list. The header STATION represents the code for the weather station that recorded this data. The position of this header tells us that the first value in each line will be the weather station code. The NAME header indicates that the second value in each line is the name of the weather station that made the recording. The rest of the headers specify what kinds of information were recorded in each reading. The data we're most interested in for now are the date (DATE), the high temperature (TMAX), and the low temperature (TMIN). This is a simple dataset that contains only temperature-related data. When you download your own weather data, you can choose to include a number of other measurements relating to wind speed, wind direction, and precipitation data.

Printing the Headers and Their Positions

To make it easier to understand the file header data, let's print each header and its position in the list:

sitka_highs.py

```
--snip--
reader = csv.reader(lines)
header_row = next(reader)

for index, column_header in enumerate(header_row):
    print(index, column_header)
```

The enumerate() function returns both the index of each item and the value of each item as you loop through a list. (Note that we've removed the line print(header_row) in favor of this more detailed version.)

Here's the output showing the index of each header:

```
0 STATION
1 NAME
2 DATE
3 TAVG
4 TMAX
5 TMIN
```

We can see that the dates and their high temperatures are stored in columns 2 and 4. To explore this data, we'll process each row of data in *sitka_weather_07-2021_simple.csv* and extract the values with the indexes 2 and 4.

Extracting and Reading Data

Now that we know which columns of data we need, let's read in some of that data. First, we'll read in the high temperature for each day:

sitka_highs.py

```
--snip--
reader = csv.reader(lines)
header_row = next(reader)

# Extract high temperatures.
❶ highs = []
❷ for row in reader:
❸     high = int(row[4])
      highs.append(high)

print(highs)
```

We make an empty list called `highs` ❶ and then loop through the remaining rows in the file ❷. The `reader` object continues from where it left off in the CSV file and automatically returns each line following its current position. Because we've already read the header row, the loop will begin at the second line where the actual data begins. On each pass through the loop we pull the data from index 4, corresponding to the header `TMAX`, and assign it to the variable `high` ❸. We use the `int()` function to convert the data, which is stored as a string, to a numerical format so we can use it. We then append this value to `highs`.

The following listing shows the data now stored in `highs`:

```
[61, 60, 66, 60, 65, 59, 58, 58, 57, 60, 60, 60, 57, 58, 60, 61, 63, 63, 70,
 64, 59, 63, 61, 58, 59, 64, 62, 70, 70, 73, 66]
```

We've extracted the high temperature for each date and stored each value in a list. Now let's create a visualization of this data.

Plotting Data in a Temperature Chart

To visualize the temperature data we have, we'll first create a simple plot of the daily highs using Matplotlib, as shown here:

sitka_highs.py

```
from pathlib import Path
import csv

import matplotlib.pyplot as plt

path = Path('weather_data/sitka_weather_07-2021_simple.csv')
lines = path.read_text(encoding='utf-8').splitlines()
    --snip--
```

```
    # Plot the high temperatures.
    plt.style.use('seaborn-v0_8')
    fig, ax = plt.subplots()
❶ ax.plot(highs, color='red')

    # Format plot.
❷ ax.set_title("Daily High Temperatures, July 2021", fontsize=24)
❸ ax.set_xlabel('', fontsize=16)
    ax.set_ylabel("Temperature (F)", fontsize=16)
    ax.tick_params(labelsize=16)

    plt.show()
```

We pass the list of highs to `plot()` and pass `color='red'` to plot the points in red ❶. (We'll plot the highs in red and the lows in blue.) We then specify a few other formatting details, such as the title, font size, and labels ❷, just as we did in Chapter 15. Because we have yet to add the dates, we won't label the *x*-axis, but `ax.set_xlabel()` does modify the font size to make the default labels more readable ❸. Figure 16-1 shows the resulting plot: a simple line graph of the high temperatures for July 2021 in Sitka, Alaska.

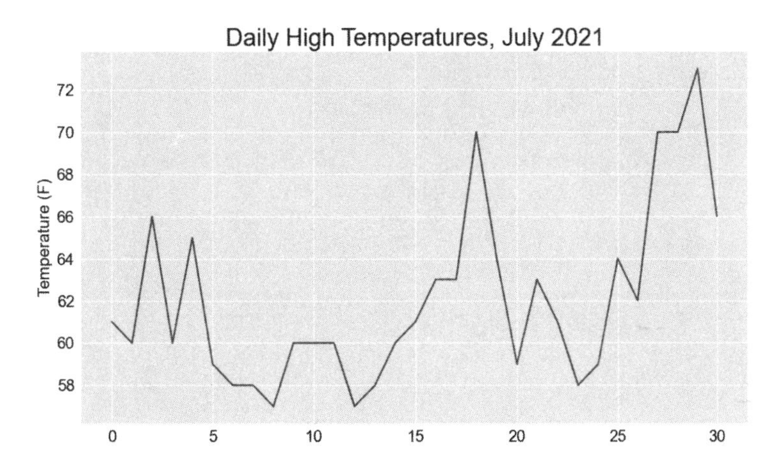

Figure 16-1: A line graph showing daily high temperatures for July 2021 in Sitka, Alaska

The datetime Module

Let's add dates to our graph to make it more useful. The first date from the weather data file is in the second row of the file:

```
"USW00025333","SITKA AIRPORT, AK US","2021-07-01",,"61","53"
```

The data will be read in as a string, so we need a way to convert the string `"2021-07-01"` to an object representing this date. We can construct an object representing July 1, 2021, using the strptime() method from the datetime module. Let's see how strptime() works in a terminal session:

```
>>> from datetime import datetime
>>> first_date = datetime.strptime('2021-07-01', '%Y-%m-%d')
>>> print(first_date)
2021-07-01 00:00:00
```

We first import the datetime class from the datetime module. Then we call the method strptime() with the string containing the date we want to process as its first argument. The second argument tells Python how the date is formatted. In this example, '%Y-' tells Python to look for a four-digit year before the first dash; '%m-' indicates a two-digit month before the second dash; and '%d' means the last part of the string is the day of the month, from 1 to 31.

The strptime() method can take a variety of arguments to determine how to interpret the date. Table 16-1 shows some of these arguments.

Table 16-1: Date and Time Formatting Arguments from the datetime Module

Argument	Meaning
%A	Weekday name, such as Monday
%B	Month name, such as January
%m	Month, as a number (01 to 12)
%d	Day of the month, as a number (01 to 31)
%Y	Four-digit year, such as 2019
%y	Two-digit year, such as 19
%H	Hour, in 24-hour format (00 to 23)
%I	Hour, in 12-hour format (01 to 12)
%p	AM or PM
%M	Minutes (00 to 59)
%S	Seconds (00 to 61)

Plotting Dates

We can improve our plot by extracting dates for the daily high temperature readings, and using these dates on the *x*-axis:

sitka_highs.py

```
from pathlib import Path
import csv
from datetime import datetime

import matplotlib.pyplot as plt

path = Path('weather_data/sitka_weather_07-2021_simple.csv')
lines = path.read_text(encoding='utf-8').splitlines()
```

```
    reader = csv.reader(lines)
    header_row = next(reader)

    # Extract dates and high temperatures.
❶ dates, highs = [], []
    for row in reader:
❷       current_date = datetime.strptime(row[2], '%Y-%m-%d')
        high = int(row[4])
        dates.append(current_date)
        highs.append(high)

    # Plot the high temperatures.
    plt.style.use('seaborn-v0_8')
    fig, ax = plt.subplots()
❸ ax.plot(dates, highs, color='red')

    # Format plot.
    ax.set_title("Daily High Temperatures, July 2021", fontsize=24)
    ax.set_xlabel('', fontsize=16)
❹ fig.autofmt_xdate()
    ax.set_ylabel("Temperature (F)", fontsize=16)
    ax.tick_params(labelsize=16)

    plt.show()
```

We create two empty lists to store the dates and high temperatures from the file ❶. We then convert the data containing the date information (`row[2]`) to a `datetime` object ❷ and append it to `dates`. We pass the dates and the high temperature values to `plot()` ❸. The call to `fig.autofmt_xdate()` ❹ draws the date labels diagonally to prevent them from overlapping. Figure 16-2 shows the improved graph.

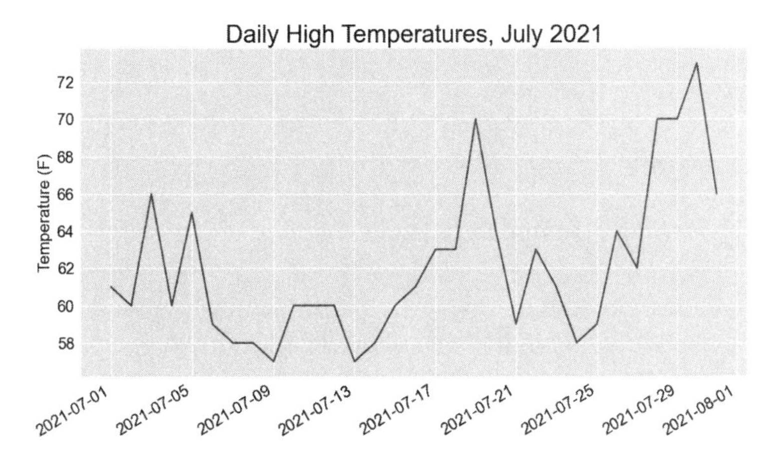

Figure 16-2: The graph is more meaningful, now that it has dates on the x-axis.

Plotting a Longer Timeframe

With our graph set up, let's include additional data to get a more complete picture of the weather in Sitka. Copy the file *sitka_weather_2021_simple.csv*, which contains a full year's worth of weather data for Sitka, to the folder where you're storing the data for this chapter's programs.

Now we can generate a graph for the entire year's weather:

sitka_highs.py

```
--snip--
path = Path('weather_data/sitka_weather_2021_simple.csv')
lines = path.read_text(encoding='utf-8').splitlines()
--snip--
# Format plot.
ax.set_title("Daily High Temperatures, 2021", fontsize=24)
ax.set_xlabel('', fontsize=16)
--snip--
```

We modify the filename to use the new data file *sitka_weather_2021_simple.csv*, and we update the title of our plot to reflect the change in its content. Figure 16-3 shows the resulting plot.

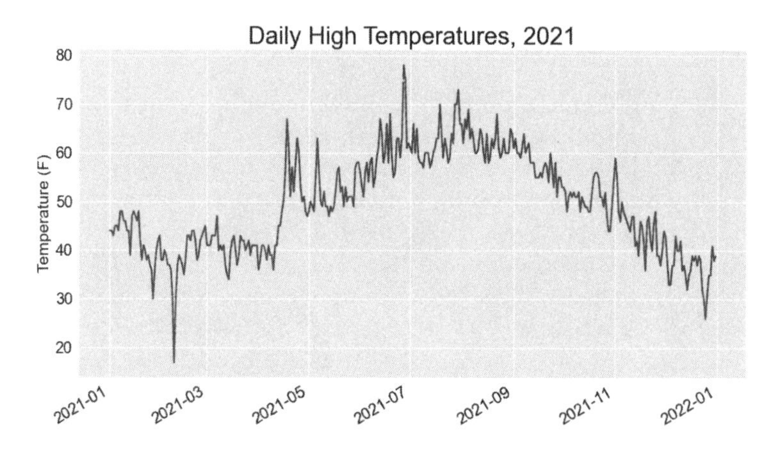

Figure 16-3: A year's worth of data

Plotting a Second Data Series

We can make our graph even more useful by including the low temperatures. We need to extract the low temperatures from the data file and then add them to our graph, as shown here:

sitka_highs_lows.py

```
--snip--
reader = csv.reader(lines)
header_row = next(reader)

# Extract dates, and high and low temperatures.
❶ dates, highs, lows = [], [], []
for row in reader:
```

```
        current_date = datetime.strptime(row[2], '%Y-%m-%d')
        high = int(row[4])
❷       low = int(row[5])
        dates.append(current_date)
        highs.append(high)
        lows.append(low)

# Plot the high and low temperatures.
plt.style.use('seaborn-v0_8')
fig, ax = plt.subplots()
ax.plot(dates, highs, color='red')
❸ ax.plot(dates, lows, color='blue')

# Format plot.
❹ ax.set_title("Daily High and Low Temperatures, 2021", fontsize=24)
--snip--
```

We add the empty list `lows` to hold low temperatures ❶, and then we extract and store the low temperature for each date from the sixth position in each row (`row[5]`) ❷. We add a call to `plot()` for the low temperatures and color these values blue ❸. Finally, we update the title ❹. Figure 16-4 shows the resulting chart.

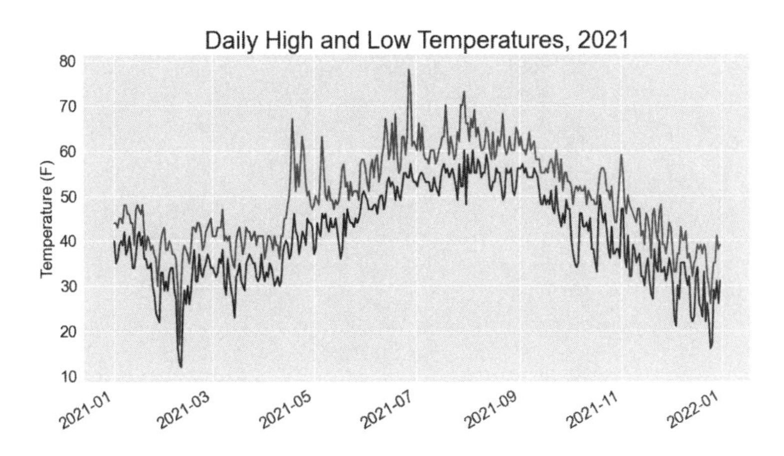

Figure 16-4: Two data series on the same plot

Shading an Area in the Chart

Having added two data series, we can now examine the range of temperatures for each day. Let's add a finishing touch to the graph by using shading to show the range between each day's high and low temperatures. To do so, we'll use the `fill_between()` method, which takes a series of *x*-values and two series of *y*-values and fills the space between the two series of *y*-values:

sitka_highs_lows.py

```
--snip--
# Plot the high and low temperatures.
plt.style.use('seaborn-v0_8')
```

```
fig, ax = plt.subplots()
❶ ax.plot(dates, highs, color='red', alpha=0.5)
ax.plot(dates, lows, color='blue', alpha=0.5)
❷ ax.fill_between(dates, highs, lows, facecolor='blue', alpha=0.1)
--snip--
```

The `alpha` argument controls a color's transparency ❶. An `alpha` value of 0 is completely transparent, and a value of 1 (the default) is completely opaque. By setting `alpha` to 0.5, we make the red and blue plot lines appear lighter.

We pass `fill_between()` the list `dates` for the *x*-values and then the two *y*-value series `highs` and `lows` ❷. The `facecolor` argument determines the color of the shaded region; we give it a low `alpha` value of 0.1 so the filled region connects the two data series without distracting from the information they represent. Figure 16-5 shows the plot with the shaded region between the highs and lows.

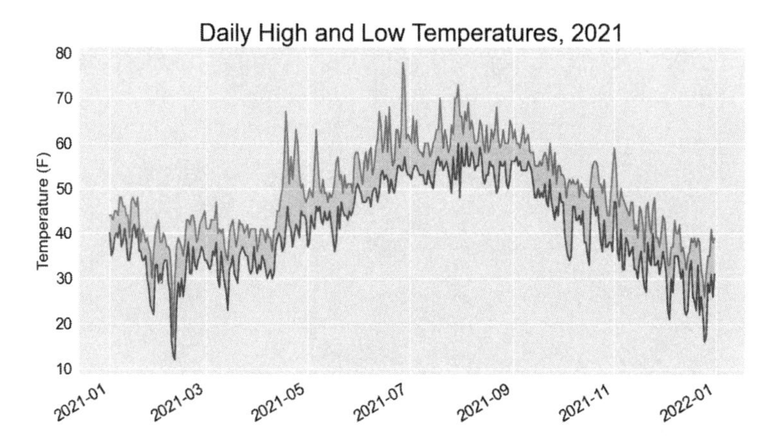

Figure 16-5: The region between the two datasets is shaded.

The shading helps make the range between the two datasets immediately apparent.

Error Checking

We should be able to run the *sitka_highs_lows.py* code using data for any location. But some weather stations collect different data than others, and some occasionally malfunction and fail to collect some of the data they're supposed to. Missing data can result in exceptions that crash our programs, unless we handle them properly.

For example, let's see what happens when we attempt to generate a temperature plot for Death Valley, California. Copy the file *death_valley_2021_simple.csv* to the folder where you're storing the data for this chapter's programs.

First, let's run the code to see the headers that are included in this data file:

death_valley_highs_lows.py

```
from pathlib import Path
import csv

path = Path('weather_data/death_valley_2021_simple.csv')
lines = path.read_text(encoding='utf-8').splitlines()

reader = csv.reader(lines)
header_row = next(reader)

for index, column_header in enumerate(header_row):
    print(index, column_header)
```

Here's the output:

```
0 STATION
1 NAME
2 DATE
3 TMAX
4 TMIN
5 TOBS
```

The date is in the same position, at index 2. But the high and low temperatures are at indexes 3 and 4, so we'll need to change the indexes in our code to reflect these new positions. Instead of including an average temperature reading for the day, this station includes `TOBS`, a reading for a specific observation time.

Change *sitka_highs_lows.py* to generate a graph for Death Valley using the indexes we just noted, and see what happens:

death_valley_highs_lows.py

```
--snip--
path = Path('weather_data/death_valley_2021_simple.csv')
lines = path.read_text(encoding='utf-8').splitlines()
    --snip--
# Extract dates, and high and low temperatures.
dates, highs, lows = [], [], []
for row in reader:
    current_date = datetime.strptime(row[2], '%Y-%m-%d')
    high = int(row[3])
    low = int(row[4])
    dates.append(current_date)
--snip--
```

We update the program to read from the Death Valley data file, and we change the indexes to correspond to this file's `TMAX` and `TMIN` positions.

When we run the program, we get an error:

```
Traceback (most recent call last):
  File "death_valley_highs_lows.py", line 17, in <module>
    high = int(row[3])
❶ ValueError: invalid literal for int() with base 10: ''
```

The traceback tells us that Python can't process the high temperature for one of the dates because it can't turn an empty string ('') into an integer ❶. Rather than looking through the data to find out which reading is missing, we'll just handle cases of missing data directly.

We'll run error-checking code when the values are being read from the CSV file to handle exceptions that might arise. Here's how to do this:

death_valley_highs_lows.py

```
--snip--
for row in reader:
    current_date = datetime.strptime(row[2], '%Y-%m-%d')
❶   try:
        high = int(row[3])
        low = int(row[4])
    except ValueError:
❷       print(f"Missing data for {current_date}")
❸   else:
        dates.append(current_date)
        highs.append(high)
        lows.append(low)

# Plot the high and low temperatures.
--snip--

# Format plot.
❹ title = "Daily High and Low Temperatures, 2021\nDeath Valley, CA"
ax.set_title(title, fontsize=20)
ax.set_xlabel('', fontsize=16)
--snip--
```

Each time we examine a row, we try to extract the date and the high and low temperature ❶. If any data is missing, Python will raise a `ValueError` and we handle it by printing an error message that includes the date of the missing data ❷. After printing the error, the loop will continue processing the next row. If all data for a date is retrieved without error, the `else` block will run and the data will be appended to the appropriate lists ❸. Because we're plotting information for a new location, we update the title to include the location on the plot, and we use a smaller font size to accommodate the longer title ❹.

When you run *death_valley_highs_lows.py* now, you'll see that only one date had missing data:

```
Missing data for 2021-05-04 00:00:00
```

Because the error is handled appropriately, our code is able to generate a plot, which skips over the missing data. Figure 16-6 shows the resulting plot.

Comparing this graph to the Sitka graph, we can see that Death Valley is warmer overall than southeast Alaska, as we expect. Also, the range of temperatures each day is greater in the desert. The height of the shaded region makes this clear.

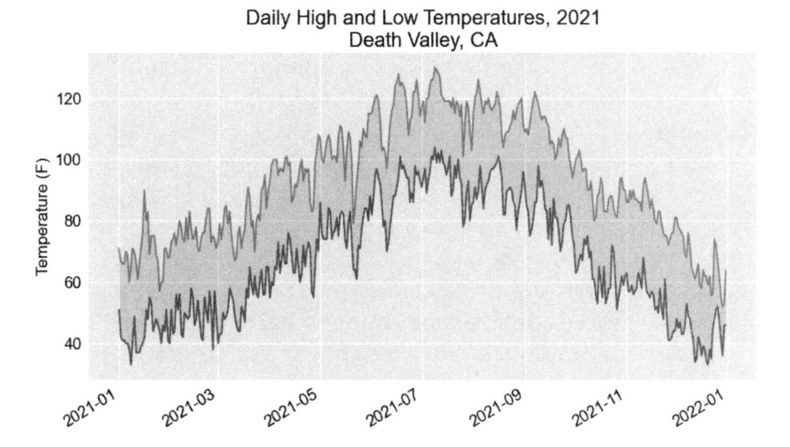

Figure 16-6: Daily high and low temperatures for Death Valley

Many datasets you work with will have missing, improperly formatted, or incorrect data. You can use the tools you learned in the first half of this book to handle these situations. Here we used a `try`-`except`-`else` block to handle missing data. Sometimes you'll use `continue` to skip over some data, or use `remove()` or `del` to eliminate some data after it's been extracted. Use any approach that works, as long as the result is a meaningful, accurate visualization.

Downloading Your Own Data

To download your own weather data, follow these steps:

1. Visit the NOAA Climate Data Online site at *https://www.ncdc.noaa.gov/cdo-web*. In the Discover Data By section, click **Search Tool**. In the Select a Dataset box, choose **Daily Summaries**.
2. Select a date range, and in the Search For section, choose **ZIP Codes**. Enter the ZIP code you're interested in and click **Search**.
3. On the next page, you'll see a map and some information about the area you're focusing on. Below the location name, click **View Full Details**, or click the map and then click **Full Details**.
4. Scroll down and click **Station List** to see the weather stations that are available in this area. Click one of the station names and then click **Add to Cart**. This data is free, even though the site uses a shopping cart icon. In the upper-right corner, click the cart.
5. In Select the Output Format, choose **Custom GHCN-Daily CSV**. Make sure the date range is correct and click **Continue**.

6. On the next page, you can select the kinds of data you want. You can download one kind of data (for example, focusing on air temperature) or you can download all the data available from this station. Make your choices and then click **Continue**.
7. On the last page, you'll see a summary of your order. Enter your email address and click **Submit Order**. You'll receive a confirmation that your order was received, and in a few minutes, you should receive another email with a link to download your data.

The data you download should be structured just like the data we worked with in this section. It might have different headers than those you saw in this section, but if you follow the same steps we used here, you should be able to generate visualizations of the data you're interested in.

TRY IT YOURSELF

16-1. Sitka Rainfall: Sitka is located in a temperate rainforest, so it gets a fair amount of rainfall. In the data file *sitka_weather_2021_full.csv* is a header called `PRCP`, which represents daily rainfall amounts. Make a visualization focusing on the data in this column. You can repeat the exercise for Death Valley if you're curious how little rainfall occurs in a desert.

16-2. Sitka–Death Valley Comparison: The temperature scales on the Sitka and Death Valley graphs reflect the different data ranges. To accurately compare the temperature range in Sitka to that of Death Valley, you need identical scales on the *y*-axis. Change the settings for the *y*-axis on one or both of the charts in Figures 16-5 and 16-6. Then make a direct comparison between temperature ranges in Sitka and Death Valley (or any two places you want to compare).

16-3. San Francisco: Are temperatures in San Francisco more like temperatures in Sitka or temperatures in Death Valley? Download some data for San Francisco, and generate a high-low temperature plot for San Francisco to make a comparison.

16-4. Automatic Indexes: In this section, we hardcoded the indexes corresponding to the `TMIN` and `TMAX` columns. Use the header row to determine the indexes for these values, so your program can work for Sitka or Death Valley. Use the station name to automatically generate an appropriate title for your graph as well.

16-5. Explore: Generate a few more visualizations that examine any other weather aspect you're interested in for any locations you're curious about.

Mapping Global Datasets: GeoJSON Format

In this section, you'll download a dataset representing all the earthquakes that have occurred in the world during the previous month. Then you'll make a map showing the location of these earthquakes and how significant each one was. Because the data is stored in the GeoJSON format, we'll work with it using the json module. Using Plotly's scatter_geo() plot, you'll create visualizations that clearly show the global distribution of earthquakes.

Downloading Earthquake Data

Make a folder called *eq_data* inside the folder where you're saving this chapter's programs. Copy the file *eq_1_day_m1.geojson* into this new folder. Earthquakes are categorized by their magnitude on the Richter scale. This file includes data for all earthquakes with a magnitude M1 or greater that took place in the last 24 hours (at the time of this writing). This data comes from one of the United States Geological Survey's earthquake data feeds, at *https://earthquake.usgs.gov/earthquakes/feed.*

Examining GeoJSON Data

When you open *eq_1_day_m1.geojson*, you'll see that it's very dense and hard to read:


```
{"type":"FeatureCollection","metadata":{"generated":1649052296000,...
{"type":"Feature","properties":{"mag":1.6,"place":"63 km SE of Ped...
{"type":"Feature","properties":{"mag":2.2,"place":"27 km SSE of Ca...
{"type":"Feature","properties":{"mag":3.7,"place":"102 km SSE of S...
{"type":"Feature","properties":{"mag":2.92000008,"place":"49 km SE...
{"type":"Feature","properties":{"mag":1.4,"place":"44 km NE of Sus...
--snip--
```


This file is formatted more for machines than humans. But we can see that the file contains some dictionaries, as well as information that we're interested in, such as earthquake magnitudes and locations.

The json module provides a variety of tools for exploring and working with JSON data. Some of these tools will help us reformat the file so we can look at the raw data more easily before we work with it programmatically.

Let's start by loading the data and displaying it in a format that's easier to read. This is a long data file, so instead of printing it, we'll rewrite the data to a new file. Then we can open that file and scroll back and forth through the data more easily:

eq_explore_data.py

```
from pathlib import Path
import json

# Read data as a string and convert to a Python object.
path = Path('eq_data/eq_data_1_day_m1.geojson')
contents = path.read_text(encoding='utf-8')
❶ all_eq_data = json.loads(contents)

# Create a more readable version of the data file.
❷ path = Path('eq_data/readable_eq_data.geojson')
❸ readable_contents = json.dumps(all_eq_data, indent=4)
path.write_text(readable_contents)
```

We read the data file as a string, and use `json.loads()` to convert the string representation of the file to a Python object ❶. This is the same approach we used in Chapter 10. In this case, the entire dataset is converted to a single dictionary, which we assign to `all_eq_data`. We then define a new `path` where we can write this same data in a more readable format ❷. The `json.dumps()` function that you saw in Chapter 10 can take an optional `indent` argument ❸, which tells it how much to indent nested elements in the data structure.

When you look in your *eq_data* directory and open the file *readable_eq_data.json*, here's the first part of what you'll see:

readable_eq_data.json

```
{
    "type": "FeatureCollection",
❶   "metadata": {
        "generated": 1649052296000,
        "url": "https://earthquake.usgs.gov/earthquakes/.../1.0_day.geojson",
        "title": "USGS Magnitude 1.0+ Earthquakes, Past Day",
        "status": 200,
        "api": "1.10.3",
        "count": 160
    },
❷   "features": [
    --snip--
```

The first part of the file includes a section with the key `"metadata"` ❶. This tells us when the data file was generated and where we can find the data online. It also gives us a human-readable title and the number of earthquakes included in this file. In this 24-hour period, `160` earthquakes were recorded.

This GeoJSON file has a structure that's helpful for location-based data. The information is stored in a list associated with the key `"features"` ❷. Because this file contains earthquake data, the data is in list form where every item in the list corresponds to a single earthquake. This structure might look confusing, but it's quite powerful. It allows geologists to store as much information as they need to in a dictionary about each earthquake, and then stuff all those dictionaries into one big list.

Let's look at a dictionary representing a single earthquake:

readable_eq_data.json

```
    --snip--
        {
            "type": "Feature",
❶           "properties": {
                "mag": 1.6,
                --snip--
❷               "title": "M 1.6 - 27 km NNW of Susitna, Alaska"
            },
❸           "geometry": {
                "type": "Point",
                "coordinates": [
❹                   -150.7585,
❺                   61.7591,
                    56.3
```

```
            ]
        },
        "id": "ak0224bju1jx"
    },
```

The key "properties" contains a lot of information about each earthquake ❶. We're mainly interested in the magnitude of each earthquake, associated with the key "mag". We're also interested in the "title" of each event, which provides a nice summary of its magnitude and location ❷.

The key "geometry" helps us understand where the earthquake occurred ❸. We'll need this information to map each event. We can find the longitude ❹ and the latitude ❺ for each earthquake in a list associated with the key "coordinates".

This file contains way more nesting than we'd use in the code we write, so if it looks confusing, don't worry: Python will handle most of the complexity. We'll only be working with one or two nesting levels at a time. We'll start by pulling out a dictionary for each earthquake that was recorded in the 24-hour time period.

NOTE *When we talk about locations, we often say the location's latitude first, followed by its longitude. This convention probably arose because humans discovered latitude long before we developed the concept of longitude. However, many geospatial frameworks list the longitude first and then the latitude, because this corresponds to the* (x, y) *convention we use in mathematical representations. The GeoJSON format follows the (longitude, latitude) convention. If you use a different framework, it's important to learn what convention that framework follows.*

Making a List of All Earthquakes

First, we'll make a list that contains all the information about every earthquake that occurred.

eq_explore_data.py

```
from pathlib import Path
import json

# Read data as a string and convert to a Python object.
path = Path('eq_data/eq_data_1_day_m1.geojson')
contents = path.read_text(encoding='utf-8')
all_eq_data = json.loads(contents)

# Examine all earthquakes in the dataset.
all_eq_dicts = all_eq_data['features']
print(len(all_eq_dicts))
```

We take the data associated with the key 'features' in the all_eq_data dictionary, and assign it to all_eq_dicts. We know this file contains records of 160 earthquakes, and the output verifies that we've captured all the earthquakes in the file:

```
160
```

Notice how short this code is. The neatly formatted file *readable_eq_data.json* has over 6,000 lines. But in just a few lines, we can read through all that data and store it in a Python list. Next, we'll pull the magnitudes from each earthquake.

Extracting Magnitudes

We can loop through the list containing data about each earthquake, and extract any information we want. Let's pull out the magnitude of each earthquake:

eq_explore_data.py

```
--snip--
all_eq_dicts = all_eq_data['features']

❶ mags = []
for eq_dict in all_eq_dicts:
❷     mag = eq_dict['properties']['mag']
    mags.append(mag)

print(mags[:10])
```

We make an empty list to store the magnitudes, and then loop through the list `all_eq_dicts` ❶. Inside this loop, each earthquake is represented by the dictionary `eq_dict`. Each earthquake's magnitude is stored in the `'properties'` section of this dictionary, under the key `'mag'` ❷. We store each magnitude in the variable `mag` and then append it to the list `mags`.

We print the first `10` magnitudes, so we can see whether we're getting the correct data:

```
[1.6, 1.6, 2.2, 3.7, 2.92000008, 1.4, 4.6, 4.5, 1.9, 1.8]
```

Next, we'll pull the location data for each earthquake, and then we can make a map of the earthquakes.

Extracting Location Data

The location data for each earthquake is stored under the key `"geometry"`. Inside the geometry dictionary is a `"coordinates"` key, and the first two values in this list are the longitude and latitude. Here's how we'll pull this data:

eq_explore_data.py

```
--snip--
all_eq_dicts = all_eq_data['features']

mags, lons, lats = [], [], []
for eq_dict in all_eq_dicts:
    mag = eq_dict['properties']['mag']
❶     lon = eq_dict['geometry']['coordinates'][0]
    lat = eq_dict['geometry']['coordinates'][1]
    mags.append(mag)
    lons.append(lon)
```

```
    lats.append(lat)

print(mags[:10])
print(lons[:5])
print(lats[:5])
```

We make empty lists for the longitudes and latitudes. The code `eq_dict['geometry']` accesses the dictionary representing the geometry element of the earthquake ❶. The second key, `'coordinates'`, pulls the list of values associated with `'coordinates'`. Finally, the `0` index asks for the first value in the list of coordinates, which corresponds to an earthquake's longitude.

When we print the first `5` longitudes and latitudes, the output shows that we're pulling the correct data:

```
[1.6, 1.6, 2.2, 3.7, 2.92000008, 1.4, 4.6, 4.5, 1.9, 1.8]
[-150.7585, -153.4716, -148.7531, -159.6267, -155.248336791992]
[61.7591, 59.3152, 63.1633, 54.5612, 18.7551670074463]
```

With this data, we can move on to mapping each earthquake.

Building a World Map

Using the information we've pulled so far, we can build a simple world map. Although it won't look presentable yet, we want to make sure the information is displayed correctly before focusing on style and presentation issues. Here's the initial map:

eq_world_map.py

```
from pathlib import Path
import json

import plotly.express as px

--snip--
for eq_dict in all_eq_dicts:
    --snip--

title = 'Global Earthquakes'
❶ fig = px.scatter_geo(lat=lats, lon=lons, title=title)
fig.show()
```

We import `plotly.express` with the alias `px`, just as we did in Chapter 15. The `scatter_geo()` function ❶ allows you to overlay a scatterplot of geographic data on a map. In the simplest use of this chart type, you only need to provide a list of latitudes and a list of longitudes. We pass the list `lats` to the `lat` argument, and `lons` to the `lon` argument.

When you run this file, you should see a map that looks like the one in Figure 16-7. This again shows the power of the Plotly Express library; in just three lines of code, we have a map of global earthquake activity.

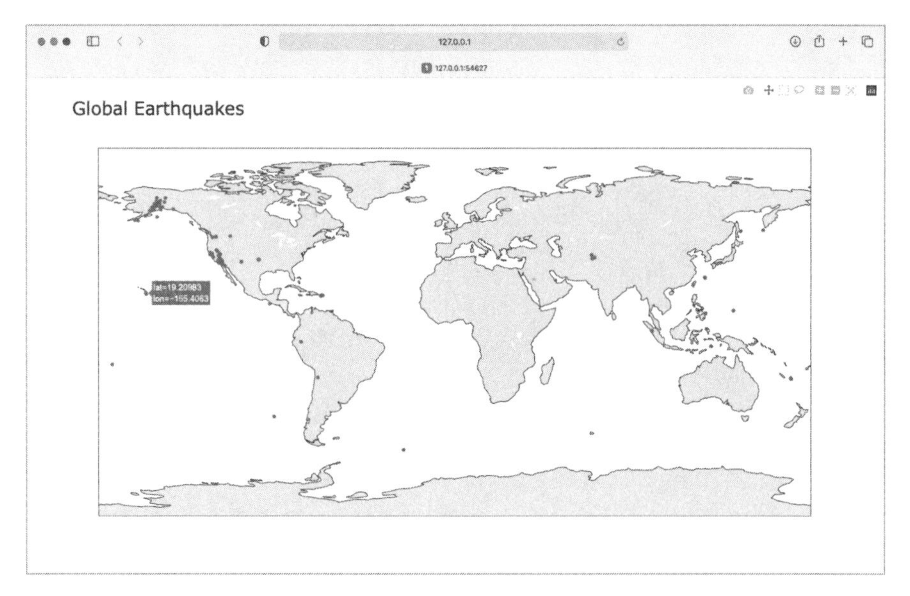

Figure 16-7: A simple map showing where all the earthquakes in the last 24 hours occurred

Now that we know the information in our dataset is being plotted correctly, we can make a few changes to make the map more meaningful and easier to read.

Representing Magnitudes

A map of earthquake activity should show the magnitude of each earthquake. We can also include more data, now that we know the data is being plotted correctly.

```
--snip--
# Read data as a string and convert to a Python object.
path = Path('eq_data/eq_data_30_day_m1.geojson')
contents = path.read_text(encoding='utf-8')
--snip--

title = 'Global Earthquakes'
fig = px.scatter_geo(lat=lats, lon=lons, size=mags, title=title)
fig.show()
```

We load the file *eq_data_30_day_m1.geojson*, to include a full 30 days' worth of earthquake activity. We also use the size argument in the `px.scatter_geo()` call, which specifies how the points on the map will be sized. We pass the list `mags` to `size`, so earthquakes with a higher magnitude will show up as larger points on the map.

The resulting map is shown in Figure 16-8. Earthquakes usually occur near tectonic plate boundaries, and the longer period of earthquake activity included in this map reveals the exact locations of these boundaries.

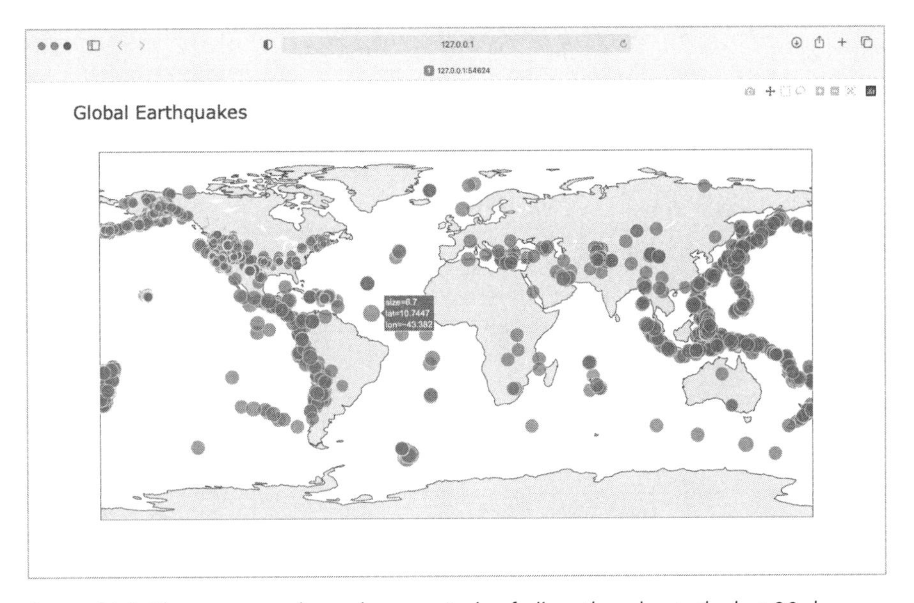

Figure 16-8: The map now shows the magnitude of all earthquakes in the last 30 days.

This map is better, but it's still difficult to pick out which points represent the most significant earthquakes. We can improve this further by using color to represent magnitudes as well.

Customizing Marker Colors

We can use Plotly's color scales to customize each marker's color, according to the severity of the corresponding earthquake. We'll also use a different projection for the base map.

eq_world_map.py

```
--snip--
fig = px.scatter_geo(lat=lats, lon=lons, size=mags, title=title,
❶       color=mags,
❷       color_continuous_scale='Viridis',
❸       labels={'color':'Magnitude'},
❹       projection='natural earth',
    )
fig.show()
```

All the significant changes here occur in the `px.scatter_geo()` function call. The `color` argument tells Plotly what values it should use to determine where each marker falls on the color scale ❶. We use the `mags` list to determine the color for each point, just as we did with the `size` argument.

The `color_continuous_scale` argument tells Plotly which color scale to use ❷. *Viridis* is a color scale that ranges from dark blue to bright yellow, and it works well for this dataset. By default, the color scale on the right of the map is labeled *color*; this is not representative of what the colors actually mean. The `labels` argument, shown in Chapter 15, takes a dictionary as a value ❸. We only need to set one custom label on this chart, making sure the color scale is labeled *Magnitude* instead of *color*.

We add one more argument, to modify the base map over which the earthquakes are plotted. The `projection` argument accepts a number of common map projections ❹. Here we use the `'natural earth'` projection, which rounds the ends of the map. Also, note the trailing comma after this last argument. When a function call has a long list of arguments spanning multiple lines like this, it's common practice to add a trailing comma so you're always ready to add another argument on the next line.

When you run the program now, you'll see a much nicer-looking map. In Figure 16-9, the color scale shows the severity of individual earthquakes; the most severe earthquakes stand out as light-yellow points, in contrast to many darker points. You can also tell which regions of the world have more significant earthquake activity.

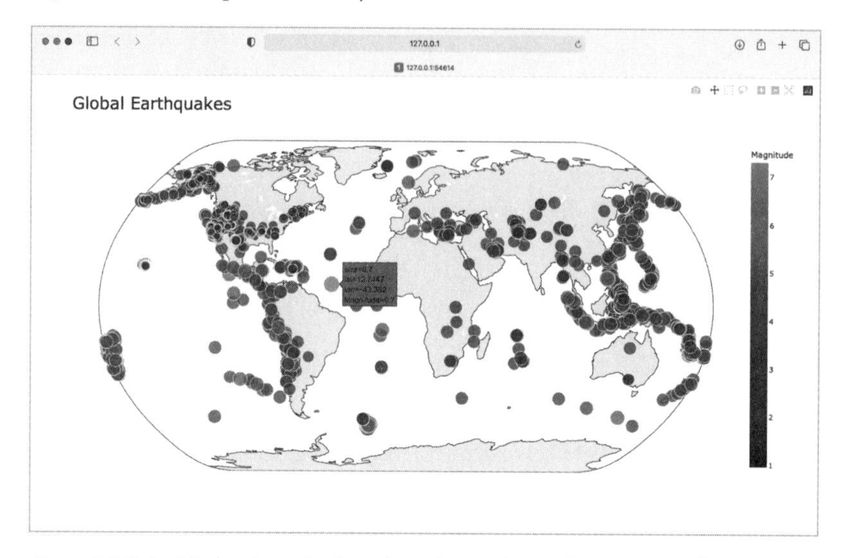

Figure 16-9: In 30 days' worth of earthquakes, color and size are used to represent the magnitude of each earthquake.

Other Color Scales

You can choose from a number of other color scales. To see the available color scales, enter the following two lines in a Python terminal session:

```
>>> import plotly.express as px
>>> px.colors.named_colorscales()
['aggrnyl', 'agsunset', 'blackbody', ..., 'mygbm']
```

Feel free to try out these color scales in the earthquake map, or with any dataset where continuously varying colors can help show patterns in the data.

Adding Hover Text

To finish this map, we'll add some informative text that appears when you hover over the marker representing an earthquake. In addition to showing the longitude and latitude, which appear by default, we'll show the magnitude and provide a description of the approximate location as well.

To make this change, we need to pull a little more data from the file:

eq_world_map.py

```
--snip--
❶ mags, lons, lats, eq_titles = [], [], [], []
      mag = eq_dict['properties']['mag']
      lon = eq_dict['geometry']['coordinates'][0]
      lat = eq_dict['geometry']['coordinates'][1]
❷     eq_title = eq_dict['properties']['title']
      mags.append(mag)
      lons.append(lon)
      lats.append(lat)
      eq_titles.append(eq_title)

  title = 'Global Earthquakes'
  fig = px.scatter_geo(lat=lats, lon=lons, size=mags, title=title,
          --snip--
          projection='natural earth',
❸         hover_name=eq_titles,
      )
  fig.show()
```

We first make a list called `eq_titles` to store the title of each earthquake ❶. The `'title'` section of the data contains a descriptive name of the magnitude and location of each earthquake, in addition to its longitude and latitude. We pull this information and assign it to the variable `eq_title` ❷, and then append it to the list `eq_titles`.

In the `px.scatter_geo()` call, we pass `eq_titles` to the `hover_name` argument ❸. Plotly will now add the information from the title of each earthquake to the hover text on each point. When you run this program, you should be able to hover over any marker, see a description of where that earthquake took place, and read its exact magnitude. An example of this information is shown in Figure 16-10.

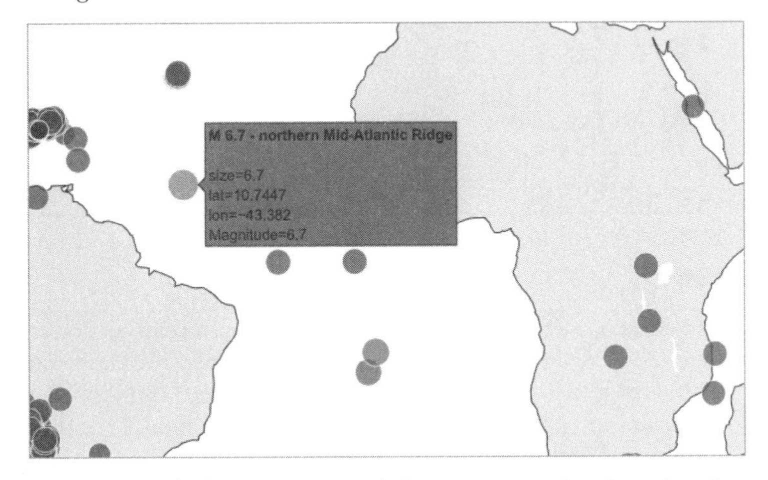

Figure 16-10: The hover text now includes a summary of each earthquake.

This is impressive! In less than 30 lines of code, we've created a visually appealing and meaningful map of global earthquake activity that also illustrates the geological structure of the planet. Plotly offers a wide range of ways you can customize the appearance and behavior of your visualizations. Using Plotly's many options, you can make charts and maps that show exactly what you want them to.

> **TRY IT YOURSELF**
>
> **16-6. Refactoring:** The loop that pulls data from `all_eq_dicts` uses variables for the magnitude, longitude, latitude, and title of each earthquake before appending these values to their appropriate lists. This approach was chosen for clarity in how to pull data from a GeoJSON file, but it's not necessary in your code. Instead of using these temporary variables, pull each value from `eq_dict` and append it to the appropriate list in one line. Doing so should shorten the body of this loop to just four lines.
>
> **16-7. Automated Title:** In this section, we used the generic title *Global Earthquakes*. Instead, you can use the title for the dataset in the metadata part of the GeoJSON file. Pull this value and assign it to the variable `title`.
>
> **16-8. Recent Earthquakes:** You can find online data files containing information about the most recent earthquakes over 1-hour, 1-day, 7-day, and 30-day periods. Go to *https://earthquake.usgs.gov/earthquakes/feed/v1.0/geojson.php* and you'll see a list of links to datasets for various time periods, focusing on earthquakes of different magnitudes. Download one of these datasets and create a visualization of the most recent earthquake activity.
>
> **16-9. World Fires:** In the resources for this chapter, you'll find a file called *world_fires_1_day.csv*. This file contains information about fires burning in different locations around the globe, including the latitude, longitude, and brightness of each fire. Using the data-processing work from the first part of this chapter and the mapping work from this section, make a map that shows which parts of the world are affected by fires.
>
> You can download more recent versions of this data at *https://earthdata.nasa.gov/earth-observation-data/near-real-time/firms/active-fire-data*. You can find links to the data in CSV format in the *SHP, KML, and TXT Files* section.

Summary

In this chapter, you learned how to work with real-world datasets. You processed CSV and GeoJSON files, and extracted the data you want to focus on. Using historical weather data, you learned more about working with Matplotlib, including how to use the `datetime` module and how to plot multiple data series on one chart. You plotted geographical data on a world map in Plotly, and learned to customize the style of the map.

As you gain experience working with CSV and JSON files, you'll be able to process almost any data you want to analyze. You can download most online datasets in either or both of these formats. By working with these formats, you'll be able to learn how to work with other data formats more easily as well.

In the next chapter, you'll write programs that automatically gather their own data from online sources, and then you'll create visualizations of that data. These are fun skills to have if you want to program as a hobby and are critical skills if you're interested in programming professionally.

17

WORKING WITH APIS

In this chapter, you'll learn how to write a self-contained program that generates a visualization based on data it retrieves. Your program will use an *application programming interface (API)* to automatically request specific information from a website and then use that information to generate a visualization. Because programs written like this will always use current data to generate a visualization, even when that data might be rapidly changing, the visualization will always be up to date.

Using an API

An API is a part of a website designed to interact with programs. Those programs use very specific URLs to request certain information. This kind of request is called an *API call*. The requested data will be returned in an easily processed format, such as JSON or CSV. Most apps that use external data sources, such as apps that integrate with social media sites, rely on API calls.

Git and GitHub

We'll base our visualization on information from GitHub (*https://github.com*), a site that allows programmers to collaborate on coding projects. We'll use GitHub's API to request information about Python projects on the site, and then generate an interactive visualization of the relative popularity of these projects using Plotly.

GitHub takes its name from Git, a distributed version control system. Git helps people manage their work on a project in a way that prevents changes made by one person from interfering with changes other people are making. When you implement a new feature in a project, Git tracks the changes you make to each file. When your new code works, you *commit* the changes you've made, and Git records the new state of your project. If you make a mistake and want to revert your changes, you can easily return to any previously working state. (To learn more about version control using Git, see Appendix D.) Projects on GitHub are stored in *repositories*, which contain everything associated with the project: its code, information on its collaborators, any issues or bug reports, and so on.

When users on GitHub like a project, they can "star" it to show their support and keep track of projects they might want to use. In this chapter, we'll write a program to automatically download information about the most-starred Python projects on GitHub, and then we'll create an informative visualization of these projects.

Requesting Data Using an API Call

GitHub's API lets you request a wide range of information through API calls. To see what an API call looks like, enter the following into your browser's address bar and press ENTER:

```
https://api.github.com/search/repositories?q=language:python+sort:stars
```

This call returns the number of Python projects currently hosted on GitHub, as well as information about the most popular Python repositories. Let's examine the call. The first part, `https://api.github.com/`, directs the request to the part of GitHub that responds to API calls. The next part, `search/repositories`, tells the API to conduct a search through all the repositories on GitHub.

The question mark after `repositories` signals that we're about to pass an argument. The `q` stands for *query*, and the equal sign (`=`) lets us begin specifying a query (`q=`). By using `language:python`, we indicate that we want information only on repositories that have Python as the primary language. The final part, `+sort:stars`, sorts the projects by the number of stars they've been given.

The following snippet shows the first few lines of the response:

```
{
❶   "total_count": 8961993,
❷   "incomplete_results": true,
❸   "items": [
      {
        "id": 54346799,
        "node_id": "MDEwOlJlcG9zaXRvcnk1NDM0Njc5OQ==",
        "name": "public-apis",
        "full_name": "public-apis/public-apis",
        --snip--
```

You can see from the response that this URL is not primarily intended to be entered by humans, because it's in a format that's meant to be processed by a program. GitHub found just under nine million Python projects as of this writing ❶. The value for `"incomplete_results"` is `true`, which tells us that GitHub didn't fully process the query ❷. GitHub limits how long each query can run, in order to keep the API responsive for all users. In this case it found some of the most popular Python repositories, but it didn't have time to find all of them; we'll fix that in a moment. The `"items"` returned are displayed in the list that follows, which contains details about the most popular Python projects on GitHub ❸.

Installing Requests

The *Requests* package allows a Python program to easily request information from a website and examine the response. Use pip to install Requests:

```
$ python -m pip install --user requests
```

If you use a command other than **python** to run programs or start a terminal session, such as **python3**, your command will look like this:

```
$ python3 -m pip install --user requests
```

Processing an API Response

Now we'll write a program to automatically issue an API call and process the results:

python_repos.py

```
import requests

# Make an API call and check the response.
❶ url = "https://api.github.com/search/repositories"
url += "?q=language:python+sort:stars+stars:>10000"

❷ headers = {"Accept": "application/vnd.github.v3+json"}
❸ r = requests.get(url, headers=headers)
❹ print(f"Status code: {r.status_code}")
```

```
# Convert the response object to a dictionary.
❺ response_dict = r.json()

# Process results.
print(response_dict.keys())
```

We first import the requests module. Then we assign the URL of the API call to the url variable ❶. This is a long URL, so we break it into two lines. The first line is the main part of the URL, and the second line is the query string. We've included one more condition to the original query string: stars:>10000, which tells GitHub to only look for Python repositories that have more than 10,000 stars. This should allow GitHub to return a complete, consistent set of results.

GitHub is currently on the third version of its API, so we define headers for the API call that ask explicitly to use this version of the API, and return the results in the JSON format ❷. Then we use requests to make the call to the API ❸. We call get() and pass it the URL and the header that we defined, and we assign the response object to the variable r.

The response object has an attribute called status_code, which tells us whether the request was successful. (A status code of 200 indicates a successful response.) We print the value of status_code so we can make sure the call went through successfully ❹. We asked the API to return the information in JSON format, so we use the json() method to convert the information to a Python dictionary ❺. We assign the resulting dictionary to response_dict.

Finally, we print the keys from response_dict and see the following output:

```
Status code: 200
dict_keys(['total_count', 'incomplete_results', 'items'])
```

Because the status code is 200, we know that the request was successful. The response dictionary contains only three keys: 'total_count', 'incomplete_results', and 'items'. Let's take a look inside the response dictionary.

Working with the Response Dictionary

With the information from the API call represented as a dictionary, we can work with the data stored there. Let's generate some output that summarizes the information. This is a good way to make sure we received the information we expected, and to start examining the information we're interested in:

python_repos.py

```
import requests

# Make an API call and store the response.
--snip--

# Convert the response object to a dictionary.
response_dict = r.json()
```

```
❶ print(f"Total repositories: {response_dict['total_count']}")
  print(f"Complete results: {not response_dict['incomplete_results']}")

  # Explore information about the repositories.
❷ repo_dicts = response_dict['items']
  print(f"Repositories returned: {len(repo_dicts)}")

  # Examine the first repository.
❸ repo_dict = repo_dicts[0]
❹ print(f"\nKeys: {len(repo_dict)}")
❺ for key in sorted(repo_dict.keys()):
      print(key)
```

We start exploring the response dictionary by printing the value associated with `'total_count'`, which represents the total number of Python repositories returned by this API call ❶. We also use the value associated with `'incomplete_results'`, so we'll know if GitHub was able to fully process the query. Rather than printing this value directly, we print its opposite: a value of `True` will indicate that we received a complete set of results.

The value associated with `'items'` is a list containing a number of dictionaries, each of which contains data about an individual Python repository. We assign this list of dictionaries to `repo_dicts` ❷. We then print the length of `repo_dicts` to see how many repositories we have information for.

To look closer at the information returned about each repository, we pull out the first item from `repo_dicts` and assign it to `repo_dict` ❸. We then print the number of keys in the dictionary to see how much information we have ❹. Finally, we print all the dictionary's keys to see what kind of information is included ❺.

The results give us a clearer picture of the actual data:

```
  Status code: 200
❶ Total repositories: 248
❷ Complete results: True
  Repositories returned: 30

❸ Keys: 78
  allow_forking
  archive_url
  archived
  --snip--
  url
  visiblity
  watchers
  watchers_count
```

At the time of this writing, there are only `248` Python repositories with over 10,000 stars ❶. We can see that GitHub was able to fully process the API call ❷. In this response, GitHub returned information about the first `30` repositories that match the conditions of our query. If we want more repositories, we can request additional pages of data.

GitHub's API returns a lot of information about each repository: there are 78 keys in `repo_dict` ❸. When you look through these keys, you'll get a sense of the kind of information you can extract about a project. (The only way to know what information is available through an API is to read the documentation or to examine the information through code, as we're doing here.)

Let's pull out the values for some of the keys in `repo_dict`:

python_repos.py

```
--snip--
# Examine the first repository.
repo_dict = repo_dicts[0]

print("\nSelected information about first repository:")
❶ print(f"Name: {repo_dict['name']}")
❷ print(f"Owner: {repo_dict['owner']['login']}")
❸ print(f"Stars: {repo_dict['stargazers_count']}")
print(f"Repository: {repo_dict['html_url']}")
❹ print(f"Created: {repo_dict['created_at']}")
❺ print(f"Updated: {repo_dict['updated_at']}")
print(f"Description: {repo_dict['description']}")
```

Here, we print the values for a number of keys from the first repository's dictionary. We start with the name of the project ❶. An entire dictionary represents the project's owner, so we use the key `owner` to access the dictionary representing the owner, and then use the key `login` to get the owner's login name ❷. Next, we print how many stars the project has earned ❸ and the URL for the project's GitHub repository. We then show when it was created ❹ and when it was last updated ❺. Finally, we print the repository's description.

The output should look something like this:

```
Status code: 200
Total repositories: 248
Complete results: True
Repositories returned: 30

Selected information about first repository:
Name: public-apis
Owner: public-apis
Stars: 191493
Repository: https://github.com/public-apis/public-apis
Created: 2016-03-20T23:49:42Z
Updated: 2022-05-12T06:37:11Z
Description: A collective list of free APIs
```

We can see that the most-starred Python project on GitHub as of this writing is *public-apis*. Its owner is an organization with the same name, and it has been starred by almost 200,000 GitHub users. We can see the URL for the project's repository, its creation date of March 2016, and that it was updated recently. Additionally, the description tells us that *public-apis* contains a list of free APIs that programmers might be interested in.

Summarizing the Top Repositories

When we make a visualization for this data, we'll want to include more than one repository. Let's write a loop to print selected information about each repository the API call returns so we can include them all in the visualization:

python_repos.py

```
--snip--
# Explore information about the repositories.
repo_dicts = response_dict['items']
print(f"Repositories returned: {len(repo_dicts)}")

❶ print("\nSelected information about each repository:")
❷ for repo_dict in repo_dicts:
    print(f"\nName: {repo_dict['name']}")
    print(f"Owner: {repo_dict['owner']['login']}")
    print(f"Stars: {repo_dict['stargazers_count']}")
    print(f"Repository: {repo_dict['html_url']}")
    print(f"Description: {repo_dict['description']}")
```

We first print an introductory message ❶. Then we loop through all the dictionaries in `repo_dicts` ❷. Inside the loop, we print the name of each project, its owner, how many stars it has, its URL on GitHub, and the project's description:

```
Status code: 200
Total repositories: 248
Complete results: True
Repositories returned: 30

Selected information about each repository:

Name: public-apis
Owner: public-apis
Stars: 191494
Repository: https://github.com/public-apis/public-apis
Description: A collective list of free APIs

Name: system-design-primer
Owner: donnemartin
Stars: 179952
Repository: https://github.com/donnemartin/system-design-primer
Description: Learn how to design large-scale systems. Prep for the system
  design interview.  Includes Anki flashcards.
--snip--

Name: PayloadsAllTheThings
Owner: swisskyrepo
Stars: 37227
Repository: https://github.com/swisskyrepo/PayloadsAllTheThings
Description: A list of useful payloads and bypass for Web Application Security
  and Pentest/CTF
```

Some interesting projects appear in these results, and it might be worth looking at a few. But don't spend too much time here, because we're about to create a visualization that will make the results much easier to read.

Monitoring API Rate Limits

Most APIs have *rate limits*, which means there's a limit to how many requests you can make in a certain amount of time. To see if you're approaching GitHub's limits, enter *https://api.github.com/rate_limit* into a web browser. You should see a response that begins like this:

```
{
  "resources": {
    --snip--
❶   "search": {
❷     "limit": 10,
❸     "remaining": 9,
❹     "reset": 1652338832,
      "used": 1,
      "resource": "search"
    },
    --snip--
```

The information we're interested in is the rate limit for the search API ❶. We see that the limit is 10 requests per minute ❷ and that we have 9 requests remaining for the current minute ❸. The value associated with the key "reset" represents the time in *Unix* or *epoch time* (the number of seconds since midnight on January 1, 1970) when our quota will reset ❹. If you reach your quota, you'll get a short response that lets you know you've reached the API limit. If you reach the limit, just wait until your quota resets.

NOTE *Many APIs require you to register and obtain an API key or access token to make API calls. As of this writing, GitHub has no such requirement, but if you obtain an access token, your limits will be much higher.*

Visualizing Repositories Using Plotly

Let's make a visualization using the data we've gathered to show the relative popularity of Python projects on GitHub. We'll make an interactive bar chart: the height of each bar will represent the number of stars the project has acquired, and you'll be able to click the bar's label to go to that project's home on GitHub.

Save a copy of the program we've been working on as *python_repos_visual.py*, then modify it so it reads as follows:

python_repos_visual.py

```
import requests
import plotly.express as px

# Make an API call and check the response.
url = "https://api.github.com/search/repositories"
url += "?q=language:python+sort:stars+stars:>10000"

headers = {"Accept": "application/vnd.github.v3+json"}
r = requests.get(url, headers=headers)
❶ print(f"Status code: {r.status_code}")

# Process overall results.
response_dict = r.json()
❷ print(f"Complete results: {not response_dict['incomplete_results']}")

# Process repository information.
repo_dicts = response_dict['items']
❸ repo_names, stars = [], []
for repo_dict in repo_dicts:
    repo_names.append(repo_dict['name'])
    stars.append(repo_dict['stargazers_count'])

# Make visualization.
❹ fig = px.bar(x=repo_names, y=stars)
fig.show()
```

We import Plotly Express and then make the API call as we have been doing. We continue to print the status of the API call response so we'll know if there is a problem ❶. When we process the overall results, we continue to print the message confirming that we got a complete set of results ❷. We remove the rest of the `print()` calls because we're no longer in the exploratory phase; we know we have the data we want.

We then create two empty lists ❸ to store the data we'll include in the initial chart. We'll need the name of each project to label the bars (`repo_names`) and the number of stars to determine the height of the bars (`stars`). In the loop, we append the name of each project and the number of stars it has to these lists.

We make the initial visualization with just two lines of code ❹. This is consistent with Plotly Express's philosophy that you should be able to see your visualization as quickly as possible before refining its appearance. Here we use the `px.bar()` function to create a bar chart. We pass the list `repo_names` as the `x` argument and `stars` as the `y` argument.

Figure 17-1 shows the resulting chart. We can see that the first few projects are significantly more popular than the rest, but all of them are important projects in the Python ecosystem.

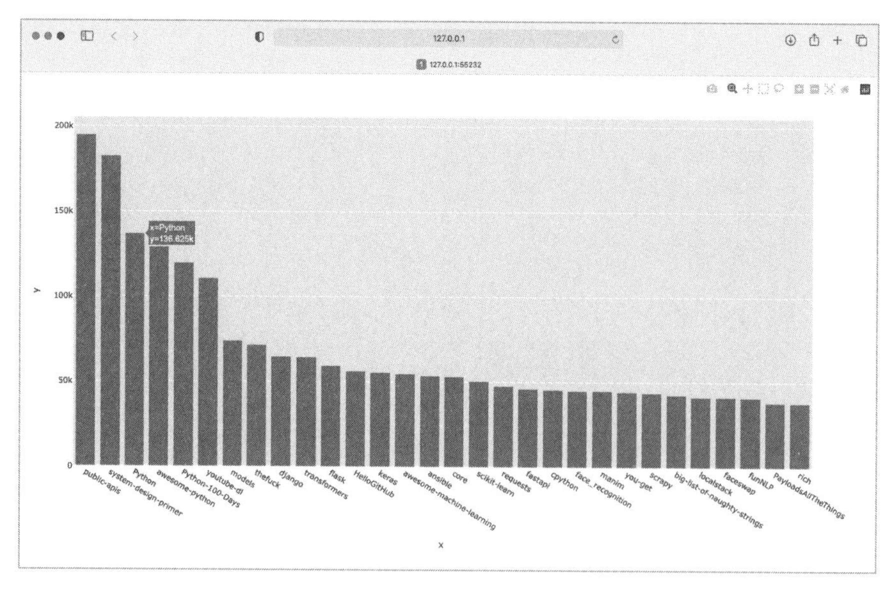

Figure 17-1: The most-starred Python projects on GitHub

Styling the Chart

Plotly supports a number of ways to style and customize the plots, once you know the information in the plot is correct. We'll make some changes in the initial `px.bar()` call and then make some further adjustments to the `fig` object after it's been created.

We'll start styling the chart by adding a title and labels for each axis:

python_repos_visual.py

```
--snip--
# Make visualization.
title = "Most-Starred Python Projects on GitHub"
labels = {'x': 'Repository', 'y': 'Stars'}
fig = px.bar(x=repo_names, y=stars, title=title, labels=labels)

❶ fig.update_layout(title_font_size=28, xaxis_title_font_size=20,
        yaxis_title_font_size=20)

fig.show()
```

We first add a title and labels for each axis, as we did in Chapters 15 and 16. We then use the `fig.update_layout()` method to modify specific elements of the chart ❶. Plotly uses a convention where aspects of a chart element are connected by underscores. As you become familiar with Plotly's documentation, you'll start to see consistent patterns in how different elements of a chart are named and modified. Here we set the title font size to `28` and the font size for each axis title to `20`. The result is shown in Figure 17-2.

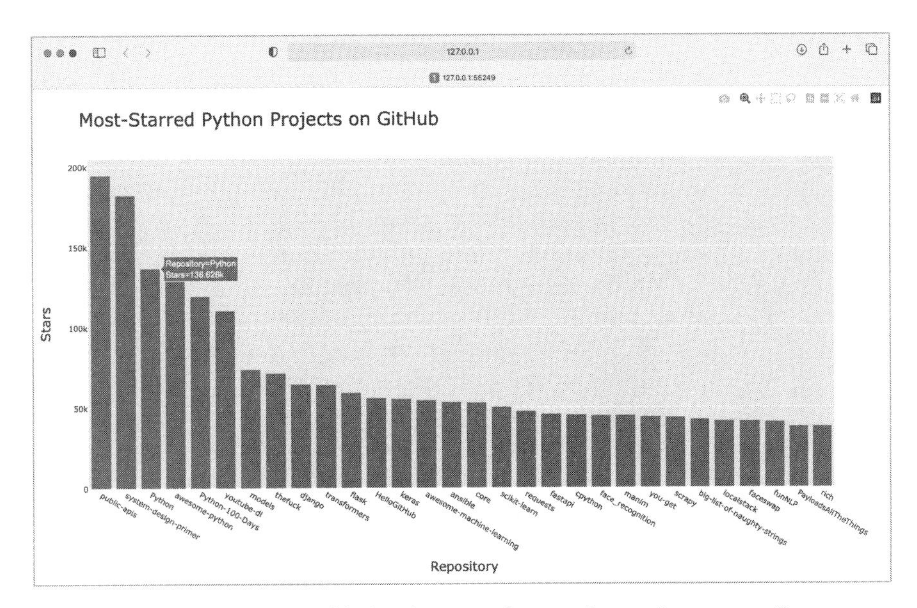

Figure 17-2: A title has been added to the main chart, and to each axis as well.

Adding Custom Tooltips

In Plotly, you can hover the cursor over an individual bar to show the information the bar represents. This is commonly called a *tooltip*, and in this case, it currently shows the number of stars a project has. Let's create a custom tooltip to show each project's description as well as the project's owner.

We need to pull some additional data to generate the tooltips:

python_repos_visual.py

```
--snip--
# Process repository information.
repo_dicts = response_dict['items']
❶ repo_names, stars, hover_texts = [], [], []
for repo_dict in repo_dicts:
    repo_names.append(repo_dict['name'])
    stars.append(repo_dict['stargazers_count'])

    # Build hover texts.
❷     owner = repo_dict['owner']['login']
    description = repo_dict['description']
❸     hover_text = f"{owner}<br />{description}"
    hover_texts.append(hover_text)

# Make visualization.
title = "Most-Starred Python Projects on GitHub"
labels = {'x': 'Repository', 'y': 'Stars'}
❹ fig = px.bar(x=repo_names, y=stars, title=title, labels=labels,
        hover_name=hover_texts)
```

```
fig.update_layout(title_font_size=28, xaxis_title_font_size=20,
        yaxis_title_font_size=20)

fig.show()
```

We first define a new empty list, `hover_texts`, to hold the text we want to display for each project ❶. In the loop where we process the data, we pull the owner and the description for each project ❷. Plotly allows you to use HTML code within text elements, so we generate a string for the label with a line break (`<br />`) between the project owner's username and the description ❸. We then append this label to the list `hover_texts`.

In the `px.bar()` call, we add the `hover_name` argument and pass it `hover_texts` ❹. This is the same approach we used to customize the label for each dot in the map of global earthquake activity. As Plotly creates each bar, it will pull labels from this list and only display them when the viewer hovers over a bar. Figure 17-3 shows one of these custom tooltips.

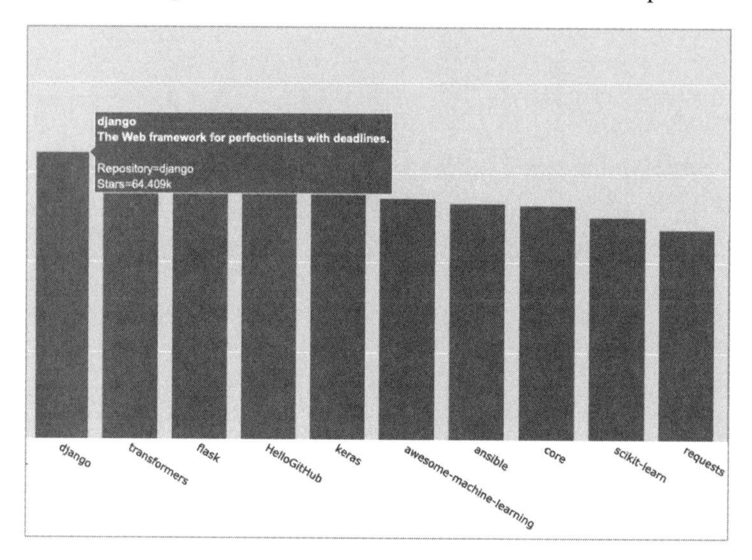

Figure 17-3: Hovering over a bar shows the project's owner and description.

Adding Clickable Links

Because Plotly allows you to use HTML on text elements, we can easily add links to a chart. Let's use the *x*-axis labels as a way to let the viewer visit any project's home page on GitHub. We need to pull the URLs from the data and use them when generating the *x*-axis labels:

python_repos_visual.py

```
--snip--
# Process repository information.
repo_dicts = response_dict['items']
❶ repo_links, stars, hover_texts = [], [], []
for repo_dict in repo_dicts:
    # Turn repo names into active links.
    repo_name = repo_dict['name']
❷   repo_url = repo_dict['html_url']
```

```
❸       repo_link = f"<a href='{repo_url}'>{repo_name}</a>"
        repo_links.append(repo_link)

        stars.append(repo_dict['stargazers_count'])
        --snip--

# Make visualization.
title = "Most-Starred Python Projects on GitHub"
labels = {'x': 'Repository', 'y': 'Stars'}
fig = px.bar(x=repo_links, y=stars, title=title, labels=labels,
        hover_name=hover_texts)

fig.update_layout(title_font_size=28, xaxis_title_font_size=20,
        yaxis_title_font_size=20)

fig.show()
```

We update the name of the list we're creating from `repo_names` to `repo_links` to more accurately communicate the kind of information we're putting together for the chart ❶. We then pull the URL for the project from `repo_dict` and assign it to the temporary variable `repo_url` ❷. Next, we generate a link to the project ❸. We use the HTML anchor tag, which has the form `<a href='URL'>link text</a>`, to generate the link. We then append this link to `repo_links`.

When we call `px.bar()`, we use `repo_links` for the *x*-values in the chart. The result looks the same as before, but now the viewer can click any of the project names at the bottom of the chart to visit that project's home page on GitHub. Now we have an interactive, informative visualization of data retrieved through an API!

Customizing Marker Colors

Once a chart has been created, almost any aspect of the chart can be customized through an update method. We've used the `update_layout()` method previously. Another method, `update_traces()`, can be used to customize the data that's represented on a chart.

Let's change the bars to a darker blue, with some transparency:

```
--snip--
fig.update_layout(title_font_size=28, xaxis_title_font_size=20,
        yaxis_title_font_size=20)

fig.update_traces(marker_color='SteelBlue', marker_opacity=0.6)

fig.show()
```

In Plotly, a *trace* refers to a collection of data on a chart. The `update_traces()` method can take a number of different arguments; any argument that starts with `marker_` affects the markers on the chart. Here we set each marker's color to `'SteelBlue'`; any named CSS color will work here. We also set the opacity of each marker to `0.6`. An opacity of 1.0 will be entirely opaque, and an opacity of 0 will be entirely invisible.

More About Plotly and the GitHub API

Plotly's documentation is extensive and well organized; however, it can be hard to know where to start reading. A good place to start is with the article "Plotly Express in Python," at *https://plotly.com/python/plotly-express*. This is an overview of all the plots you can make with Plotly Express, and you can find links to longer articles about each individual chart type.

If you want to understand how to customize Plotly charts better, the article "Styling Plotly Express Figures in Python" will expand on what you've seen in Chapters 15–17. You can find this article at *https://plotly.com/python/styling-plotly-express*.

For more about the GitHub API, refer to its documentation at *https://docs.github.com/en/rest*. Here you'll learn how to pull a wide variety of information from GitHub. To expand on what you saw in this project, look for the Search section of the reference in the sidebar. If you have a GitHub account, you can work with your own data as well as the publicly available data from other users' repositories.

The Hacker News API

To explore how to use API calls on other sites, let's take a quick look at Hacker News (*https://news.ycombinator.com*). On Hacker News, people share articles about programming and technology and engage in lively discussions about those articles. The Hacker News API provides access to data about all submissions and comments on the site, and you can use the API without having to register for a key.

The following call returns information about the current top article as of this writing:

```
https://hacker-news.firebaseio.com/v0/item/31353677.json
```

When you enter this URL in a browser, you'll see that the text on the page is enclosed by braces, meaning it's a dictionary. But the response is difficult to examine without some better formatting. Let's run this URL through the `json.dumps()` method, like we did in the earthquake project in Chapter 16, so we can explore the kind of information that's returned about an article:

hn_article.py

```
import requests
import json

# Make an API call, and store the response.
url = "https://hacker-news.firebaseio.com/v0/item/31353677.json"
r = requests.get(url)
print(f"Status code: {r.status_code}")

# Explore the structure of the data.
response_dict = r.json()
response_string = json.dumps(response_dict, indent=4)
❶ print(response_string)
```

Everything in this program should look familiar, because we've used it all in the previous two chapters. The main difference here is that we can print the formatted response string ❶ instead of writing it to a file, because the output is not particularly long.

The output is a dictionary of information about the article with the ID 31353677:

```
{
    "by": "sohkamyung",
❶   "descendants": 302,
    "id": 31353677,
❷   "kids": [
        31354987,
        31354235,
        --snip--
    ],
    "score": 785,
    "time": 1652361401,
❸   "title": "Astronomers reveal first image of the black hole
        at the heart of our galaxy",
    "type": "story",
❹   "url": "https://public.nrao.edu/news/.../"
}
```

The dictionary contains a number of keys we can work with. The key "descendants" tells us the number of comments the article has received ❶. The key "kids" provides the IDs of all comments made directly in response to this submission ❷. Each of these comments might have comments of their own as well, so the number of descendants a submission has is usually greater than its number of kids. We can see the title of the article being discussed ❸ and a URL for the article being discussed as well ❹.

The following URL returns a simple list of all the IDs of the current top articles on Hacker News:

```
https://hacker-news.firebaseio.com/v0/topstories.json
```

We can use this call to find out which articles are on the home page right now, and then generate a series of API calls similar to the one we just examined. With this approach, we can print a summary of all the articles on the front page of Hacker News at the moment:

hn_submissions.py

```
from operator import itemgetter

import requests

# Make an API call and check the response.
❶ url = "https://hacker-news.firebaseio.com/v0/topstories.json"
r = requests.get(url)
print(f"Status code: {r.status_code}")

# Process information about each submission.
❷ submission_ids = r.json()
```

```
❸ submission_dicts = []
  for submission_id in submission_ids[:30]:
      # Make a new API call for each submission.
❹     url = f"https://hacker-news.firebaseio.com/v0/item/{submission_id}.json"
      r = requests.get(url)
      print(f"id: {submission_id}\tstatus: {r.status_code}")
      response_dict = r.json()

      # Build a dictionary for each article.
❺     submission_dict = {
          'title': response_dict['title'],
          'hn_link': f"https://news.ycombinator.com/item?id={submission_id}",
          'comments': response_dict['descendants'],
      }
❻     submission_dicts.append(submission_dict)

❼ submission_dicts = sorted(submission_dicts, key=itemgetter('comments'),
                              reverse=True)

❽ for submission_dict in submission_dicts:
      print(f"\nTitle: {submission_dict['title']}")
      print(f"Discussion link: {submission_dict['hn_link']}")
      print(f"Comments: {submission_dict['comments']}")
```

First, we make an API call and print the status of the response ❶. This API call returns a list containing the IDs of up to 500 of the most popular articles on Hacker News at the time the call is issued. We then convert the response object to a Python list ❷, which we assign to `submission_ids`. We'll use these IDs to build a set of dictionaries, each of which contains information about one of the current submissions.

We set up an empty list called `submission_dicts` to store these dictionaries ❸. We then loop through the IDs of the top 30 submissions. We make a new API call for each submission by generating a URL that includes the current value of `submission_id` ❹. We print the status of each request along with its ID, so we can see whether it's successful.

Next, we create a dictionary for the submission currently being processed ❺. We store the title of the submission, a link to the discussion page for that item, and the number of comments the article has received so far. Then we append each `submission_dict` to the list `submission_dicts` ❻.

Each submission on Hacker News is ranked according to an overall score based on a number of factors, including how many times it's been voted on, how many comments it's received, and how recent the submission is. We want to sort the list of dictionaries by the number of comments. To do this, we use a function called `itemgetter()` ❼, which comes from the `operator` module. We pass this function the key `'comments'`, and it pulls the value associated with that key from each dictionary in the list. The `sorted()` function then uses this value as its basis for sorting the list. We sort the list in reverse order, to place the most-commented stories first.

Once the list is sorted, we loop through the list ❽ and print out three pieces of information about each of the top submissions: the title, a link to the discussion page, and the number of comments the submission currently has:

```
Status code: 200
id: 31390506    status: 200
id: 31389893    status: 200
id: 31390742    status: 200
--snip--

Title: Fly.io: The reclaimer of Heroku's magic
Discussion link: https://news.ycombinator.com/item?id=31390506
Comments: 134

Title: The weird Hewlett Packard FreeDOS option
Discussion link: https://news.ycombinator.com/item?id=31389893
Comments: 64

Title: Modern JavaScript Tutorial
Discussion link: https://news.ycombinator.com/item?id=31390742
Comments: 20
--snip--
```

You would use a similar process to access and analyze information with any API. With this data, you could make a visualization showing which submissions have inspired the most active recent discussions. This is also the basis for apps that provide a customized reading experience for sites like Hacker News. To learn more about what kind of information you can access through the Hacker News API, visit the documentation page at *https://github.com/HackerNews/API*.

NOTE *Hacker News sometimes allows companies it supports to make special hiring posts, and comments are disabled on these posts. If you run this program while one of these posts is present, you'll get a* `KeyError`*. If this causes an issue, you can wrap the code that builds* `submission_dict` *in a* `try-except` *block and skip over these posts.*

TRY IT YOURSELF

17-1. Other Languages: Modify the API call in *python_repos.py* so it generates a chart showing the most popular projects in other languages. Try languages such as *JavaScript, Ruby, C, Java, Perl, Haskell,* and *Go.*

17-2. Active Discussions: Using the data from *hn_submissions.py*, make a bar chart showing the most active discussions currently happening on Hacker News. The height of each bar should correspond to the number of comments each submission has. The label for each bar should include the submission's title and act as a link to the discussion page for that submission. If you get a `KeyError` when creating a chart, use a `try-except` block to skip over the promotional posts.

(continued)

17-3. Testing python_repos.py: In *python_repos.py*, we printed the value of `status_code` to make sure the API call was successful. Write a program called *test_python_repos.py* that uses `pytest` to assert that the value of `status_code` is 200. Figure out some other assertions you can make: for example, that the number of items returned is expected and that the total number of repositories is greater than a certain amount.

17-4. Further Exploration: Visit the documentation for Plotly and either the GitHub API or the Hacker News API. Use some of the information you find there to either customize the style of the plots we've already made or pull some different information and create your own visualizations. If you're curious about exploring other APIs, take a look at the APIs mentioned in the GitHub repository at *https://github.com/public-apis*.

Summary

In this chapter, you learned how to use APIs to write self-contained programs that automatically gather the data they need and use that data to create a visualization. You used the GitHub API to explore the most-starred Python projects on GitHub, and you also looked briefly at the Hacker News API. You learned how to use the Requests package to automatically issue an API call and how to process the results of that call. We also introduced some Plotly settings that further customize the appearance of the charts you generate.

In the next chapter, you'll use Django to build a web application as your final project.

18

GETTING STARTED WITH DJANGO

As the internet has evolved, the line between websites and mobile apps has blurred. Websites and apps both help users interact with data in a variety of ways. Fortunately, you can use Django to build a single project that serves a dynamic website as well as a set of mobile apps. *Django* is Python's most popular *web framework,* a set of tools designed for building interactive web applications. In this chapter, you'll learn how to use Django to build a project called Learning Log, an online journal system that lets you keep track of information you've learned about different topics.

We'll write a specification for this project, and then define models for the data the app will work with. We'll use Django's admin system to enter some initial data, and then write views and templates so Django can build the site's pages.

Django can respond to page requests and make it easier to read and write to a database, manage users, and much more. In Chapters 19 and 20, you'll refine the Learning Log project, and then deploy it to a live server so you (and everyone else in the world) can use it.

Setting Up a Project

When starting work on something as significant as a web app, you first need to describe the project's goals in a specification, or *spec*. Once you have a clear set of goals, you can start to identify manageable tasks to achieve those goals.

In this section, we'll write a spec for Learning Log and start working on the first phase of the project. This will involve setting up a virtual environment and building out the initial aspects of a Django project.

Writing a Spec

A full spec details the project goals, describes the project's functionality, and discusses its appearance and user interface. Like any good project or business plan, a spec should keep you focused and help keep your project on track. We won't write a full project spec here, but we'll lay out a few clear goals to keep the development process focused. Here's the spec we'll use:

> We'll write a web app called Learning Log that allows users to log the topics they're interested in and make journal entries as they learn about each topic. The Learning Log home page will describe the site and invite users to either register or log in. Once logged in, a user can create new topics, add new entries, and read and edit existing entries.

When you're researching a new topic, maintaining a journal of what you've learned can help you keep track of new information and information you've already found. This is especially true when studying technical subjects. A good app, like the one we'll be creating, can help make this process more efficient.

Creating a Virtual Environment

To work with Django, we'll first set up a virtual environment. A *virtual environment* is a place on your system where you can install packages and isolate them from all other Python packages. Separating one project's libraries from other projects is beneficial and will be necessary when we deploy Learning Log to a server in Chapter 20.

Create a new directory for your project called *learning_log*, switch to that directory in a terminal, and enter the following code to create a virtual environment:

```
learning_log$ python -m venv ll_env
learning_log$
```

Here we're running the venv virtual environment module and using it to create an environment named *ll_env* (note that this name starts with two lowercase *Ls*, not two ones). If you use a command such as python3 when running programs or installing packages, make sure to use that command here.

Activating the Virtual Environment

Now we need to activate the virtual environment, using the following command:

```
learning_log$ source ll_env/bin/activate
(ll_env)learning_log$
```

This command runs the script *activate* in *ll_env/bin/*. When the environment is active, you'll see the name of the environment in parentheses. This indicates that you can install new packages to the environment and use packages that have already been installed. Packages you install in *ll_env* will not be available when the environment is inactive.

NOTE *If you're using Windows, use the command* **ll_env\Scripts\activate** *(without the word* source*) to activate the virtual environment. If you're using PowerShell, you might need to capitalize* Activate.

To stop using a virtual environment, enter **deactivate**:

```
(ll_env)learning_log$ deactivate
learning_log$
```

The environment will also become inactive when you close the terminal it's running in.

Installing Django

With the virtual environment activated, enter the following to update pip and install Django:

```
(ll_env)learning_log$ pip install --upgrade pip
(ll_env)learning_log$ pip install django
Collecting django
--snip--
Installing collected packages: sqlparse, asgiref, django
Successfully installed asgiref-3.5.2 django-4.1 sqlparse-0.4.2
(ll_env)learning_log$
```

Because it downloads resources from a variety of sources, pip is upgraded fairly often. It's a good idea to upgrade pip whenever you make a new virtual environment.

We're working in a virtual environment now, so the command to install Django is the same on all systems. There's no need to use longer commands, such as python -m pip install *package_name*, or to include the --user flag. Keep in mind that Django will be available only when the *ll_env* environment is active.

NOTE *Django releases a new version about every eight months, so you may see a newer version when you install Django. This project will most likely work as it's written here, even on newer versions of Django. If you want to make sure to use the same version of Django you see here, use the command* **`pip install django==4.1.*`**. *This will install the latest release of Django 4.1. If you have any issues related to the version you're using, see the online resources for this book at* https://ehmatthes.github.io/pcc_3e.

Creating a Project in Django

Without leaving the active virtual environment (remember to look for *ll_env* in parentheses in the terminal prompt), enter the following commands to create a new project:

```
❶ (ll_env)learning_log$ django-admin startproject ll_project .
❷ (ll_env)learning_log$ ls
  ll_env ll_project manage.py
❸ (ll_env)learning_log$ ls ll_project
  __init__.py asgi.py settings.py urls.py wsgi.py
```

The `startproject` command ❶ tells Django to set up a new project called *ll_project*. The dot (`.`) at the end of the command creates the new project with a directory structure that will make it easy to deploy the app to a server when we're finished developing it.

NOTE *Don't forget this dot, or you might run into some configuration issues when you deploy the app. If you forget the dot, delete the files and folders that were created (except* ll_env*) and run the command again.*

Running the `ls` command (`dir` on Windows) ❷ shows that Django has created a new directory called *ll_project*. It also created a *manage.py* file, which is a short program that takes in commands and feeds them to the relevant part of Django. We'll use these commands to manage tasks, such as working with databases and running servers.

The *ll_project* directory contains four files ❸; the most important are *settings.py*, *urls.py*, and *wsgi.py*. The *settings.py* file controls how Django interacts with your system and manages your project. We'll modify a few of these settings and add some settings of our own as the project evolves. The *urls.py* file tells Django which pages to build in response to browser requests. The *wsgi.py* file helps Django serve the files it creates. The filename is an acronym for "web server gateway interface."

Creating the Database

Django stores most of the information for a project in a database, so next we need to create a database that Django can work with. Enter the following command (still in an active environment):

```
  (ll_env)learning_log$ python manage.py migrate
❶ Operations to perform:
```

```
  Apply all migrations: admin, auth, contenttypes, sessions
Running migrations:
  Applying contenttypes.0001_initial... OK
  Applying auth.0001_initial... OK
  --snip--
  Applying sessions.0001_initial... OK
❷ (ll_env)learning_log$ ls
db.sqlite3 ll_env ll_project manage.py
```

Anytime we modify a database, we say we're *migrating* the database. Issuing the `migrate` command for the first time tells Django to make sure the database matches the current state of the project. The first time we run this command in a new project using SQLite (more about SQLite in a moment), Django will create a new database for us. Here, Django reports that it will prepare the database to store information it needs to handle administrative and authentication tasks ❶.

Running the `ls` command shows that Django created another file called *db.sqlite3* ❷. *SQLite* is a database that runs off a single file; it's ideal for writing simple apps because you won't have to pay much attention to managing the database.

NOTE *In an active virtual environment, use the command* `python` *to run* `manage.py` *commands, even if you use something different, like* `python3`*, to run other programs. In a virtual environment, the command* `python` *refers to the version of Python that was used to create the virtual environment.*

Viewing the Project

Let's make sure that Django has set up the project properly. Enter the **`runserver`** command to view the project in its current state:

```
(ll_env)learning_log$ python manage.py runserver
Watching for file changes with StatReloader
Performing system checks...

❶ System check identified no issues (0 silenced).
May 19, 2022 - 21:52:35
❷ Django version 4.1, using settings 'll_project.settings'
❸ Starting development server at http://127.0.0.1:8000/
Quit the server with CONTROL-C.
```

Django should start a server called the *development server*, so you can view the project on your system to see how well it works. When you request a page by entering a URL in a browser, the Django server responds to that request by building the appropriate page and sending it to the browser.

Django first checks to make sure the project is set up properly ❶; it then reports the version of Django in use and the name of the settings file in use ❷. Finally, it reports the URL where the project is being served ❸. The URL *http://127.0.0.1:8000/* indicates that the project is listening for requests on port 8000 on your computer, which is called a localhost. The term *localhost* refers to a server that only processes requests on your system; it doesn't allow anyone else to see the pages you're developing.

Open a web browser and enter the URL *http://localhost:8000/*, or *http://127.0.0.1:8000/* if the first one doesn't work. You should see something like Figure 18-1: a page that Django creates to let you know everything is working properly so far. Keep the server running for now, but when you want to stop the server, press CTRL-C in the terminal where the `runserver` command was issued.

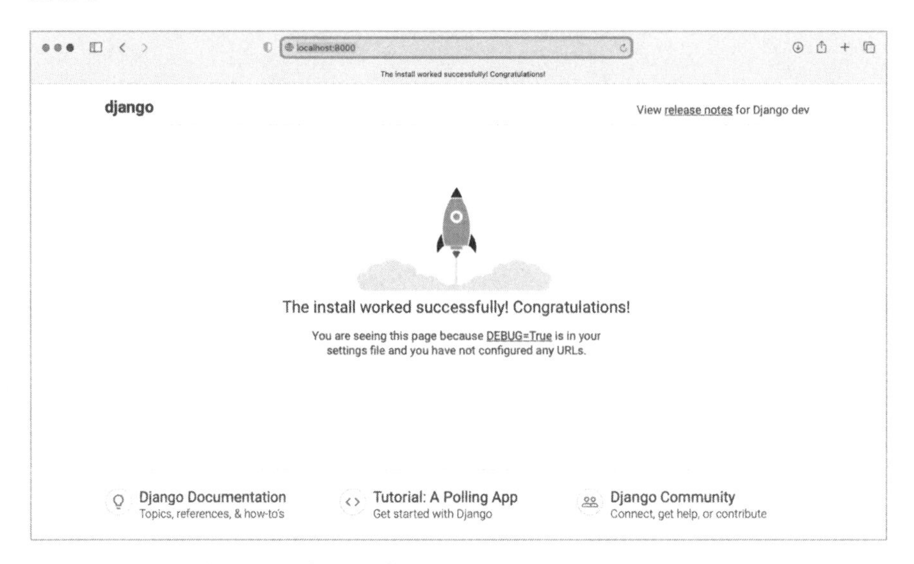

Figure 18-1: Everything is working so far.

NOTE *If you receive the error message "That port is already in use," tell Django to use a different port by entering* **`python manage.py runserver 8001`** *and then cycling through higher numbers until you find an open port.*

TRY IT YOURSELF

18-1. New Projects: To get a better idea of what Django does, build a couple empty projects and look at what Django creates. Make a new folder with a simple name, like *tik_gram* or *insta_tok* (outside of your *learning_log* directory), navigate to that folder in a terminal, and create a virtual environment. Install Django and run the command **`django-admin startproject tg_project .`** (making sure to include the dot at the end of the command).

Look at the files and folders this command creates, and compare them to Learning Log. Do this a few times, until you're familiar with what Django creates when starting a new project. Then delete the project directories if you wish.

Starting an App

A Django *project* is organized as a group of individual *apps* that work together to make the project work as a whole. For now, we'll create one app to do most of our project's work. We'll add another app in Chapter 19 to manage user accounts.

You should leave the development server running in the terminal window you opened earlier. Open a new terminal window (or tab) and navigate to the directory that contains *manage.py*. Activate the virtual environment, and then run the `startapp` command:

```
learning_log$ source ll_env/bin/activate
(ll_env)learning_log$ python manage.py startapp learning_logs
❶ (ll_env)learning_log$ ls
db.sqlite3 learning_logs ll_env ll_project manage.py
❷ (ll_env)learning_log$ ls learning_logs/
__init__.py admin.py apps.py migrations models.py tests.py views.py
```

The command `startapp` *appname* tells Django to create the infrastructure needed to build an app. When you look in the project directory now, you'll see a new folder called *learning_logs* ❶. Use the `ls` command to see what Django has created ❷. The most important files are *models.py*, *admin.py*, and *views.py*. We'll use *models.py* to define the data we want to manage in our app. We'll look at *admin.py* and *views.py* a little later.

Defining Models

Let's think about our data for a moment. Each user will need to create a number of topics in their learning log. Each entry they make will be tied to a topic, and these entries will be displayed as text. We'll also need to store the timestamp of each entry so we can show users when they made each one.

Open the file *models.py* and look at its existing content:

models.py

```
from django.db import models

# Create your models here.
```

A module called `models` is being imported, and we're being invited to create models of our own. A *model* tells Django how to work with the data that will be stored in the app. A model is a class; it has attributes and methods, just like every class we've discussed. Here's the model for the topics users will store:

```
from django.db import models

class Topic(models.Model):
    """A topic the user is learning about."""
❶     text = models.CharField(max_length=200)
❷     date_added = models.DateTimeField(auto_now_add=True)
```

```
❸     def __str__(self):
          """Return a string representation of the model."""
          return self.text
```

We've created a class called `Topic`, which inherits from `Model`—a parent class included in Django that defines a model's basic functionality. We add two attributes to the `Topic` class: `text` and `date_added`.

The `text` attribute is a `CharField`, a piece of data that's made up of characters or text ❶. You use `CharField` when you want to store a small amount of text, such as a name, a title, or a city. When we define a `CharField` attribute, we have to tell Django how much space it should reserve in the database. Here we give it a `max_length` of `200` characters, which should be enough to hold most topic names.

The `date_added` attribute is a `DateTimeField`, a piece of data that will record a date and time ❷. We pass the argument `auto_now_add=True`, which tells Django to automatically set this attribute to the current date and time whenever the user creates a new topic.

It's a good idea to tell Django how you want it to represent an instance of a model. If a model has a `__str__()` method, Django calls that method whenever it needs to generate output referring to an instance of that model. Here we've written a `__str__()` method that returns the value assigned to the `text` attribute ❸.

To see the different kinds of fields you can use in a model, see the "Model Field Reference" page at *https://docs.djangoproject.com/en/4.1/ref/models/fields*. You won't need all the information right now, but it will be extremely useful when you're developing your own Django projects.

Activating Models

To use our models, we have to tell Django to include our app in the overall project. Open *settings.py* (in the *ll_project* directory); you'll see a section that tells Django which apps are installed in the project:

settings.py

```
--snip--
INSTALLED_APPS = [
    'django.contrib.admin',
    'django.contrib.auth',
    'django.contrib.contenttypes',
    'django.contrib.sessions',
    'django.contrib.messages',
    'django.contrib.staticfiles',
]
--snip--
```

Add our app to this list by modifying `INSTALLED_APPS` so it looks like this:

```
--snip--
INSTALLED_APPS = [
    # My apps.
    'learning_logs',
```

```
    # Default django apps.
    'django.contrib.admin',
    --snip--
]
--snip--
```

Grouping apps together in a project helps keep track of them as the project grows to include more apps. Here we start a section called `My apps`, which includes only `'learning_logs'` for now. It's important to place your own apps before the default apps, in case you need to override any behavior of the default apps with your own custom behavior.

Next, we need to tell Django to modify the database so it can store information related to the model `Topic`. From the terminal, run the following command:

```
(ll_env)learning_log$ python manage.py makemigrations learning_logs
Migrations for 'learning_logs':
  learning_logs/migrations/0001_initial.py
    - Create model Topic
(ll_env)learning_log$
```

The command `makemigrations` tells Django to figure out how to modify the database so it can store the data associated with any new models we've defined. The output here shows that Django has created a migration file called *0001_initial.py*. This migration will create a table for the model `Topic` in the database.

Now we'll apply this migration and have Django modify the database for us:

```
(ll_env)learning_log$ python manage.py migrate
Operations to perform:
  Apply all migrations: admin, auth, contenttypes, learning_logs, sessions
Running migrations:
  Applying learning_logs.0001_initial... OK
```

Most of the output from this command is identical to the output from the first time we issued the `migrate` command. We need to check the last line in this output, where Django confirms that the migration for `learning_logs` worked `OK`.

Whenever we want to modify the data that Learning Log manages, we'll follow these three steps: modify *models.py*, call `makemigrations` on `learning_logs`, and tell Django to `migrate` the project.

The Django Admin Site

Django makes it easy to work with your models through its admin site. Django's *admin site* is only meant to be used by the site's administrators; it's not meant for regular users. In this section, we'll set up the admin site and use it to add some topics through the `Topic` model.

Setting Up a Superuser

Django allows you to create a *superuser*, a user who has all privileges available on the site. A user's *privileges* control the actions they can take. The most restrictive privilege settings allow a user to only read public information on the site. Registered users typically have the privilege of reading their own private data and some selected information available only to members. To effectively administer a project, the site owner usually needs access to all information stored on the site. A good administrator is careful with their users' sensitive information, because users put a lot of trust into the apps they access.

To create a superuser in Django, enter the following command and respond to the prompts:

```
(ll_env)learning_log$ python manage.py createsuperuser
❶ Username (leave blank to use 'eric'): ll_admin
❷ Email address:
❸ Password:
Password (again):
Superuser created successfully.
(ll_env)learning_log$
```

When you issue the command `createsuperuser`, Django prompts you to enter a username for the superuser ❶. Here I'm using `ll_admin`, but you can enter any username you want. You can enter an email address or just leave this field blank ❷. You'll need to enter your password twice ❸.

NOTE *Some sensitive information can be hidden from a site's administrators. For example, Django doesn't store the password you enter; instead, it stores a string derived from the password, called a* hash. *Each time you enter your password, Django hashes your entry and compares it to the stored hash. If the two hashes match, you're authenticated. By requiring hashes to match, Django ensures that if an attacker gains access to a site's database, they'll be able to read the stored hashes but not the passwords. When a site is set up properly, it's almost impossible to get the original passwords from the hashes.*

Registering a Model with the Admin Site

Django includes some models in the admin site automatically, such as `User` and `Group`, but the models we create need to be added manually.

When we started the `learning_logs` app, Django created an *admin.py* file in the same directory as *models.py*. Open the *admin.py* file:

admin.py

```
from django.contrib import admin

# Register your models here.
```

To register Topic with the admin site, enter the following:

```
from django.contrib import admin

from .models import Topic

admin.site.register(Topic)
```

This code first imports the model we want to register, `Topic`. The dot in front of `models` tells Django to look for *models.py* in the same directory as *admin.py*. The code `admin.site.register()` tells Django to manage our model through the admin site.

Now use the superuser account to access the admin site. Go to *http://localhost:8000/admin/* and enter the username and password for the superuser you just created. You should see a screen similar to the one shown in Figure 18-2. This page allows you to add new users and groups, and change existing ones. You can also work with data related to the `Topic` model that we just defined.

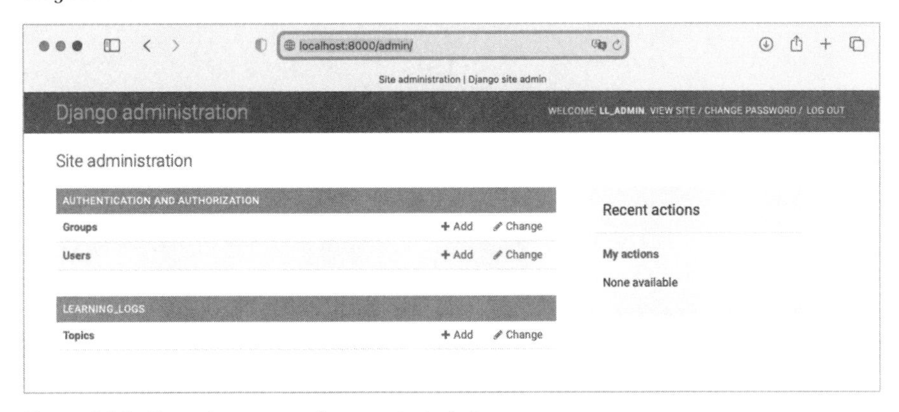

Figure 18-2: The admin site with `Topic` *included*

NOTE *If you see a message in your browser that the web page is not available, make sure you still have the Django server running in a terminal window. If you don't, activate a virtual environment and reissue the command* **`python manage.py runserver`**. *If you're having trouble viewing your project at any point in the development process, closing any open terminals and reissuing the* **`runserver`** *command is a good first troubleshooting step.*

Adding Topics

Now that `Topic` has been registered with the admin site, let's add our first topic. Click **Topics** to go to the Topics page, which is mostly empty, because we have no topics to manage yet. Click **Add Topic**, and a form for adding a new topic appears. Enter **`Chess`** in the first box and click **Save**. You'll be sent back to the Topics admin page, and you'll see the topic you just created.

Let's create a second topic so we'll have more data to work with. Click **Add Topic** again, and enter **`Rock Climbing`**. Click **Save**, and you'll be sent back to the main Topics page again. Now you'll see Chess and Rock Climbing listed.

Defining the Entry Model

For a user to record what they've been learning about chess and rock climbing, we need to define a model for the kinds of entries users can make in their learning logs. Each entry needs to be associated with a particular topic. This relationship is called a *many-to-one relationship*, meaning many entries can be associated with one topic.

Here's the code for the Entry model. Place it in your *models.py* file:

models.py

```
from django.db import models

class Topic(models.Model):
    --snip--

❶ class Entry(models.Model):
    """Something specific learned about a topic."""
❷    topic = models.ForeignKey(Topic, on_delete=models.CASCADE)
❸    text = models.TextField()
    date_added = models.DateTimeField(auto_now_add=True)

❹    class Meta:
        verbose_name_plural = 'entries'

    def __str__(self):
        """Return a simple string representing the entry."""
❺        return f"{self.text[:50]}..."
```

The Entry class inherits from Django's base Model class, just as Topic did ❶. The first attribute, topic, is a ForeignKey instance ❷. A *foreign key* is a database term; it's a reference to another record in the database. This is the code that connects each entry to a specific topic. Each topic is assigned a *key*, or ID, when it's created. When Django needs to establish a connection between two pieces of data, it uses the keys associated with each piece of information. We'll use these connections shortly to retrieve all the entries associated with a certain topic. The on_delete=models.CASCADE argument tells Django that when a topic is deleted, all the entries associated with that topic should be deleted as well. This is known as a *cascading delete*.

Next is an attribute called text, which is an instance of TextField ❸. This kind of field doesn't need a size limit, because we don't want to limit the size of individual entries. The date_added attribute allows us to present entries in the order they were created, and to place a timestamp next to each entry.

The Meta class is nested inside the Entry class ❹. The Meta class holds extra information for managing a model; here, it lets us set a special attribute telling Django to use Entries when it needs to refer to more than one entry. Without this, Django would refer to multiple entries as Entrys.

The __str__() method tells Django which information to show when it refers to individual entries. Because an entry can be a long body of text, __str__() returns just the first 50 characters of text ❺. We also add an ellipsis to clarify that we're not always displaying the entire entry.

Migrating the Entry Model

Because we've added a new model, we need to migrate the database again. This process will become quite familiar: you modify *models.py*, run the command **`python manage.py makemigrations`** ***`app_name`***, and then run the command **`python manage.py migrate`**.

Migrate the database and check the output by entering the following commands:

```
(ll_env)learning_log$ python manage.py makemigrations learning_logs
Migrations for 'learning_logs':
❶  learning_logs/migrations/0002_entry.py
    - Create model Entry
(ll_env)learning_log$ python manage.py migrate
Operations to perform:
  --snip--
❷  Applying learning_logs.0002_entry... OK
```

A new migration called *0002_entry.py* is generated, which tells Django how to modify the database to store information related to the model `Entry` ❶. When we issue the `migrate` command, we see that Django applied this migration and everything worked properly ❷.

Registering Entry with the Admin Site

We also need to register the `Entry` model. Here's what *admin.py* should look like now:

admin.py

```
from django.contrib import admin

from .models import Topic, Entry

admin.site.register(Topic)
admin.site.register(Entry)
```

Go back to *http://localhost/admin/*, and you should see Entries listed under *Learning_Logs*. Click the **Add** link for Entries, or click **Entries** and then choose **Add entry**. You should see a drop-down list to select the topic you're creating an entry for and a text box for adding an entry. Select **Chess** from the drop-down list, and add an entry. Here's the first entry I made:

> The opening is the first part of the game, roughly the first ten moves or so. In the opening, it's a good idea to do three things—bring out your bishops and knights, try to control the center of the board, and castle your king.
>
> Of course, these are just guidelines. It will be important to learn when to follow these guidelines and when to disregard these suggestions.

When you click **Save**, you'll be brought back to the main admin page for entries. Here, you'll see the benefit of using `text[:50]` as the string representation for each entry; it's much easier to work with multiple entries in the admin interface if you see only the first part of an entry, rather than the entire text of each entry.

Make a second entry for Chess and one entry for Rock Climbing so we have some initial data. Here's a second entry for Chess:

> In the opening phase of the game, it's important to bring out your bishops and knights. These pieces are powerful and maneuverable enough to play a significant role in the beginning moves of a game.

And here's a first entry for Rock Climbing:

> One of the most important concepts in climbing is to keep your weight on your feet as much as possible. There's a myth that climbers can hang all day on their arms. In reality, good climbers have practiced specific ways of keeping their weight over their feet whenever possible.

These three entries will give us something to work with as we continue to develop Learning Log.

The Django Shell

Now that we've entered some data, we can examine it programmatically through an interactive terminal session. This interactive environment is called the Django *shell*, and it's a great environment for testing and troubleshooting your project. Here's an example of an interactive shell session:

```
(ll_env)learning_log$ python manage.py shell
❶ >>> from learning_logs.models import Topic
>>> Topic.objects.all()
<QuerySet [<Topic: Chess>, <Topic: Rock Climbing>]>
```

The command `python manage.py shell`, run in an active virtual environment, launches a Python interpreter that you can use to explore the data stored in your project's database. Here, we import the model `Topic` from the `learning_logs.models` module ❶. We then use the method `Topic.objects.all()` to get all instances of the model `Topic`; the list that's returned is called a *queryset*.

We can loop over a queryset just as we'd loop over a list. Here's how you can see the ID that's been assigned to each topic object:

```
>>> topics = Topic.objects.all()
>>> for topic in topics:
...     print(topic.id, topic)
...
1 Chess
2 Rock Climbing
```

We assign the queryset to `topics` and then print each topic's `id` attribute and the string representation of each topic. We can see that `Chess` has an ID of `1` and `Rock Climbing` has an ID of `2`.

If you know the ID of a particular object, you can use the method `Topic.objects.get()` to retrieve that object and examine any attribute the object has. Let's look at the `text` and `date_added` values for `Chess`:

```
>>> t = Topic.objects.get(id=1)
>>> t.text
'Chess'
>>> t.date_added
datetime.datetime(2022, 5, 20, 3, 33, 36, 928759,
    tzinfo=datetime.timezone.utc)
```

We can also look at the entries related to a certain topic. Earlier, we defined the `topic` attribute for the `Entry` model. This was a `ForeignKey`, a connection between each entry and a topic. Django can use this connection to get every entry related to a certain topic, like this:

```
❶ >>> t.entry_set.all()
<QuerySet [<Entry: The opening is the first part of the game, roughly...>,
<Entry:
In the opening phase of the game, it's important t...>]>
```

To get data through a foreign key relationship, you use the lowercase name of the related model followed by an underscore and the word `set` ❶. For example, say you have the models `Pizza` and `Topping`, and `Topping` is related to `Pizza` through a foreign key. If your object is called `my_pizza`, representing a single pizza, you can get all of the pizza's toppings using the code `my_pizza.topping_set.all()`.

We'll use this syntax when we begin to code the pages users can request. The shell is really useful for making sure your code retrieves the data you want it to. If your code works as you expect it to in the shell, it should also work properly in the files within your project. If your code generates errors or doesn't retrieve the data you expect it to, it's much easier to troubleshoot your code in the simple shell environment than within the files that generate web pages. We won't refer to the shell much, but you should continue using it to practice working with Django's syntax for accessing the data stored in the project.

Each time you modify your models, you'll need to restart the shell to see the effects of those changes. To exit a shell session, press CTRL-D; on Windows, press CTRL-Z and then press ENTER.

TRY IT YOURSELF

18-2. Short Entries: The `__str__()` method in the `Entry` model currently appends an ellipsis to every instance of `Entry` when Django shows it in the admin site or the shell. Add an `if` statement to the `__str__()` method that adds an ellipsis

(continued)

only if the entry is longer than 50 characters. Use the admin site to add an entry that's fewer than 50 characters in length, and check that it doesn't have an ellipsis when viewed.

18-3. The Django API: When you write code to access the data in your project, you're writing a *query*. Skim through the documentation for querying your data at *https://docs.djangoproject.com/en/4.1/topics/db/queries*. Much of what you see will look new to you, but it will be quite useful as you start to work on your own projects.

18-4. Pizzeria: Start a new project called `pizzeria_project` with an app called `pizzas`. Define a model `Pizza` with a field called `name`, which will hold name values, such as `Hawaiian` and `Meat Lovers`. Define a model called `Topping` with fields called `pizza` and `name`. The `pizza` field should be a foreign key to `Pizza`, and `name` should be able to hold values such as `pineapple`, `Canadian bacon`, and `sausage`.

Register both models with the admin site, and use the site to enter some pizza names and toppings. Use the shell to explore the data you entered.

Making Pages: The Learning Log Home Page

Making web pages with Django consists of three stages: defining URLs, writing views, and writing templates. You can do these in any order, but in this project we'll always start by defining the URL pattern. A *URL pattern* describes the way the URL is laid out. It also tells Django what to look for when matching a browser request with a site URL, so it knows which page to return.

Each URL then maps to a particular view. The *view* function retrieves and processes the data needed for that page. The view function often renders the page using a *template*, which contains the overall structure of the page. To see how this works, let's make the home page for Learning Log. We'll define the URL for the home page, write its view function, and create a simple template.

Because we just want to ensure that Learning Log works as it's supposed to, we'll make a simple page for now. A functioning web app is fun to style when it's complete; an app that looks good but doesn't work well is pointless. For now, the home page will display only a title and a brief description.

Mapping a URL

Users request pages by entering URLs into a browser and clicking links, so we'll need to decide what URLs are needed. The home page URL is first: it's the base URL people use to access the project. At the moment, the base URL, *http://localhost:8000/*, returns the default Django site that lets us know the project was set up correctly. We'll change this by mapping the base URL to Learning Log's home page.

In the main *ll_project* folder, open the file *urls.py*. You should see the following code:

ll_project/urls.py

```
❶ from django.contrib import admin
  from django.urls import path

❷ urlpatterns = [
❸     path('admin/', admin.site.urls),
  ]
```

The first two lines import the admin module and a function to build URL paths ❶. The body of the file defines the urlpatterns variable ❷. In this *urls.py* file, which defines URLs for the project as a whole, the urlpatterns variable includes sets of URLs from the apps in the project. The list includes the module admin.site.urls, which defines all the URLs that can be requested from the admin site ❸.

We need to include the URLs for learning_logs, so add the following:

```
from django.contrib import admin
from django.urls import path, include

urlpatterns = [
    path('admin/', admin.site.urls),
    path('', include('learning_logs.urls')),
]
```

We've imported the include() function, and we've also added a line to include the module learning_logs.urls.

The default *urls.py* is in the *ll_project* folder; now we need to make a second *urls.py* file in the *learning_logs* folder. Create a new Python file, save it as *urls.py* in *learning_logs*, and enter this code into it:

learning_logs/urls.py

```
❶ """Defines URL patterns for learning_logs."""

❷ from django.urls import path

❸ from . import views

❹ app_name = 'learning_logs'
❺ urlpatterns = [
      # Home page
❻     path('', views.index, name='index'),
  ]
```

To make it clear which *urls.py* we're working in, we add a docstring at the beginning of the file ❶. We then import the path function, which is needed when mapping URLs to views ❷. We also import the views module ❸; the dot tells Python to import the *views.py* module from the same directory as the current *urls.py* module. The variable app_name helps Django distinguish this *urls.py* file from files of the same name in other apps within the project ❹. The variable urlpatterns in this module is a list of individual pages that can be requested from the learning_logs app ❺.

The actual URL pattern is a call to the `path()` function, which takes three arguments ❺. The first argument is a string that helps Django route the current request properly. Django receives the requested URL and tries to route the request to a view. It does this by searching all the URL patterns we've defined to find one that matches the current request. Django ignores the base URL for the project (*http://localhost:8000/*), so the empty string (`''`) matches the base URL. Any other URL won't match this pattern, and Django will return an error page if the URL requested doesn't match any existing URL patterns.

The second argument in `path()` ❻ specifies which function to call in *views.py*. When a requested URL matches the pattern we're defining, Django calls the `index()` function from *views.py*. (We'll write this view function in the next section.) The third argument provides the name *index* for this URL pattern so we can refer to it more easily in other files throughout the project. Whenever we want to provide a link to the home page, we'll use this name instead of writing out a URL.

Writing a View

A view function takes in information from a request, prepares the data needed to generate a page, and then sends the data back to the browser. It often does this by using a template that defines what the page will look like.

The file *views.py* in *learning_logs* was generated automatically when we ran the command `python manage.py startapp`. Here's what's in *views.py* right now:

views.py

```
from django.shortcuts import render

# Create your views here.
```

Currently, this file just imports the `render()` function, which renders the response based on the data provided by views. Open *views.py* and add the following code for the home page:

```
from django.shortcuts import render

def index(request):
    """The home page for Learning Log."""
    return render(request, 'learning_logs/index.html')
```

When a URL request matches the pattern we just defined, Django looks for a function called `index()` in the *views.py* file. Django then passes the `request` object to this view function. In this case, we don't need to process any data for the page, so the only code in the function is a call to `render()`. The `render()` function here passes two arguments: the original `request` object and a template it can use to build the page. Let's write this template.

Writing a Template

The template defines what the page should look like, and Django fills in the relevant data each time the page is requested. A template allows you to access any data provided by the view. Because our view for the home page provides no data, this template is fairly simple.

Inside the *learning_logs* folder, make a new folder called *templates*. Inside the *templates* folder, make another folder called *learning_logs*. This might seem a little redundant (we have a folder named *learning_logs* inside a folder named *templates* inside a folder named *learning_logs*), but it sets up a structure that Django can interpret unambiguously, even in the context of a large project containing many individual apps. Inside the inner *learning_logs* folder, make a new file called *index.html*. The path to the file will be *learning_log/learning_logs/templates/learning_logs/index.html*. Enter the following code into that file:

index.html

```
<p>Learning Log</p>

<p>Learning Log helps you keep track of your learning, for any topic you're
interested in.</p>
```

This is a very simple file. If you're not familiar with HTML, the `<p></p>` tags signify paragraphs. The `<p>` tag opens a paragraph, and the `</p>` tag closes a paragraph. We have two paragraphs: the first acts as a title, and the second describes what users can do with Learning Log.

Now when you request the project's base URL, *http://localhost:8000/*, you should see the page we just built instead of the default Django page. Django will take the requested URL, and that URL will match the pattern `''`; then Django will call the function `views.index()`, which will render the page using the template contained in *index.html*. Figure 18-3 shows the resulting page.

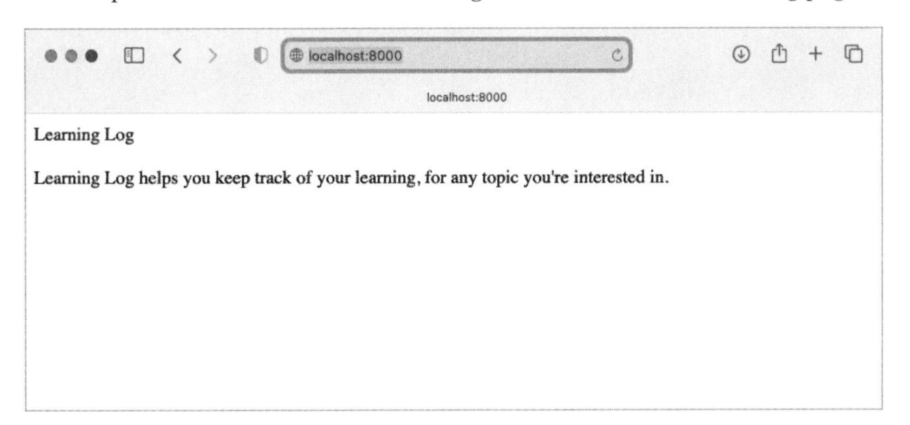

Figure 18-3: The home page for Learning Log

Although it might seem like a complicated process for creating one page, this separation between URLs, views, and templates works quite well. It allows you to think about each aspect of a project separately. In larger projects, it allows individuals working on the project to focus on the areas in which they're strongest. For example, a database specialist can focus on the models, a programmer can focus on the view code, and a frontend specialist can focus on the templates.

NOTE *You might see the following error message:*

```
ModuleNotFoundError: No module named 'learning_logs.urls'
```

If you do, stop the development server by pressing CTRL-C in the terminal window where you issued the `runserver` *command. Then reissue the command* **`python manage.py runserver`**. *You should be able to see the home page. Anytime you run into an error like this, try stopping and restarting the server.*

> **TRY IT YOURSELF**
>
> **18-5. Meal Planner:** Consider an app that helps people plan their meals throughout the week. Make a new folder called *meal_planner*, and start a new Django project inside this folder. Then make a new app called `meal_plans`. Make a simple home page for this project.
>
> **18-6. Pizzeria Home Page:** Add a home page to the Pizzeria project you started in Exercise 18-4 (page 388).

Building Additional Pages

Now that we've established a routine for building a page, we can start to build out the Learning Log project. We'll build two pages that display data: a page that lists all topics and a page that shows all the entries for a particular topic. For each page, we'll specify a URL pattern, write a view function, and write a template. But before we do this, we'll create a base template that all templates in the project can inherit from.

Template Inheritance

When building a website, some elements will need to be repeated on each page. Rather than writing these elements directly into each page, you can write a base template containing the repeated elements and then have each page inherit from the base. This approach lets you focus on developing the unique aspects of each page, and makes it much easier to change the overall look and feel of the project.

The Parent Template

We'll create a template called *base.html* in the same directory as *index.html*. This file will contain elements common to all pages; every other template will inherit from *base.html*. The only element we want to repeat on each page right now is the title at the top. Because we'll include this template on every page, let's make the title a link to the home page:

base.html

```
<p>
❶ <a href="{% url 'learning_logs:index' %}">Learning Log</a>
```

```
</p>

❷ {% block content %}{% endblock content %}
```

The first part of this file creates a paragraph containing the name of the project, which also acts as a home page link. To generate a link, we use a *template tag*, which is indicated by braces and percent signs (`{% %}`). A template tag generates information to be displayed on a page. The template tag `{% url 'learning_logs:index' %}` shown here generates a URL matching the URL pattern defined in *learning_logs/urls.py* with the name `'index'` ❶. In this example, `learning_logs` is the *namespace* and `index` is a uniquely named URL pattern in that namespace. The namespace comes from the value we assigned to `app_name` in the *learning_logs/urls.py* file.

In a simple HTML page, a link is surrounded by the *anchor tag* `<a>`:

```
<a href="link_url">link text</a>
```

Having the template tag generate the URL for us makes it much easier to keep our links up to date. We only need to change the URL pattern in *urls.py*, and Django will automatically insert the updated URL the next time the page is requested. Every page in our project will inherit from *base.html*, so from now on, every page will have a link back to the home page.

On the last line, we insert a pair of `block` tags ❷. This block, named `content`, is a placeholder; the child template will define the kind of information that goes in the `content` block.

A child template doesn't have to define every block from its parent, so you can reserve space in parent templates for as many blocks as you like; the child template uses only as many as it needs.

NOTE *In Python code, we almost always use four spaces when we indent. Template files tend to have more levels of nesting than Python files, so it's common to use only two spaces for each indentation level.*

The Child Template

Now we need to rewrite *index.html* to inherit from *base.html*. Add the following code to *index.html*:

index.html

```
❶ {% extends 'learning_logs/base.html' %}

❷ {% block content %}
    <p>Learning Log helps you keep track of your learning, for any topic you're
    interested in.</p>
❸ {% endblock content %}
```

If you compare this to the original *index.html*, you can see that we've replaced the Learning Log title with the code for inheriting from a parent template ❶. A child template must have an `{% extends %}` tag on the first line to tell Django which parent template to inherit from. The file *base.html* is part of `learning_logs`, so we include *learning_logs* in the path to the parent template. This line pulls in everything contained in the *base.html* template and allows *index.html* to define what goes in the space reserved by the `content` block.

We define the content block by inserting a `{% block %}` tag with the name `content` ❷. Everything that we aren't inheriting from the parent template goes inside the `content` block. Here, that's the paragraph describing the Learning Log project. We indicate that we're finished defining the content by using an `{% endblock content %}` tag ❸. The `{% endblock %}` tag doesn't require a name, but if a template grows to contain multiple blocks, it can be helpful to know exactly which block is ending.

You can start to see the benefit of template inheritance: in a child template, we only need to include content that's unique to that page. This not only simplifies each template, but also makes it much easier to modify the site. To modify an element common to many pages, you only need to modify the parent template. Your changes are then carried over to every page that inherits from that template. In a project that includes tens or hundreds of pages, this structure can make it much easier and faster to improve your site.

In a large project, it's common to have one parent template called *base.html* for the entire site and parent templates for each major section of the site. All the section templates inherit from *base.html*, and each page in the site inherits from a section template. This way you can easily modify the look and feel of the site as a whole, any section in the site, or any individual page. This configuration provides a very efficient way to work, and encourages you to steadily update your project over time.

The Topics Page

Now that we have an efficient approach to building pages, we can focus on our next two pages: the general topics page and the page to display entries for a single topic. The topics page will show all topics that users have created, and it's the first page that will involve working with data.

The Topics URL Pattern

First, we define the URL for the topics page. It's common to choose a simple URL fragment that reflects the kind of information presented on the page. We'll use the word *topics*, so the URL *http://localhost:8000/topics/* will return this page. Here's how we modify *learning_logs/urls.py*:

learning_logs/urls.py

```
"""Defines URL patterns for learning_logs."""
--snip--
urlpatterns = [
    # Home page
    path('', views.index, name='index'),
    # Page that shows all topics.
    path('topics/', views.topics, name='topics'),
]
```

The new URL pattern is the word *topics*, followed by a forward slash. When Django examines a requested URL, this pattern will match any URL that has the base URL followed by *topics*. You can include or omit a forward slash at the end, but there can't be anything else after the word *topics*, or the pattern won't match. Any request with a URL that matches this pattern will then be passed to the function `topics()` in *views.py*.

The Topics View

The `topics()` function needs to retrieve some data from the database and send it to the template. Add the following to *views.py*:

views.py

```
from django.shortcuts import render

❶ from .models import Topic

def index(request):
    --snip--

❷ def topics(request):
    """Show all topics."""
❸     topics = Topic.objects.order_by('date_added')
❹     context = {'topics': topics}
❺     return render(request, 'learning_logs/topics.html', context)
```

We first import the model associated with the data we need ❶. The `topics()` function needs one parameter: the `request` object Django received from the server ❷. We query the database by asking for the `Topic` objects, sorted by the `date_added` attribute ❸. We assign the resulting queryset to `topics`.

We then define a context that we'll send to the template ❹. A *context* is a dictionary in which the keys are names we'll use in the template to access the data we want, and the values are the data we need to send to the template. In this case, there's one key-value pair, which contains the set of topics we'll display on the page. When building a page that uses data, we call `render()` with the `request` object, the template we want to use, and the `context` dictionary ❺.

The Topics Template

The template for the topics page receives the `context` dictionary, so the template can use the data that `topics()` provides. Make a file called *topics.html* in the same directory as *index.html*. Here's how we can display the topics in the template:

topics.html

```
{% extends 'learning_logs/base.html' %}

{% block content %}

  <p>Topics</p>
```

```
❶ <ul>
❷   {% for topic in topics %}
❸     <li>{{ topic.text }}</li>
❹   {% empty %}
      <li>No topics have been added yet.</li>
❺   {% endfor %}
❻ </ul>

{% endblock content %}
```

We use the {% extends %} tag to inherit from *base.html*, just as we did on the home page, and then we open a content block. The body of this page contains a bulleted list of the topics that have been entered. In standard HTML, a bulleted list is called an *unordered list* and is indicated by the tags <ul></ul>. The opening tag <ul> begins the bulleted list of topics ❶.

Next we use a template tag that's equivalent to a for loop, which loops through the list topics from the context dictionary ❷. The code used in templates differs from Python in some important ways. Python uses indentation to indicate which lines of a for statement are part of a loop. In a template, every for loop needs an explicit {% endfor %} tag indicating where the end of the loop occurs. So in a template, you'll see loops written like this:

```
{% for item in list %}
  do something with each item
{% endfor %}
```

Inside the loop, we want to turn each topic into an item in the bulleted list. To print a variable in a template, wrap the variable name in double braces. The braces won't appear on the page; they just indicate to Django that we're using a template variable. So the code {{ topic.text }} ❸ will be replaced by the value of the current topic's text attribute on each pass through the loop. The HTML tag <li></li> indicates a *list item*. Anything between these tags, inside a pair of <ul></ul> tags, will appear as a bulleted item in the list.

We also use the {% empty %} template tag ❹, which tells Django what to do if there are no items in the list. In this case, we print a message informing the user that no topics have been added yet. The last two lines close out the for loop ❺ and then close out the bulleted list ❻.

Now we need to modify the base template to include a link to the topics page. Add the following code to *base.html*:

base.html

```
<p>
❶ <a href="{% url 'learning_logs:index' %}">Learning Log</a> -
❷ <a href="{% url 'learning_logs:topics' %}">Topics</a>
</p>

{% block content %}{% endblock content %}
```

We add a dash after the link to the home page ❶, and then add a link to the topics page using the {% url %} template tag again ❷. This line tells

Django to generate a link matching the URL pattern with the name `'topics'` in *learning_logs/urls.py*.

Now when you refresh the home page in your browser, you'll see a Topics link. When you click the link, you'll see a page that looks similar to Figure 18-4.

Figure 18-4: The topics page

Individual Topic Pages

Next, we need to create a page that can focus on a single topic, showing the topic name and all the entries for that topic. We'll define a new URL pattern, write a view, and create a template. We'll also modify the topics page so each item in the bulleted list links to its corresponding topic page.

The Topic URL Pattern

The URL pattern for the topic page is a little different from the prior URL patterns because it will use the topic's `id` attribute to indicate which topic was requested. For example, if the user wants to see the detail page for the Chess topic (where the `id` is 1), the URL will be *http://localhost:8000/topics/1/*. Here's a pattern to match this URL, which you should place in *learning_logs/urls.py*:

learning_logs/urls.py

```
--snip--
urlpatterns = [
    --snip--
    # Detail page for a single topic.
    path('topics/<int:topic_id>/', views.topic, name='topic'),
]
```

Let's examine the string `'topics/<int:topic_id>/'` in this URL pattern. The first part of the string tells Django to look for URLs that have the word *topics* after the base URL. The second part of the string, `/<int:topic_id>/`, matches an integer between two forward slashes and assigns the integer value to an argument called `topic_id`.

When Django finds a URL that matches this pattern, it calls the view function `topic()` with the value assigned to `topic_id` as an argument. We'll use the value of `topic_id` to get the correct topic inside the function.

The Topic View

The `topic()` function needs to get the topic and all associated entries from the database, much like what we did earlier in the Django shell:

views.py

```
--snip--
❶ def topic(request, topic_id):
      """Show a single topic and all its entries."""
❷     topic = Topic.objects.get(id=topic_id)
❸     entries = topic.entry_set.order_by('-date_added')
❹     context = {'topic': topic, 'entries': entries}
❺     return render(request, 'learning_logs/topic.html', context)
```

This is the first view function that requires a parameter other than the `request` object. The function accepts the value captured by the expression `/<int:topic_id>/` and assigns it to `topic_id` ❶. Then we use `get()` to retrieve the topic, just as we did in the Django shell ❷. Next, we get all of the entries associated with this topic and order them according to `date_added` ❸. The minus sign in front of `date_added` sorts the results in reverse order, which will display the most recent entries first. We store the topic and entries in the `context` dictionary ❹ and call `render()` with the `request` object, the *topic.html* template, and the `context` dictionary ❺.

NOTE *The code phrases at ❷ and ❸ are called* queries, *because they query the database for specific information. When you're writing queries like these in your own projects, it's helpful to try them out in the Django shell first. You'll get much quicker feedback in the shell than you would by writing a view and template and then checking the results in a browser.*

The Topic Template

The template needs to display the name of the topic and the entries. We also need to inform the user if no entries have been made yet for this topic.

topic.html

```
{% extends 'learning_logs/base.html' %}

{% block content %}

❶   <p>Topic: {{ topic.text }}</p>

    <p>Entries:</p>
❷   <ul>
❸     {% for entry in entries %}
        <li>
❹         <p>{{ entry.date_added|date:'M d, Y H:i' }}</p>
❺         <p>{{ entry.text|linebreaks }}</p>
        </li>
❻     {% empty %}
```

```
      <li>There are no entries for this topic yet.</li>
    {% endfor %}
  </ul>

{% endblock content %}
```

We extend *base.html*, as we'll do for all pages in the project. Next, we show the text attribute of the topic that's been requested ❶. The variable `topic` is available because it's included in the `context` dictionary. We then start a bulleted list ❷ to show each of the entries and loop through them ❸, as we did with the topics earlier.

Each bullet lists two pieces of information: the timestamp and the full text of each entry. For the timestamp ❹, we display the value of the attribute `date_added`. In Django templates, a vertical line (`|`) represents a template *filter*—a function that modifies the value in a template variable during the rendering process. The filter `date:'M d, Y H:i'` displays timestamps in the format *January 1, 2022 23:00*. The next line displays the value of the current entry's `text` attribute. The filter `linebreaks` ❺ ensures that long text entries include line breaks in a format understood by browsers, rather than showing a block of uninterrupted text. We again use the `{% empty %}` template tag ❻ to print a message informing the user that no entries have been made.

Links from the Topics Page

Before we look at the topic page in a browser, we need to modify the topics template so each topic links to the appropriate page. Here's the change you need to make to *topics.html*:

topics.html

```
--snip--
    {% for topic in topics %}
      <li>
        <a href="{% url 'learning_logs:topic' topic.id %}">
          {{ topic.text }}</a>
      </li>
    {% empty %}
--snip--
```

We use the URL template tag to generate the proper link, based on the URL pattern in `learning_logs` with the name `'topic'`. This URL pattern requires a `topic_id` argument, so we add the attribute `topic.id` to the URL template tag. Now each topic in the list of topics is a link to a topic page, such as *http://localhost:8000/topics/1/*.

When you refresh the topics page and click a topic, you should see a page that looks like Figure 18-5.

NOTE *There's a subtle but important difference between* `topic.id` *and* `topic_id`. *The expression* `topic.id` *examines a topic and retrieves the value of the corresponding ID. The variable* `topic_id` *is a reference to that ID in the code. If you run into errors when working with IDs, make sure you're using these expressions in the appropriate ways.*

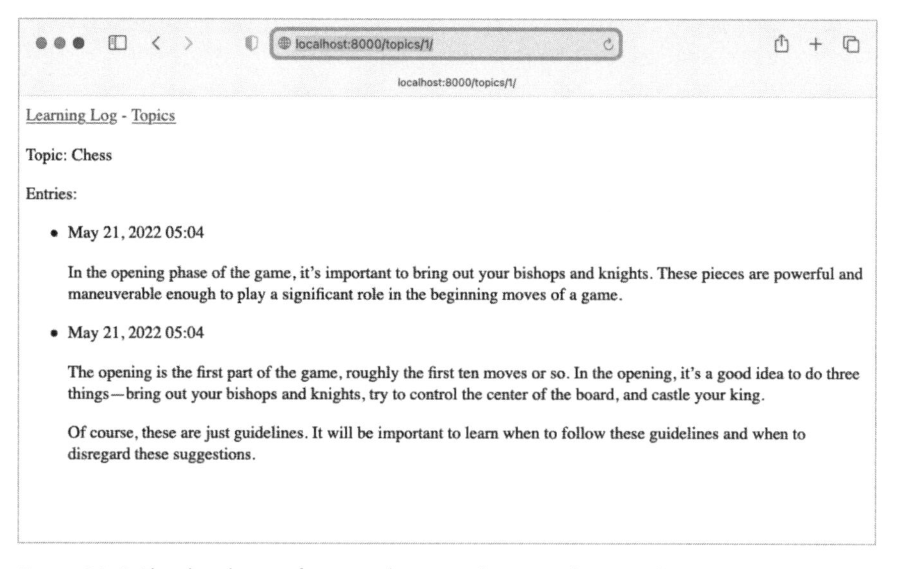

Figure 18-5: The detail page for a single topic, showing all entries for a topic

> **TRY IT YOURSELF**
>
> **18-7. Template Documentation:** Skim the Django template documentation at *https://docs.djangoproject.com/en/4.1/ref/templates*. You can refer back to it when you're working on your own projects.
>
> **18-8. Pizzeria Pages:** Add a page to the Pizzeria project from Exercise 18-6 (page 392) that shows the names of available pizzas. Then link each pizza name to a page displaying the pizza's toppings. Make sure you use template inheritance to build your pages efficiently.

Summary

In this chapter, you learned how to start building a simple web app using the Django framework. You saw a brief project specification, installed Django to a virtual environment, set up a project, and checked that the project was set up correctly. You set up an app and defined models to represent the data for your app. You learned about databases and how Django helps you migrate your database after you make a change to your models. You created a superuser for the admin site, and you used the admin site to enter some initial data.

You also explored the Django shell, which allows you to work with your project's data in a terminal session. You learned how to define URLs, create view functions, and write templates to make pages for your site. You also used template inheritance to simplify the structure of individual templates and make it easier to modify the site as the project evolves.

In Chapter 19, you'll make intuitive, user-friendly pages that allow users to add new topics and entries and edit existing entries without going through the admin site. You'll also add a user registration system, allowing users to create an account and make their own learning log. This is the heart of a web app—the ability to create something that any number of users can interact with.

19

USER ACCOUNTS

At the heart of a web application is the ability for any user, anywhere in the world, to register an account with your app and start using it. In this chapter, you'll build forms so users can add their own topics and entries, and edit existing entries. You'll also learn how Django guards against common attacks against form-based pages, so you won't have to spend much time thinking about securing your apps.

You'll also implement a user authentication system. You'll build a registration page for users to create accounts, and then restrict access to certain pages to logged-in users only. Then you'll modify some of the view functions so users can only see their own data. You'll learn to keep your users' data safe and secure.

Allowing Users to Enter Data

Before we build an authentication system for creating accounts, we'll first add some pages that allow users to enter their own data. We'll give users the ability to add a new topic, add a new entry, and edit their previous entries.

Currently, only a superuser can enter data through the admin site. We don't want users to interact with the admin site, so we'll use Django's form-building tools to build pages that allow users to enter data.

Adding New Topics

Let's start by allowing users to add a new topic. Adding a form-based page works in much the same way as adding the pages we've already built: we define a URL, write a view function, and write a template. The one significant difference is the addition of a new module called *forms.py*, which will contain the forms.

The Topic ModelForm

Any page that lets a user enter and submit information on a web page involves an HTML element called a *form*. When users enter information, we need to *validate* that the information provided is the right kind of data and is not malicious, such as code designed to interrupt our server. We then need to process and save valid information to the appropriate place in the database. Django automates much of this work.

The simplest way to build a form in Django is to use a `ModelForm`, which uses the information from the models we defined in Chapter 18 to build a form automatically. Write your first form in the file *forms.py*, which should be created in the same directory as *models.py*:

forms.py

```
from django import forms

from .models import Topic

❶ class TopicForm(forms.ModelForm):
      class Meta:
❷         model = Topic
❸         fields = ['text']
❹         labels = {'text': ''}
```

We first import the `forms` module and the model we'll work with, `Topic`. We then define a class called `TopicForm`, which inherits from `forms.ModelForm` ❶.

The simplest version of a `ModelForm` consists of a nested `Meta` class telling Django which model to base the form on and which fields to include in the form. Here we specify that the form should be based on the `Topic` model ❷, and that it should only include the `text` field ❸. The empty string in the labels dictionary tells Django not to generate a label for the `text` field ❹.

The new_topic URL

The URL for a new page should be short and descriptive. When the user wants to add a new topic, we'll send them to *http://localhost:8000/new_topic/*. Here's the URL pattern for the `new_topic` page; add this to *learning_logs/urls.py*:

learning_logs/urls.py

```
--snip--
urlpatterns = [
    --snip--
    # Page for adding a new topic.
    path('new_topic/', views.new_topic, name='new_topic'),
]
```

This URL pattern sends requests to the view function `new_topic()`, which we'll write next.

The new_topic() View Function

The `new_topic()` function needs to handle two different situations: initial requests for the `new_topic` page, in which case it should show a blank form; and the processing of any data submitted in the form. After data from a submitted form is processed, it needs to redirect the user back to the `topics` page:

views.py

```
from django.shortcuts import render, redirect

from .models import Topic
from .forms import TopicForm

--snip--
def new_topic(request):
    """Add a new topic."""
❶   if request.method != 'POST':
        # No data submitted; create a blank form.
❷       form = TopicForm()
    else:
        # POST data submitted; process data.
❸       form = TopicForm(data=request.POST)
❹       if form.is_valid():
❺           form.save()
❻           return redirect('learning_logs:topics')

    # Display a blank or invalid form.
❼   context = {'form': form}
    return render(request, 'learning_logs/new_topic.html', context)
```

We import the function `redirect`, which we'll use to redirect the user back to the `topics` page after they submit their topic. We also import the form we just wrote, `TopicForm`.

GET and POST Requests

The two main types of requests you'll use when building apps are GET and POST. You use *GET* requests for pages that only read data from the server. You usually use *POST* requests when the user needs to submit information through a form. We'll be specifying the POST method for processing all of our forms. (A few other kinds of requests exist, but we won't use them in this project.)

The `new_topic()` function takes in the `request` object as a parameter. When the user initially requests this page, their browser will send a GET request. Once the user has filled out and submitted the form, their browser will submit a POST request. Depending on the request, we'll know whether the user is requesting a blank form (GET) or asking us to process a completed form (POST).

We use an `if` test to determine whether the request method is GET or POST ❶. If the request method isn't POST, the request is probably GET, so we need to return a blank form. (If it's another kind of request, it's still safe to return a blank form.) We make an instance of `TopicForm` ❷, assign it to the variable `form`, and send the form to the template in the `context` dictionary ❼. Because we included no arguments when instantiating `TopicForm`, Django creates a blank form that the user can fill out.

If the request method is POST, the `else` block runs and processes the data submitted in the form. We make an instance of `TopicForm` ❸ and pass it the data entered by the user, which is assigned to `request.POST`. The `form` object that's returned contains the information submitted by the user.

We can't save the submitted information in the database until we've checked that it's valid ❹. The `is_valid()` method checks that all required fields have been filled in (all fields in a form are required by default) and that the data entered matches the field types expected—for example, that the length of `text` is less than 200 characters, as we specified in *models.py* in Chapter 18. This automatic validation saves us a lot of work. If everything is valid, we can call `save()` ❺, which writes the data from the form to the database.

Once we've saved the data, we can leave this page. The `redirect()` function takes in the name of a view and redirects the user to the page associated with that view. Here we use `redirect()` to redirect the user's browser to the `topics` page ❻, where the user should see the topic they just entered in the list of topics.

The `context` variable is defined at the end of the view function, and the page is rendered using the template *new_topic.html*, which we'll create next. This code is placed outside of any `if` block; it will run if a blank form was created, and it will run if a submitted form is determined to be invalid. An invalid form will include some default error messages to help the user submit acceptable data.

The new_topic Template

Now we'll make a new template called *new_topic.html* to display the form we just created:

new_topic.html

```
{% extends "learning_logs/base.html" %}

{% block content %}
  <p>Add a new topic:</p>

❶ <form action="{% url 'learning_logs:new_topic' %}" method='post'>
❷   {% csrf_token %}
❸   {{ form.as_div }}
❹   <button name="submit">Add topic</button>
  </form>

{% endblock content %}
```

This template extends *base.html*, so it has the same base structure as the rest of the pages in Learning Log. We use the `<form></form>` tags to define an HTML form ❶. The `action` argument tells the browser where to send the data submitted in the form; in this case, we send it back to the view function `new_topic()`. The `method` argument tells the browser to submit the data as a POST request.

Django uses the template tag `{% csrf_token %}` ❷ to prevent attackers from using the form to gain unauthorized access to the server. (This kind of attack is called a *cross-site request forgery*.) Next, we display the form; here you can see how simple Django can make certain tasks, such as displaying a form. We only need to include the template variable `{{ form.as_div }}` for Django to create all the fields necessary to display the form automatically ❸. The `as_div` modifier tells Django to render all the form elements as HTML `<div></div>` elements; this is a simple way to display the form neatly.

Django doesn't create a submit button for forms, so we define one before closing the form ❹.

Linking to the new_topic Page

Next, we include a link to the `new_topic` page on the `topics` page:

topics.html

```
{% extends "learning_logs/base.html" %}

{% block content %}

  <p>Topics</p>

  <ul>
    --snip--
  </ul>

  <a href="{% url 'learning_logs:new_topic' %}">Add a new topic</a>

{% endblock content %}
```

Place the link after the list of existing topics. Figure 19-1 shows the resulting form; try using the form to add a few new topics of your own.

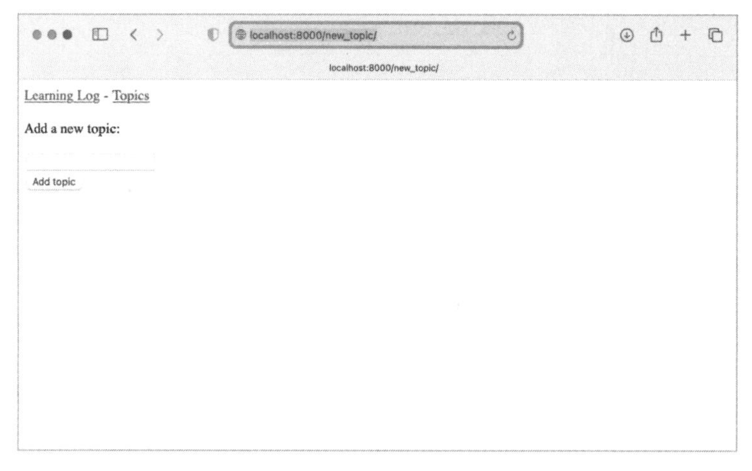

Figure 19-1: The page for adding a new topic

Adding New Entries

Now that the user can add a new topic, they'll want to add new entries too. We'll again define a URL, write a view function and a template, and link to the page. But first, we'll add another class to *forms.py*.

The Entry ModelForm

We need to create a form associated with the `Entry` model, but this time, with a bit more customization than `TopicForm`:

forms.py

```
from django import forms

from .models import Topic, Entry

class TopicForm(forms.ModelForm):
    --snip--

class EntryForm(forms.ModelForm):
    class Meta:
        model = Entry
        fields = ['text']
❶       labels = {'text': ''}
❷       widgets = {'text': forms.Textarea(attrs={'cols': 80})}
```

We update the `import` statement to include `Entry` as well as `Topic`. We make a new class called `EntryForm` that inherits from `forms.ModelForm`. The `EntryForm` class has a nested `Meta` class listing the model it's based on, and the field to include in the form. We again give the field `'text'` a blank label ❶.

For `EntryForm`, we include the `widgets` attribute ❷. A *widget* is an HTML form element, such as a single-line text box, multiline text area, or drop-down list. By including the `widgets` attribute, you can override Django's default widget choices. Here we're telling Django to use a `forms.Textarea` element with a width of 80 columns, instead of the default 40 columns. This gives users enough room to write a meaningful entry.

The new_entry URL

New entries must be associated with a particular topic, so we need to include a `topic_id` argument in the URL for adding a new entry. Here's the URL, which you add to *learning_logs/urls.py*:

learning_logs/urls.py

```
--snip--
urlpatterns = [
    --snip--
    # Page for adding a new entry.
    path('new_entry/<int:topic_id>/', views.new_entry, name='new_entry'),
]
```

This URL pattern matches any URL with the form *http://localhost:8000/new_entry/id/*, where *id* is a number matching the topic ID. The code `<int:topic_id>` captures a numerical value and assigns it to the variable `topic_id`. When a URL matching this pattern is requested, Django sends the request and the topic's ID to the `new_entry()` view function.

The new_entry() View Function

The view function for `new_entry` is much like the function for adding a new topic. Add the following code to your *views.py* file:

views.py

```
from django.shortcuts import render, redirect

from .models import Topic
from .forms import TopicForm, EntryForm

--snip--
def new_entry(request, topic_id):
    """Add a new entry for a particular topic."""
❶   topic = Topic.objects.get(id=topic_id)

❷   if request.method != 'POST':
        # No data submitted; create a blank form.
❸       form = EntryForm()
    else:
        # POST data submitted; process data.
❹       form = EntryForm(data=request.POST)
        if form.is_valid():
❺           new_entry = form.save(commit=False)
❻           new_entry.topic = topic
            new_entry.save()
❼           return redirect('learning_logs:topic', topic_id=topic_id)

    # Display a blank or invalid form.
    context = {'topic': topic, 'form': form}
    return render(request, 'learning_logs/new_entry.html', context)
```

We update the import statement to include the EntryForm we just made. The definition of new_entry() has a topic_id parameter to store the value it receives from the URL. We'll need the topic to render the page and process the form's data, so we use topic_id to get the correct topic object ❶.

Next, we check whether the request method is POST or GET ❷. The if block executes if it's a GET request, and we create a blank instance of EntryForm ❸.

If the request method is POST, we process the data by making an instance of EntryForm, populated with the POST data from the request object ❹. We then check whether the form is valid. If it is, we need to set the entry object's topic attribute before saving it to the database. When we call save(), we include the argument commit=False ❺ to tell Django to create a new entry object and assign it to new_entry, without saving it to the database yet. We set the topic attribute of new_entry to the topic we pulled from the database at the beginning of the function ❻. Then we call save() with no arguments, saving the entry to the database with the correct associated topic.

The redirect() call requires two arguments: the name of the view we want to redirect to and the argument that view function requires ❼. Here, we're redirecting to topic(), which needs the argument topic_id. This view then renders the topic page that the user made an entry for, and they should see their new entry in the list of entries.

At the end of the function, we create a context dictionary and render the page using the *new_entry.html* template. This code will execute for a blank form, or for a form that's been submitted but turns out to be invalid.

The new_entry Template

As you can see in the following code, the template for new_entry is similar to the template for new_topic:

new_entry.html

```
{% extends "learning_logs/base.html" %}

{% block content %}

❶ <p><a href="{% url 'learning_logs:topic' topic.id %}">{{ topic }}</a></p>

  <p>Add a new entry:</p>
❷ <form action="{% url 'learning_logs:new_entry' topic.id %}" method='post'>
    {% csrf_token %}
    {{ form.as_div }}
    <button name='submit'>Add entry</button>
  </form>

{% endblock content %}
```

We show the topic at the top of the page ❶, so the user can see which topic they're adding an entry to. The topic also acts as a link back to the main page for that topic.

The form's action argument includes the topic.id value in the URL, so the view function can associate the new entry with the correct topic ❷. Other than that, this template looks just like *new_topic.html*.

Linking to the new_entry Page

Next, we need to include a link to the new_entry page from each topic page, in the topic template:

topic.html

```
{% extends "learning_logs/base.html" %}

{% block content %}

  <p>Topic: {{ topic }}</p>

  <p>Entries:</p>
  <p>
    <a href="{% url 'learning_logs:new_entry' topic.id %}">Add new entry</a>
  </p>

  <ul>
  --snip--
  </ul>

{% endblock content %}
```

We place the link to add entries just before showing the entries, because adding a new entry will be the most common action on this page. Figure 19-2 shows the new_entry page. Now users can add new topics and as many entries as they want for each topic. Try out the new_entry page by adding a few entries to some of the topics you've created.

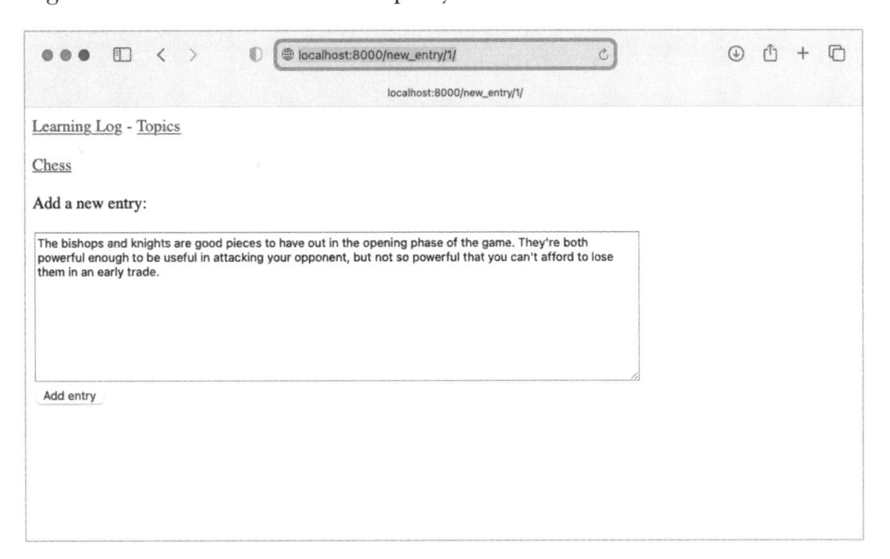

Figure 19-2: The new_entry page

Editing Entries

Now we'll make a page so users can edit the entries they've added.

The edit_entry URL

The URL for the page needs to pass the ID of the entry to be edited. Here's *learning_logs/urls.py*:

urls.py

```
--snip--
urlpatterns = [
    --snip--
    # Page for editing an entry.
    path('edit_entry/<int:entry_id>/', views.edit_entry, name='edit_entry'),
]
```

This URL pattern matches URLs like *http://localhost:8000/edit_entry/id/*. Here the value of *id* is assigned to the parameter `entry_id`. Django sends requests that match this format to the view function `edit_entry()`.

The edit_entry() View Function

When the `edit_entry` page receives a GET request, the `edit_entry()` function returns a form for editing the entry. When the page receives a POST request with revised entry text, it saves the modified text into the database:

views.py

```
from django.shortcuts import render, redirect

from .models import Topic, Entry
from .forms import TopicForm, EntryForm
--snip--

def edit_entry(request, entry_id):
    """Edit an existing entry."""
❶   entry = Entry.objects.get(id=entry_id)
    topic = entry.topic

    if request.method != 'POST':
        # Initial request; pre-fill form with the current entry.
❷       form = EntryForm(instance=entry)
    else:
        # POST data submitted; process data.
❸       form = EntryForm(instance=entry, data=request.POST)
        if form.is_valid():
❹           form.save()
❺           return redirect('learning_logs:topic', topic_id=topic.id)

    context = {'entry': entry, 'topic': topic, 'form': form}
    return render(request, 'learning_logs/edit_entry.html', context)
```

We first import the `Entry` model. We then get the entry object that the user wants to edit ❶ and the topic associated with this entry. In the `if` block, which runs for a GET request, we make an instance of `EntryForm` with the argument `instance=entry` ❷. This argument tells Django to create the form, prefilled with information from the existing entry object. The user will see their existing data and be able to edit that data.

When processing a POST request, we pass both the `instance=entry` and the `data=request.POST` arguments ❸. These arguments tell Django to create a form instance based on the information associated with the existing entry object, updated with any relevant data from `request.POST`. We then check whether the form is valid; if it is, we call `save()` with no arguments because the entry is already associated with the correct topic ❹. We then redirect to the `topic` page, where the user should see the updated version of the entry they edited ❺.

If we're showing an initial form for editing the entry or if the submitted form is invalid, we create the `context` dictionary and render the page using the *edit_entry.html* template.

The edit_entry Template

Next, we create an *edit_entry.html* template, which is similar to *new_entry.html*:

edit_entry.html

```
{% extends "learning_logs/base.html" %}

{% block content %}

  <p><a href="{% url 'learning_logs:topic' topic.id %}">{{ topic }}</a></p>

  <p>Edit entry:</p>

❶ <form action="{% url 'learning_logs:edit_entry' entry.id %}" method='post'>
    {% csrf_token %}
    {{ form.as_div }}
❷   <button name="submit">Save changes</button>
  </form>

{% endblock content %}
```

The `action` argument sends the form back to the `edit_entry()` function for processing ❶. We include the `entry.id` as an argument in the `{% url %}` tag, so the view function can modify the correct entry object. We label the submit button as `Save changes` to remind the user they're saving edits, not creating a new entry ❷.

Linking to the edit_entry Page

Now we need to include a link to the `edit_entry` page for each entry on the topic page:

topic.html

```
--snip--
    {% for entry in entries %}
      <li>
```

```
            <p>{{ entry.date_added|date:'M d, Y H:i' }}</p>
            <p>{{ entry.text|linebreaks }}</p>
            <p>
              <a href="{% url 'learning_logs:edit_entry' entry.id %}">
               Edit entry</a></p>
          </li>
--snip--
```

We include the edit link after each entry's date and text has been displayed. We use the {% url %} template tag to determine the URL for the named URL pattern edit_entry, along with the ID attribute of the current entry in the loop (entry.id). The link text Edit entry appears after each entry on the page. Figure 19-3 shows what the topic page looks like with these links.

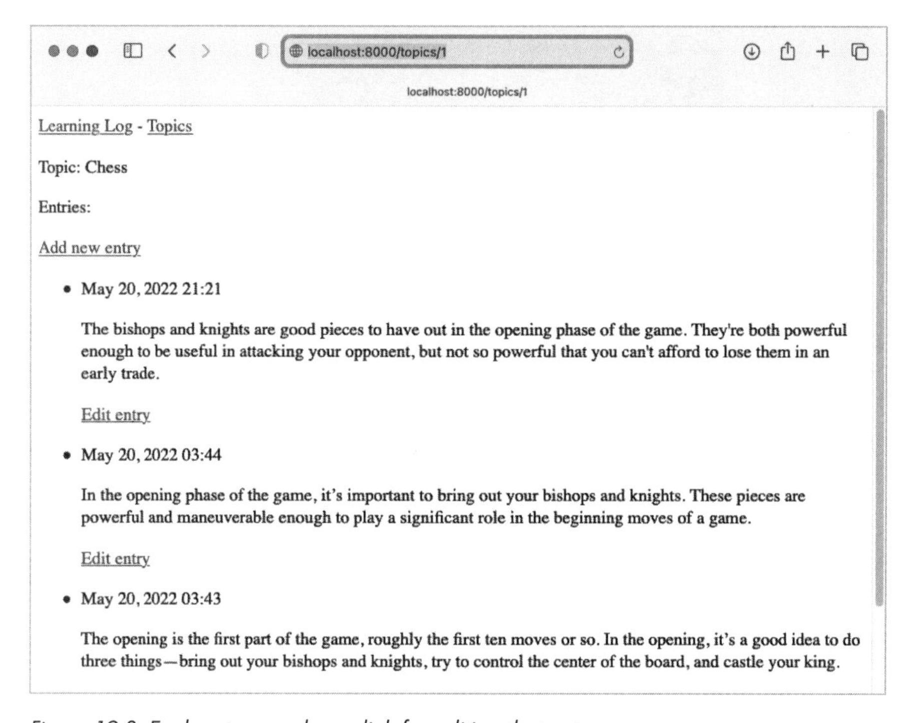

Figure 19-3: Each entry now has a link for editing that entry.

Learning Log now has most of the functionality it needs. Users can add topics and entries, and they can read through any set of entries they want. In the next section, we'll implement a user registration system so anyone can make an account with Learning Log and create their own set of topics and entries.

TRY IT YOURSELF

19-1. Blog: Start a new Django project called *Blog*. Create an app called *blogs*, with one model that represents an overall blog, and one model that represents an individual blog post. Give each model an appropriate set of fields. Create a superuser for the project, and use the admin site to make a blog and a couple of short posts. Make a home page that shows all posts in an appropriate order.

Create pages for making a blog, for making new posts, and for editing existing posts. Use your pages to make sure they work.

Setting Up User Accounts

In this section, we'll set up a user registration and authorization system so people can register an account, log in, and log out. We'll create a new app to contain all the functionality related to working with users. We'll use the default user authentication system included with Django to do as much of the work as possible. We'll also modify the `Topic` model slightly so every topic belongs to a certain user.

The accounts App

We'll start by creating a new app called `accounts`, using the `startapp` command:

```
(ll_env)learning_log$ python manage.py startapp accounts
(ll_env)learning_log$ ls
❶ accounts db.sqlite3 learning_logs ll_env ll_project manage.py
(ll_env)learning_log$ ls accounts
❷ __init__.py admin.py apps.py migrations models.py tests.py views.py
```

The default authentication system is built around the concept of user accounts, so using the name `accounts` makes integration with the default system easier. The `startapp` command shown here makes a new directory called *accounts* ❶ with a structure identical to the `learning_logs` app ❷.

Adding accounts to settings.py

We need to add our new app to `INSTALLED_APPS` in *settings.py*, like so:

settings.py

```
--snip--
INSTALLED_APPS = [
    # My apps
    'learning_logs',
    'accounts',
```

```
    # Default django apps.
    --snip--
]
--snip--
```

Now Django will include the accounts app in the overall project.

Including the URLs from accounts

Next, we need to modify the root *urls.py* so it includes the URLs we'll write for the accounts app:

ll_project/urls.py

```
from django.contrib import admin
from django.urls import path, include

urlpatterns = [
    path('admin/', admin.site.urls),
    path('accounts/', include('accounts.urls')),
    path('', include('learning_logs.urls')),
]
```

We add a line to include the file *urls.py* from accounts. This line will match any URL that starts with the word *accounts*, such as *http://localhost:8000/accounts/login/*.

The Login Page

We'll first implement a login page. We'll use the default login view Django provides, so the URL pattern for this app looks a little different. Make a new *urls.py* file in the directory *learning_log/accounts/* and add the following to it:

accounts/urls.py

```
"""Defines URL patterns for accounts."""

from django.urls import path, include

app_name = 'accounts'
urlpatterns = [
    # Include default auth urls.
    path('', include('django.contrib.auth.urls')),
]
```

We import the path function, and then import the include function so we can include some default authentication URLs that Django has defined. These default URLs include named URL patterns, such as 'login' and 'logout'. We set the variable app_name to 'accounts' so Django can distinguish these URLs from URLs belonging to other apps. Even default URLs provided by Django, when included in the accounts app's *urls.py* file, will be accessible through the accounts namespace.

The login page's pattern matches the URL *http://localhost:8000/accounts/login/*. When Django reads this URL, the word *accounts* tells Django to look in *accounts/urls.py*, and *login* tells it to send requests to Django's default `login` view.

The login Template

When the user requests the login page, Django will use a default view function, but we still need to provide a template for the page. The default authentication views look for templates inside a folder called *registration*, so we'll need to make that folder. Inside the *ll_project/accounts/* directory, make a directory called *templates*; inside that, make another directory called *registration*. Here's the *login.html* template, which should be saved in *ll_project/accounts/templates/registration*:

login.html

```
{% extends 'learning_logs/base.html' %}

{% block content %}

❶ {% if form.errors %}
    <p>Your username and password didn't match. Please try again.</p>
  {% endif %}

❷ <form action="{% url 'accounts:login' %}" method='post'>
    {% csrf_token %}
❸   {{ form.as_div }}

❹   <button name="submit">Log in</button>
  </form>

{% endblock content %}
```

This template extends *base.html* to ensure that the login page will have the same look and feel as the rest of the site. Note that a template in one app can inherit from a template in another app.

If the form's `errors` attribute is set, we display an error message ❶, reporting that the username and password combination doesn't match anything stored in the database.

We want the login view to process the form, so we set the `action` argument as the URL of the login page ❷. The login view sends a `form` object to the template, and it's up to us to display the form ❸ and add a submit button ❹.

The LOGIN_REDIRECT_URL Setting

Once a user logs in successfully, Django needs to know where to send that user. We control this in the settings file.

Add the following code to the end of *settings.py*:

settings.py

```
--snip--
# My settings.
LOGIN_REDIRECT_URL = 'learning_logs:index'
```

With all the default settings in *settings.py*, it's helpful to mark off the section where we're adding new settings. The first new setting we'll add is `LOGIN_REDIRECT_URL`, which tells Django which URL to redirect to after a successful login attempt.

Linking to the Login Page

Let's add the login link to *base.html* so it appears on every page. We don't want the link to display when the user is already logged in, so we nest it inside an `{% if %}` tag:

base.html

```
<p>
  <a href="{% url 'learning_logs:index' %}">Learning Log</a> -
  <a href="{% url 'learning_logs:topics' %}">Topics</a> -
❶ {% if user.is_authenticated %}
❷   Hello, {{ user.username }}.
  {% else %}
❸   <a href="{% url 'accounts:login' %}">Log in</a>
  {% endif %}
</p>

{% block content %}{% endblock content %}
```

In Django's authentication system, every template has a `user` object available that always has an `is_authenticated` attribute set: the attribute is `True` if the user is logged in and `False` if they aren't. This attribute allows you to display one message to authenticated users and another to unauthenticated users.

Here we display a greeting to users currently logged in ❶. Authenticated users have an additional `username` attribute set, which we use to personalize the greeting and remind the user they're logged in ❷. For users who haven't been authenticated, we display a link to the login page ❸.

Using the Login Page

We've already set up a user account, so let's log in to see if the page works. Go to *http://localhost:8000/admin/*. If you're still logged in as an admin, look for a **logout** link in the header and click it.

When you're logged out, go to *http://localhost:8000/accounts/login/*. You should see a login page similar to the one shown in Figure 19-4. Enter the username and password you set up earlier, and you should be brought back to the home page. The header on the home page should display a greeting personalized with your username.

Figure 19-4: The login page

Logging Out

Now we need to provide a way for users to log out. Logout requests should be submitted as POST requests, so we'll add a small logout form to *base.html*. When users click the logout button, they'll go to a page confirming that they've been logged out.

Adding a Logout Form to base.html

We'll add the form for logging out to *base.html* so it's available on every page. We'll include it in another `if` block, so only users who are already logged in can see it:

base.html

```
--snip--
{% block content %}{% endblock content %}

{% if user.is_authenticated %}
❶ <hr />
❷ <form action="{% url 'accounts:logout' %}" method='post'>
    {% csrf_token %}
    <button name='submit'>Log out</button>
  </form>
{% endif %}
```

The default URL pattern for logging out is `'accounts/logout/'`. However, the request has to be sent as a POST request; otherwise, attackers can easily force logout requests. To make the logout request use POST, we define a simple form.

We place the form at the bottom of the page, below a horizontal rule element (`<hr />`) ❶. This is an easy way to always keep the logout button in a consistent position below any other content on the page. The form itself has the logout URL as its `action` argument, and `'post'` as the request method ❷. Every form in Django needs to include the `{% csrf_token %}`, even a simple form like this one. This form is empty except for the submit button.

The LOGOUT_REDIRECT_URL Setting

When the user clicks the logout button, Django needs to know where to send them. We control this behavior in *settings.py*:

settings.py

```
--snip--
# My settings.
LOGIN_REDIRECT_URL = 'learning_logs:index'
LOGOUT_REDIRECT_URL = 'learning_logs:index'
```

The `LOGOUT_REDIRECT_URL` setting shown here tells Django to redirect logged-out users back to the home page. This is a simple way to confirm that they were logged out, because they should no longer see their username after logging out.

The Registration Page

Next, we'll build a page so new users can register. We'll use Django's default `UserCreationForm`, but write our own view function and template.

The register URL

The following code provides the URL pattern for the registration page, which should be placed in *accounts/urls.py*:

accounts/urls.py

```
"""Defines URL patterns for accounts."""

from django.urls import path, include

from . import views

app_name = accounts
urlpatterns = [
    # Include default auth urls.
    path('', include('django.contrib.auth.urls')),
    # Registration page.
    path('register/', views.register, name='register'),
]
```

We import the `views` module from `accounts`, which we need because we're writing our own view for the registration page. The pattern for the registration page matches the URL *http://localhost:8000/accounts/register/* and sends requests to the `register()` function we're about to write.

The register() View Function

The register() view function needs to display a blank registration form when the registration page is first requested, and then process completed registration forms when they're submitted. When a registration is successful, the function also needs to log the new user in. Add the following code to *accounts/views.py*:

accounts/views.py

```
from django.shortcuts import render, redirect
from django.contrib.auth import login
from django.contrib.auth.forms import UserCreationForm

def register(request):
    """Register a new user."""
    if request.method != 'POST':
        # Display blank registration form.
❶       form = UserCreationForm()
    else:
        # Process completed form.
❷       form = UserCreationForm(data=request.POST)

❸       if form.is_valid():
❹           new_user = form.save()
            # Log the user in and then redirect to home page.
❺           login(request, new_user)
❻           return redirect('learning_logs:index')

    # Display a blank or invalid form.
    context = {'form': form}
    return render(request, 'registration/register.html', context)
```

We import the render() and redirect() functions, and then we import the login() function to log the user in if their registration information is correct. We also import the default UserCreationForm. In the register() function, we check whether we're responding to a POST request. If we're not, we make an instance of UserCreationForm with no initial data ❶.

If we're responding to a POST request, we make an instance of UserCreationForm based on the submitted data ❷. We check that the data is valid ❸—in this case, that the username has the appropriate characters, the passwords match, and the user isn't trying to do anything malicious in their submission.

If the submitted data is valid, we call the form's save() method to save the username and the hash of the password to the database ❹. The save() method returns the newly created user object, which we assign to new_user. When the user's information is saved, we log them in by calling the login() function with the request and new_user objects ❺, which creates a valid session for the new user. Finally, we redirect the user to the home page ❻, where a personalized greeting in the header tells them their registration was successful.

At the end of the function, we render the page, which will be either a blank form or a submitted form that's invalid.

The register Template

Now create a template for the registration page, which will be similar to the login page. Be sure to save it in the same directory as *login.html*:

register.html

```
{% extends "learning_logs/base.html" %}

{% block content %}

  <form action="{% url 'accounts:register' %}" method='post'>
    {% csrf_token %}
    {{ form.as_div }}

    <button name="submit">Register</button>
  </form>

{% endblock content %}
```

This should look like the other form-based templates we've been writing. We use the as_div method again so Django will display all the fields in the form appropriately, including any error messages if the form isn't filled out correctly.

Linking to the Registration Page

Next, we'll add code to show the registration page link to any user who isn't currently logged in:

base.html

```
--snip--
  {% if user.is_authenticated %}
    Hello, {{ user.username }}.
  {% else %}
    <a href="{% url 'accounts:register' %}">Register</a> -
    <a href="{% url 'accounts:login' %}">Log in</a>
  {% endif %}
--snip--
```

Now users who are logged in see a personalized greeting and a logout button. Users who aren't logged in see a registration link and a login link. Try out the registration page by making several user accounts with different usernames.

In the next section, we'll restrict some of the pages so they're available only to registered users, and we'll make sure every topic belongs to a specific user.

NOTE *The registration system we've set up allows anyone to make any number of accounts for Learning Log. Some systems require users to confirm their identity by sending a confirmation email that users must reply to. By doing so, the system generates fewer spam accounts than the simple system we're using here. However, when you're learning to build apps, it's perfectly appropriate to practice with a simple user registration system like the one we're using.*

> **TRY IT YOURSELF**
>
> **19-2. Blog Accounts:** Add a user authentication and registration system to the Blog project you started in Exercise 19-1 (page 415). Make sure logged-in users see their username somewhere on the screen and unregistered users see a link to the registration page.

Allowing Users to Own Their Data

Users should be able to enter private data in their learning logs, so we'll create a system to figure out which data belongs to which user. Then we'll restrict access to certain pages so users can only work with their own data.

We'll modify the `Topic` model so every topic belongs to a specific user. This will also take care of entries, because every entry belongs to a specific topic. We'll start by restricting access to certain pages.

Restricting Access with @login_required

Django makes it easy to restrict access to certain pages through the `@login_required` decorator. Recall from Chapter 11 that a *decorator* is a directive placed just before a function definition, which modifies how the function behaves. Let's look at an example.

Restricting Access to the Topics Page

Each topic will be owned by a user, so only registered users can request the topics page. Add the following code to *learning_logs/views.py*:

learning_logs/views.py

```
from django.shortcuts import render, redirect
from django.contrib.auth.decorators import login_required

from .models import Topic, Entry
--snip--

@login_required
def topics(request):
    """Show all topics."""
    --snip--
```

We first import the login_required() function. We apply login_required() as a decorator to the topics() view function by prepending login_required with the @ symbol. As a result, Python knows to run the code in login_required() before the code in topics().

The code in login_required() checks whether a user is logged in, and Django runs the code in topics() only if they are. If the user isn't logged in, they're redirected to the login page.

To make this redirect work, we need to modify *settings.py* so Django knows where to find the login page. Add the following at the end of *settings.py*:

settings.py

```
--snip--
# My settings.
LOGIN_REDIRECT_URL = 'learning_logs:index'
LOGOUT_REDIRECT_URL = 'learning_logs:index'
LOGIN_URL = 'accounts:login'
```

Now when an unauthenticated user requests a page protected by the @login_required decorator, Django will send the user to the URL defined by LOGIN_URL in *settings.py*.

You can test this setting by logging out of any user accounts and going to the home page. Click the **Topics** link, which should redirect you to the login page. Then log in to any of your accounts, and from the home page, click the **Topics** link again. You should be able to access the topics page.

Restricting Access Throughout Learning Log

Django makes it easy to restrict access to pages, but you have to decide which pages to protect. It's best to think about which pages need to be unrestricted first, and then restrict all the other pages in the project. You can easily correct over-restricted access, and it's less dangerous than leaving sensitive pages unrestricted.

In Learning Log, we'll keep the home page and the registration page unrestricted. We'll restrict access to every other page.

Here's *learning_logs/views.py* with @login_required decorators applied to every view except index():

learning_logs/views.py

```
--snip--
@login_required
def topics(request):
    --snip--

@login_required
def topic(request, topic_id):
    --snip--

@login_required
def new_topic(request):
    --snip--
```

```
@login_required
def new_entry(request, topic_id):
    --snip--

@login_required
def edit_entry(request, entry_id):
    --snip--
```

Try accessing each of these pages while logged out; you should be redirected back to the login page. You'll also be unable to click links to pages such as new_topic. But if you enter the URL *http://localhost:8000/new_topic/*, you'll be redirected to the login page. You should restrict access to any URL that's publicly accessible and relates to private user data.

Connecting Data to Certain Users

Next, we need to connect the data to the user who submitted it. We only need to connect the data highest in the hierarchy to a user, and the lower-level data will follow. In Learning Log, topics are the highest level of data in the app, and all entries are connected to a topic. As long as each topic belongs to a specific user, we can trace the ownership of each entry in the database.

We'll modify the Topic model by adding a foreign key relationship to a user. We'll then have to migrate the database. Finally, we'll modify some of the views so they only show the data associated with the currently logged-in user.

Modifying the Topic Model

The modification to *models.py* is just two lines:

models.py

```
from django.db import models
from django.contrib.auth.models import User

class Topic(models.Model):
   """A topic the user is learning about."""
    text = models.CharField(max_length=200)
    date_added = models.DateTimeField(auto_now_add=True)
    owner = models.ForeignKey(User, on_delete=models.CASCADE)

    def __str__(self):
        """Return a string representing the topic."""
        Return self.text

class Entry(models.Model):
    --snip--
```

We import the User model from django.contrib.auth. Then we add an owner field to Topic, which establishes a foreign key relationship to the User model. If a user is deleted, all the topics associated with that user will be deleted as well.

Identifying Existing Users

When we migrate the database, Django will modify the database so it can store a connection between each topic and a user. To make the migration, Django needs to know which user to associate with each existing topic. The simplest approach is to start by assigning all existing topics to one user—for example, the superuser. But first, we need to know that user's ID.

Let's look at the IDs of all users created so far. Start a Django shell session and issue the following commands:

```
(ll_env)learning_log$ python manage.py shell
❶ >>> from django.contrib.auth.models import User
❷ >>> User.objects.all()
<QuerySet [<User: ll_admin>, <User: eric>, <User: willie>]>
❸ >>> for user in User.objects.all():
...     print(user.username, user.id)
...
ll_admin 1
eric 2
willie 3
>>>
```

We first import the `User` model into the shell session ❶. We then look at all the users that have been created so far ❷. The output shows three users for my version of the project: `ll_admin`, `eric`, and `willie`.

Next, we loop through the list of users and print each user's username and ID ❸. When Django asks which user to associate the existing topics with, we'll use one of these ID values.

Migrating the Database

Now that we know the IDs, we can migrate the database. When we do this, Python will ask us to connect the `Topic` model to a particular owner temporarily or to add a default to our *models.py* file to tell it what to do. Choose option **1**:

```
❶ (ll_env)learning_log$ python manage.py makemigrations learning_logs
❷ It is impossible to add a non-nullable field 'owner' to topic without
  specifying a default. This is because...
❸ Please select a fix:
   1) Provide a one-off default now (will be set on all existing rows with a
      null value for this column)
   2) Quit and manually define a default value in models.py.
❹ Select an option: 1
❺ Please enter the default value now, as valid Python
  The datetime and django.utils.timezone modules are available...
  Type 'exit' to exit this prompt
❻ >>> 1
  Migrations for 'learning_logs':
    learning_logs/migrations/0003_topic_owner.py
  - Add field owner to topic
  (ll_env)learning_log$
```

We start by issuing the `makemigrations` command ❶. In the output, Django indicates that we're trying to add a required (*non-nullable*) field to an existing model (`topic`) with no default value specified ❷. Django gives us two options: we can provide a default right now, or we can quit and add a default value in *models.py* ❸. Here I've chosen the first option ❹. Django then asks us to enter the default value ❺.

To associate all existing topics with the original admin user, `ll_admin`, I entered the user ID of `1` ❻. You can use the ID of any user you've created; it doesn't have to be a superuser. Django then migrates the database using this value and generates the migration file *0003_topic_owner.py*, which adds the field `owner` to the `Topic` model.

Now we can execute the migration. Enter the following in an active virtual environment:

```
(ll_env)learning_log$ python manage.py migrate
Operations to perform:
  Apply all migrations: admin, auth, contenttypes, learning_logs, sessions
Running migrations:
❶   Applying learning_logs.0003_topic_owner... OK
(ll_env)learning_log$
```

Django applies the new migration, and the result is `OK` ❶.

We can verify that the migration worked as expected in a shell session, like this:

```
>>> from learning_logs.models import Topic
>>> for topic in Topic.objects.all():
...     print(topic, topic.owner)
...
Chess ll_admin
Rock Climbing ll_admin
>>>
```

We import `Topic` from `learning_logs.models` and then loop through all existing topics, printing each topic and the user it belongs to. You can see that each topic now belongs to the user `ll_admin`. (If you get an error when you run this code, try exiting the shell and starting a new shell.)

NOTE *You can simply reset the database instead of migrating, but that will lose all existing data. It's good practice to learn how to migrate a database while maintaining the integrity of users' data. If you do want to start with a fresh database, issue the command* **`python manage.py flush`** *to rebuild the database structure. You'll have to create a new superuser, and all of your data will be gone.*

Restricting Topics Access to Appropriate Users

Currently, if you're logged in, you'll be able to see all the topics, no matter which user you're logged in as. We'll change that by showing users only the topics that belong to them.

Make the following change to the topics() function in *views.py*:

learning_logs/views.py

```
--snip--
@login_required
def topics(request):
    """Show all topics."""
    topics = Topic.objects.filter(owner=request.user).order_by('date_added')
    context = {'topics': topics}
    return render(request, 'learning_logs/topics.html', context)
--snip--
```

When a user is logged in, the request object has a request.user attribute set, which contains information about the user. The query Topic.objects .filter(owner=request.user) tells Django to retrieve only the Topic objects from the database whose owner attribute matches the current user. Because we're not changing how the topics are displayed, we don't need to change the template for the topics page at all.

To see if this works, log in as the user you connected all existing topics to, and go to the topics page. You should see all the topics. Now log out and log back in as a different user. You should see the message "No topics have been added yet."

Protecting a User's Topics

We haven't restricted access to the topic pages yet, so any registered user could try a bunch of URLs (like *http://localhost:8000/topics/1/*) and retrieve topic pages that happen to match.

Try it yourself. While logged in as the user that owns all topics, copy the URL or note the ID in the URL of a topic, and then log out and log back in as a different user. Enter that topic's URL. You should be able to read the entries, even though you're logged in as a different user.

We'll fix this now by performing a check before retrieving the requested entries in the topic() view function:

learning_logs/views.py

```
from django.shortcuts import render, redirect
from django.contrib.auth.decorators import login_required
❶ from django.http import Http404

--snip--
@login_required
def topic(request, topic_id):
    """Show a single topic and all its entries."""
    topic = Topic.objects.get(id=topic_id)
    # Make sure the topic belongs to the current user.
❷   if topic.owner != request.user:
        raise Http404

    entries = topic.entry_set.order_by('-date_added')
    context = {'topic': topic, 'entries': entries}
    return render(request, 'learning_logs/topic.html', context)
--snip--
```

A 404 response is a standard error response that's returned when a requested resource doesn't exist on a server. Here we import the `Http404` exception ❶, which we'll raise if the user requests a topic they shouldn't have access to. After receiving a topic request, we make sure the topic's user matches the currently logged-in user before rendering the page. If the requested topic's owner is not the same as the current user, we raise the `Http404` exception ❷, and Django returns a 404-error page.

Now if you try to view another user's topic entries, you'll see a "Page Not Found" message from Django. In Chapter 20, we'll configure the project so users will see a proper error page instead of a debugging page.

Protecting the edit_entry Page

The `edit_entry` pages have URLs of the form *http://localhost:8000/edit_entry/entry_id/*, where the ***entry_id*** is a number. Let's protect this page so no one can use the URL to gain access to someone else's entries:

learning_logs/views.py

```
--snip--
@login_required
def edit_entry(request, entry_id):
    """Edit an existing entry."""
    entry = Entry.objects.get(id=entry_id)
    topic = entry.topic
    if topic.owner != request.user:
        raise Http404

    if request.method != 'POST':
        --snip--
```

We retrieve the entry and the topic associated with this entry. We then check whether the owner of the topic matches the currently logged-in user; if they don't match, we raise an `Http404` exception.

Associating New Topics with the Current User

Currently, the page for adding new topics is broken because it doesn't associate new topics with any particular user. If you try adding a new topic, you'll see the message `IntegrityError` along with `NOT NULL constraint failed: learning_logs_topic.owner_id`. Django is saying you can't create a new topic without specifying a value for the topic's `owner` field.

There's a straightforward fix for this problem, because we have access to the current user through the `request` object. Add the following code, which associates the new topic with the current user:

learning_logs/views.py

```
--snip--
@login_required
def new_topic(request):
    --snip--
    else:
        # POST data submitted; process data.
        form = TopicForm(data=request.POST)
        if form.is_valid():
```

```
❶            new_topic = form.save(commit=False)
❷            new_topic.owner = request.user
❸            new_topic.save()
             return redirect('learning_logs:topics')

     # Display a blank or invalid form.
     context = {'form': form}
     return render(request, 'learning_logs/new_topic.html', context)
 --snip--
```

When we first call `form.save()`, we pass the `commit=False` argument because we need to modify the new topic before saving it to the database ❶. We then set the new topic's `owner` attribute to the current user ❷. Finally, we call `save()` on the topic instance we just defined ❸. Now the topic has all the required data and will save successfully.

You should be able to add as many new topics as you want for as many different users as you want. Each user will only have access to their own data, whether they're viewing data, entering new data, or modifying old data.

TRY IT YOURSELF

19-3. Refactoring: There are two places in *views.py* where we make sure the user associated with a topic matches the currently logged-in user. Put the code for this check in a function called `check_topic_owner()`, and call this function where appropriate.

19-4. Protecting new_entry: Currently, a user can add a new entry to another user's learning log by entering a URL with the ID of a topic belonging to another user. Prevent this attack by checking that the current user owns the entry's topic before saving the new entry.

19-5. Protected Blog: In your Blog project, make sure each blog is connected to a particular user. Make sure all posts are publicly accessible but only registered users can add posts and edit existing posts. In the view that allows users to edit their posts, make sure the user is editing their own post before processing the form.

Summary

In this chapter, you learned how forms allow users to add new topics and entries, and edit existing entries. You then learned how to implement user accounts. You gave existing users the ability to log in and out, and used Django's default `UserCreationForm` to let people create new accounts.

After building a simple user authentication and registration system, you restricted access to logged-in users for certain pages using the `@login_required` decorator. You then assigned data to specific users through a foreign key relationship. You also learned to migrate the database when the migration requires you to specify some default data.

Finally, you learned how to make sure a user can only see data that belongs to them by modifying the view functions. You retrieved appropriate data using the `filter()` method, and compared the owner of the requested data to the currently logged-in user.

It might not always be immediately obvious what data you should make available and what data you should protect, but this skill will come with practice. The decisions we've made in this chapter to secure our users' data also illustrate why working with others is a good idea when building a project: having someone else look over your project makes it more likely that you'll spot vulnerable areas.

You now have a fully functioning project running on your local machine. In the final chapter, you'll style Learning Log to make it visually appealing, and you'll deploy the project to a server so anyone with internet access can register and make an account.

20

STYLING AND DEPLOYING AN APP

Learning Log is fully functional now, but it has no styling and runs only on your local machine. In this chapter, you'll style the project in a simple but professional manner and then deploy it to a live server so anyone in the world can make an account and use it.

For the styling, we'll use the *Bootstrap* library, a collection of tools for styling web applications so they look professional on all modern devices, from a small phone to a large desktop monitor. To do this, we'll use the django-bootstrap5 app, which will also give you practice using apps made by other Django developers.

We'll deploy Learning Log using *Platform.sh*, a site that lets you push your project to one of its servers, making it available to anyone with an internet connection. We'll also start using a version control system called Git to track changes to the project.

When you're finished with Learning Log, you'll be able to develop simple web applications, give them a professional look and feel, and deploy them to a live server. You'll also be able to use more advanced learning resources as you develop your skills.

Styling Learning Log

We've purposely ignored styling until now to focus on Learning Log's functionality first. This is a good way to approach development, because an app is only useful if it works. Once an app is working, its appearance is critical so people will want to use it.

In this section, we'll install the django-bootstrap5 app and add it to the project. We'll then use it to style the individual pages in the project, so all the pages have a consistent look and feel.

The django-bootstrap5 App

We'll use django-bootstrap5 to integrate Bootstrap into our project. This app downloads the required Bootstrap files, places them in an appropriate location in your project, and makes the styling directives available in your project's templates.

To install django-bootstrap5, issue the following command in an active virtual environment:

```
(ll_env)learning_log$ pip install django-bootstrap5
--snip--
Successfully installed beautifulsoup4-4.11.1 django-bootstrap5-21.3
    soupsieve-2.3.2.post1
```

Next, we need to add django-bootstrap5 to `INSTALLED_APPS` in *settings.py*:

settings.py

```
--snip--
INSTALLED_APPS = [
    # My apps.
    'learning_logs',
    'accounts',

    # Third party apps.
    'django_bootstrap5',

    # Default django apps.
    'django.contrib.admin',
    --snip--
```

Start a new section called `Third party apps`, for apps created by other developers, and add `'django_bootstrap5'` to this section. Make sure you place this section after `My apps` but before the section containing Django's default apps.

Using Bootstrap to Style Learning Log

Bootstrap is a large collection of styling tools. It also has a number of templates you can apply to your project to create an overall style. It's much easier to use these templates than to use individual styling tools. To see the templates Bootstrap offers, go to *https://getbootstrap.com* and click **Examples**. We'll use the *Navbar static* template, which provides a simple top navigation bar and a container for the page's content.

Figure 20-1 shows what the home page will look like after we apply Bootstrap's template to *base.html* and modify *index.html* slightly.

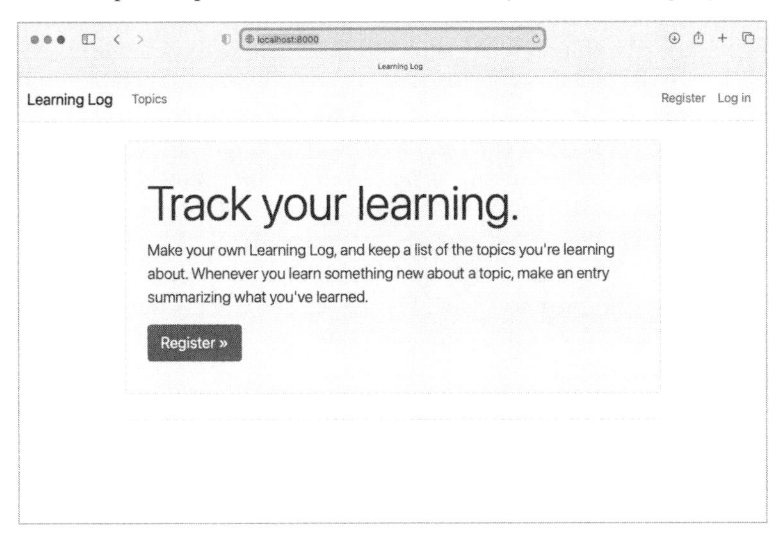

Figure 20-1: The Learning Log home page using Bootstrap

Modifying base.html

We need to rewrite *base.html* using the Bootstrap template. We'll develop the new *base.html* in sections. This is a large file; you may want to copy this file from the online resources, available at *https://ehmatthes.github.io/pcc_3e*. If you do copy the file, you should still read through the following section to understand the changes that were made.

Defining the HTML Headers

The first change we'll make to *base.html* defines the HTML headers in the file. We'll also add some requirements for using Bootstrap in our templates, and give the page a title. Delete everything in *base.html* and replace it with the following code:

base.html

```
❶ <!doctype html>
❷ <html lang="en">
❸ <head>
    <meta charset="utf-8">
    <meta name="viewport" content="width=device-width, initial-scale=1">
❹   <title>Learning Log</title>

❺   {% load django_bootstrap5 %}
    {% bootstrap_css %}
    {% bootstrap_javascript %}

  </head>
```

We first declare this file as an HTML document ❶ written in English ❷. An HTML file is divided into two main parts: the *head* and the *body*. The head of the file begins with an opening `<head>` tag ❸. The head of an HTML file doesn't hold any of the page's content; it just tells the browser what it needs to know to display the page correctly. We include a `<title>` element for the page, which will display in the browser's title bar whenever Learning Log is open ❹.

Before closing the head section, we load the collection of template tags available in django-bootstrap5 ❺. The template tag `{% bootstrap_css %}` is a custom tag from django-bootstrap5; it loads all of the CSS files required to implement Bootstrap styles. The tag that follows enables all the interactive behavior you might use on a page, such as collapsible navigation bars. The closing `</head>` tag appears on the last line.

All Bootstrap styling options are now available in any template that inherits from *base.html*. If you want to use custom template tags from django-bootstrap5, each template will need to include the `{% load django_bootstrap5 %}` tag.

Defining the Navigation Bar

The code that defines the navigation bar at the top of the page is fairly long, because it has to work equally well on narrow phone screens and wide desktop monitors. We'll work through the navigation bar in sections.

Here's the first part of the navigation bar:

base.html

```
--snip--
</head>
<body>

❶ <nav class="navbar navbar-expand-md navbar-light bg-light mb-4 border">
    <div class="container-fluid">
❷     <a class="navbar-brand" href="{% url 'learning_logs:index' %}">
          Learning Log</a>

❸     <button class="navbar-toggler" type="button" data-bs-toggle="collapse"
        data-bs-target="#navbarCollapse" aria-controls="navbarCollapse"
        aria-expanded="false" aria-label="Toggle navigation">
        <span class="navbar-toggler-icon"></span>
      </button>

❹     <div class="collapse navbar-collapse" id="navbarCollapse">
❺       <ul class="navbar-nav me-auto mb-2 mb-md-0">
❻         <li class="nav-item">
❼           <a class="nav-link" href="{% url 'learning_logs:topics' %}">
              Topics</a></li>
        </ul> <!-- End of links on left side of navbar -->
      </div> <!-- Closes collapsible parts of navbar -->

    </div> <!-- Closes navbar's container -->
  </nav> <!-- End of navbar -->
```

```
❽ {% block content %}{% endblock content %}

  </body>
  </html>
```

The first new element is the opening `<body>` tag. The *body* of an HTML file contains the content users will see on a page. Next we have a `<nav>` element, which opens the code for the navigation bar at the top of the page ❶. Everything contained in this element is styled according to the Bootstrap style rules defined by the selectors `navbar`, `navbar-expand-md`, and the rest that you see here. A *selector* determines which elements on a page a certain style rule applies to. The `navbar-light` and `bg-light` selectors style the navigation bar with a light-themed background. The `mb` in `mb-4` is short for *margin-bottom*; this selector ensures that a little space appears between the navigation bar and the rest of the page. The `border` selector provides a thin border around the light background to set it off a little from the rest of the page.

The `<div>` tag on the next line opens a resizable container that will hold the overall navigation bar. The term *div* is short for *division*; you build a web page by dividing it into sections and defining style and behavior rules that apply to that section. Any styling or behavior rules that are defined in an opening `<div>` tag affect everything you see until its corresponding closing tag, written as `</div>`.

Next we set the project's name, `Learning Log`, to appear as the first element on the navigation bar ❷. This will also serve as a link to the home page, just as it's been doing in the minimally styled version of the project we built in the previous two chapters. The `navbar-brand` selector styles this link so it stands out from the rest of the links and helps add some branding to the site.

The Bootstrap template then defines a button that appears if the browser window is too narrow to display the whole navigation bar horizontally ❸. When the user clicks the button, the navigation elements appear in a drop-down list. The `collapse` reference causes the navigation bar to collapse when the user shrinks the browser window or when the site is displayed on devices with small screens.

Next, we open a new section (`<div>`) of the navigation bar ❹. This is the part of the navigation bar that can collapse depending on the size of the browser window.

Bootstrap defines navigation elements as items in an unordered list ❺, with style rules that make it look nothing like a list. Every link or element you need on the bar can be included as an item in an unordered list ❻. Here, the only item in the list is our link to the topics page ❼. Notice the closing `</li>` tag at the end of the link; every opening tag needs a corresponding closing tag.

The rest of the lines shown here close out all of the tags that have been opened. In HTML, a comment is written like this:

```
<!-- This is an HTML comment. -->
```

Closing tags don't usually have comments, but if you're new to HTML, it can be really helpful to label some of your closing tags. A single missing tag or an extra tag can throw off the layout of an entire page. We include the content block ❽ and the closing </body> and </html> tags as well.

We're not finished with the navigation bar, but we now have a complete HTML document. If runserver is currently active, stop the current server and restart it. Go to the project's home page, and you should see a navigation bar that has some of the elements shown in Figure 20-1. Now let's add the rest of the elements to the navigation bar.

Adding User Account Links

We still need to add the links associated with user accounts. We'll start by adding all of the account-related links except the logout form.

Make the following changes to *base.html*:

base.html

```
    --snip--
    </ul> <!-- End of links on left side of navbar -->

    <!-- Account-related links -->
❶   <ul class="navbar-nav ms-auto mb-2 mb-md-0">

❷     {% if user.is_authenticated %}
        <li class="nav-item">
❸         <span class="navbar-text me-2">Hello, {{ user.username }}.
            </span></li>
❹     {% else %}
        <li class="nav-item">
          <a class="nav-link" href="{% url 'accounts:register' %}">
              Register</a></li>
        <li class="nav-item">
          <a class="nav-link" href="{% url 'accounts:login' %}">
              Log in</a></li>
      {% endif %}

    </ul> <!-- End of account-related links -->

  </div> <!-- Closes collapsible parts of navbar -->
  --snip--
```

We begin a new set of links by using another opening <ul> tag ❶. You can have as many groups of links as you need on a page. The selector ms-auto is short for *margin-start-automatic*: this selector examines the other elements in the navigation bar and works out a left (start) margin that pushes this group of links to the right side of the browser window.

The if block is the same conditional block we used earlier to display appropriate messages to users, depending on whether they're logged in ❷. The block is a little longer now because there are some styling rules inside the conditional tags. The greeting for authenticated users is wrapped in a <span> element ❸. A *span element* styles pieces of text or elements of a page that are part of a longer line. While div elements create their own divisions in a page, span elements are continuous within a larger section. This can be confusing at first, because many pages have deeply nested div elements. Here, we're using the span element to style informational text on the navigation bar: in this case, the logged-in user's name.

In the else block, which runs for unauthenticated users, we include the links for registering a new account and logging in ❹. These should look just like the link to the topics page.

If you wanted to add more links to the navigation bar, you'd add another `<li>` item to one of the `<ul>` groups that we've defined, using styling directives like the ones you've seen here.

Now let's add the logout form to the navigation bar.

Adding the Logout Form to the Navigation Bar

When we first wrote the logout form, we added it to the bottom of *base.html*. Now let's put it in a better place, in the navigation bar:

base.html

```
      --snip--
      </ul> <!-- End of account-related links -->

      {% if user.is_authenticated %}
        <form action="{% url 'accounts:logout' %}" method='post'>
          {% csrf_token %}
❶         <button name='submit' class='btn btn-outline-secondary btn-sm'>
            Log out</button>
        </form>
      {% endif %}

    </div> <!-- Closes collapsible parts of navbar -->
    --snip--
```

The logout form should be placed after the set of account-related links, but inside the collapsible section of the navigation bar. The only change in the form is the addition of a number of Bootstrap styling classes in the `<button>` element, which apply Bootstrap styling elements to the logout button ❶.

Reload the home page, and you should be able to log in and out using any of the accounts you've created.

There's still a bit more we need to add to *base.html*. We need to define two blocks that the individual pages can use to place the content specific to those pages.

Defining the Main Part of the Page

The rest of *base.html* contains the main part of the page:

base.html

```
  --snip--
  </nav> <!-- End of navbar -->

❶ <main class="container">
❷   <div class="pb-2 mb-2 border-bottom">
      {% block page_header %}{% endblock page_header %}
    </div>
```

```
❸    <div>
       {% block content %}{% endblock content %}
     </div>
   </main>

</body>
</html>
```

We first open a <main> tag ❶. The *main* element is used for the most significant part of the body of a page. Here we assign the bootstrap selector container, which is a simple way to group elements on a page. We'll place two div elements in this container.

The first div element contains a page_header block ❷. We'll use this block to title most pages. To make this section stand out from the rest of the page, we place some padding below the header. *Padding* refers to space between an element's content and its border. The selector pb-2 is a bootstrap directive that provides a moderate amount of padding at the bottom of the styled element. A *margin* is the space between an element's border and other elements on the page. The selector mb-2 provides a moderate amount of margin at the bottom of this div. We want a border on the bottom of this block, so we use the selector border-bottom, which provides a thin border at the bottom of the page_header block.

We then define one more div element that contains the block content ❸. We don't apply any specific style to this block, so we can style the content of any page as we see fit for that page. The end of the *base.html* file has closing tags for the main, body, and html elements.

When you load Learning Log's home page in a browser, you should see a professional-looking navigation bar that matches the one shown in Figure 20-1. Try resizing the window so it's really narrow; a button should replace the navigation bar. Click the button, and all the links should appear in a drop-down list.

Styling the Home Page Using a Jumbotron

To update the home page, we'll use a Bootstrap element called a *jumbotron*, a large box that stands out from the rest of the page. Typically, it's used on home pages to hold a brief description of the overall project and a call to action that invites the viewer to get involved.

Here's the revised *index.html* file:

index.html

```
{% extends "learning_logs/base.html" %}

❶ {% block page_header %}
❷   <div class="p-3 mb-4 bg-light border rounded-3">
      <div class="container-fluid py-4">
❸       <h1 class="display-3">Track your learning.</h1>

❹       <p class="lead">Make your own Learning Log, and keep a list of the
        topics you're learning about. Whenever you learn something new
        about a topic, make an entry summarizing what you've learned.</p>
```

```
❺       <a class="btn btn-primary btn-lg mt-1"
          href="{% url 'accounts:register' %}">Register &raquo;</a>
      </div>
    </div>
  {% endblock page_header %}
```

We first tell Django that we're about to define what goes in the page _header block ❶. A jumbotron is implemented as a pair of div elements with a set of styling directives applied to them ❷. The outer div has padding and margin settings, a light background color, and rounded corners. The inner div is a container that changes along with the window size and has some padding as well. The py-4 selector adds padding to the top and bottom of the div element. Feel free to adjust the numbers in these settings and see how the home page changes.

Inside the jumbotron are three elements. The first is a short message, Track your learning, that gives new visitors a sense of what Learning Log does ❸. The <h1> element is a first-level header, and the display-3 selector adds a thinner and taller look to this particular header. We also include a longer message that provides more information about what the user can do with their learning log ❹. This is formatted as a lead paragraph, which is meant to stand out from regular paragraphs.

Rather than just using a text link, we create a button that invites users to register an account on Learning Log ❺. This is the same link as in the header, but the button stands out on the page and shows the viewer what they need to do in order to start using the project. The selectors you see here style this as a large button that represents a call to action. The code » is an *HTML entity* that looks like two right angle brackets combined (>>). Finally, we provide closing div tags and close the page_header block. With only two div elements in this file, it's not particularly helpful to label the closing div tags. We aren't adding anything else to this page, so we don't need to define the content block in this template.

The home page now looks like Figure 20-1. This is a significant improvement over the unstyled version of the project!

Styling the Login Page

We've refined the overall appearance of the login page, but the login form itself doesn't have any styling yet. Let's make the form look consistent with the rest of the page by modifying *login.html*:

login.html

```
  {% extends 'learning_logs/base.html' %}
❶ {% load django_bootstrap5 %}

❷ {% block page_header %}
    <h2>Log in to your account.</h2>
  {% endblock page_header %}

  {% block content %}
```

```
    <form action="{% url 'accounts:login' %}" method='post'>
      {% csrf_token %}
❸     {% bootstrap_form form %}
❹     {% bootstrap_button button_type="submit" content="Log in" %}
    </form>

  {% endblock content %}
```

We first load the bootstrap5 template tags into this template ❶. We then define the page_header block, which tells the user what the page is for ❷. Notice that we've removed the {% if form.errors %} block from the template; django-bootstrap5 manages form errors automatically.

To display the form, we use the template tag {% bootstrap_form %} ❸; this replaces the {{ form.as_div }} element we were using in Chapter 19. The {% booststrap_form %} template tag inserts Bootstrap style rules into the form's individual elements as the form is rendered. To generate the submit button, we use the {% bootstrap_button %} tag with arguments that designate it as a submit button, and give it the label Log in ❹.

Figure 20-2 shows the login form now. The page is much cleaner, with consistent styling and a clear purpose. Try logging in with an incorrect username or password; you'll see that even the error messages are styled consistently and integrate well with the overall site.

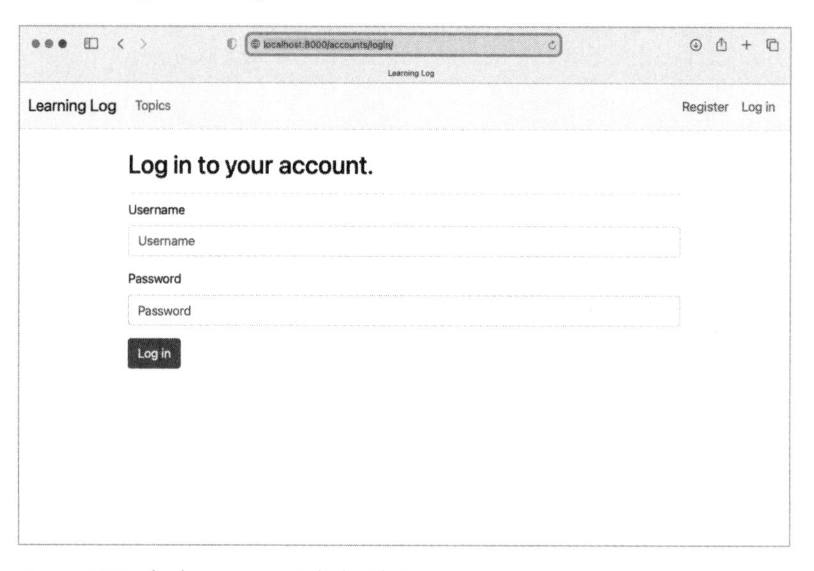

Figure 20-2: The login page styled with Bootstrap

Styling the Topics Page

Let's make sure the pages for viewing information are styled appropriately as well, starting with the topics page:

topics.html

```
{% extends 'learning_logs/base.html' %}

{% block page_header %}
```

```
❶   <h1>Topics</h1>
  {% endblock page_header %}

  {% block content %}

❷   <ul class="list-group border-bottom pb-2 mb-4">
      {% for topic in topics %}
❸       <li class="list-group-item border-0">
          <a href="{% url 'learning_logs:topic' topic.id %}">
            {{ topic.text }}</a>
        </li>
      {% empty %}
❹       <li class="list-group-item border-0">No topics have been added yet.</li>
      {% endfor %}
    </ul>

    <a href="{% url 'learning_logs:new_topic' %}">Add a new topic</a>

  {% endblock content %}
```

We don't need the `{% load bootstrap5 %}` tag, because we're not using any custom bootstrap5 template tags in this file. We move the heading `Topics` into the `page_header` block and make it an `<h1>` element instead of a simple paragraph ❶.

The main content on this page is a list of topics, so we use Bootstrap's *list group* component to render the page. This applies a simple set of styling directives to the overall list and to each item in the list. When we open the `<ul>` tag, we first include the `list-group` class to apply the default style directives to the list ❷. We further customize the list by putting a border at the bottom of the list, a little padding below the list (`pb-2`), and a margin below the bottom border (`mb-4`).

Each item in the list needs the `list-group-item` class, and we customize the default style by removing the border around individual items ❸. The message that's displayed when the list is empty needs these same classes ❹.

When you visit the topics page now, you should see a page with styling that matches the home page.

Styling the Entries on the Topic Page

On the topic page, we'll use Bootstrap's card component to make each entry stand out. A *card* is a nestable set of divs with flexible, predefined styles that are perfect for displaying a topic's entries:

topic.html

```
  {% extends 'learning_logs/base.html' %}

❶ {% block page_header %}
    <h1>{{ topic.text }}</h1>
  {% endblock page_header %}
```

```
{% block content %}
  <p>
    <a href="{% url 'learning_logs:new_entry' topic.id %}">Add new entry</a>
  </p>

  {% for entry in entries %}
❷   <div class="card mb-3">
      <!-- Card header with timestamp and edit link -->
❸     <h4 class="card-header">
        {{ entry.date_added|date:'M d, Y H:i' }}
❹       <small><a href="{% url 'learning_logs:edit_entry' entry.id %}">
          edit entry</a></small>
      </h4>
      <!-- Card body with entry text -->
❺     <div class="card-body">{{ entry.text|linebreaks }}</div>
    </div>
  {% empty %}
❻   <p>There are no entries for this topic yet.</p>
  {% endfor %}

{% endblock content %}
```

We first place the topic in the page_header block ❶. Then we delete the unordered list structure previously used in this template. Instead of making each entry a list item, we open a div element with the selector card ❷. This card has two nested elements: one to hold the timestamp and the link to edit the entry, and another to hold the body of the entry. The card selector takes care of most of the styling we need for this div; we customize the card by adding a small margin to the bottom of each card (mb-3).

The first element in the card is a header, which is an <h4> element with the selector card-header ❸. This header contains the date the entry was made and a link to edit the entry. The <small> tag around the edit_entry link makes it appear a little smaller than the timestamp ❹. The second element is a div with the selector card-body ❺, which places the text of the entry in a simple box on the card. Notice that the Django code for including the information on the page hasn't changed; only elements that affect the appearance of the page have. Since we no longer have an unordered list, we've replaced the list item tags around the empty list message with simple paragraph tags ❻.

Figure 20-3 shows the topic page with its new look. Learning Log's functionality hasn't changed, but it looks significantly more professional and inviting to users.

If you want to use a different Bootstrap template for a project, follow a process that's similar to what we've done so far in this chapter. Copy the template you want to use into *base.html*, and modify the elements that contain actual content so the template displays your project's information. Then use Bootstrap's individual styling tools to style the content on each page.

NOTE *The Bootstrap project has excellent documentation. Visit the home page at* https://getbootstrap.com *and click* ***Docs*** *to learn more about what Bootstrap offers.*

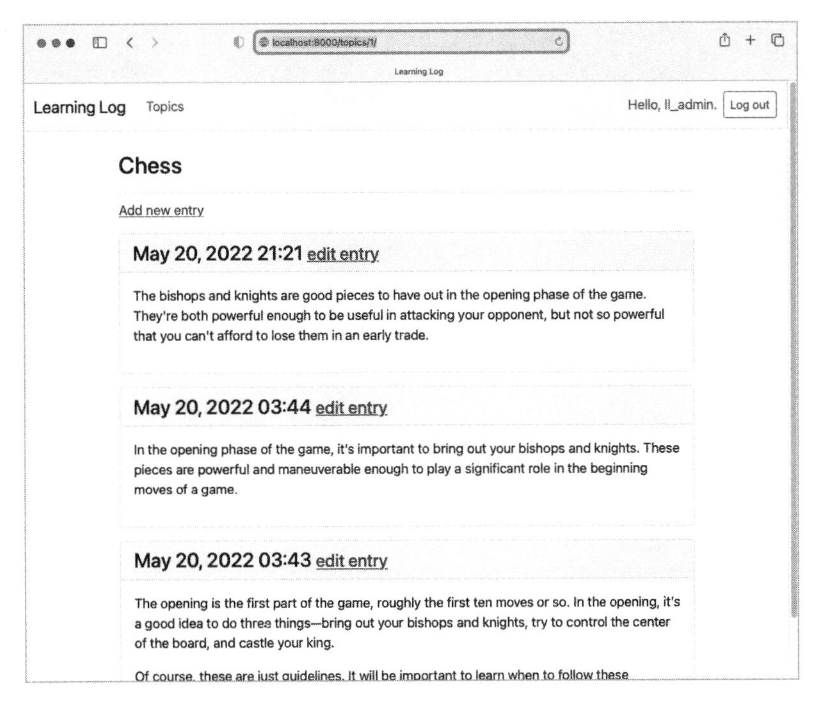

Figure 20-3: The topic page with Bootstrap styling

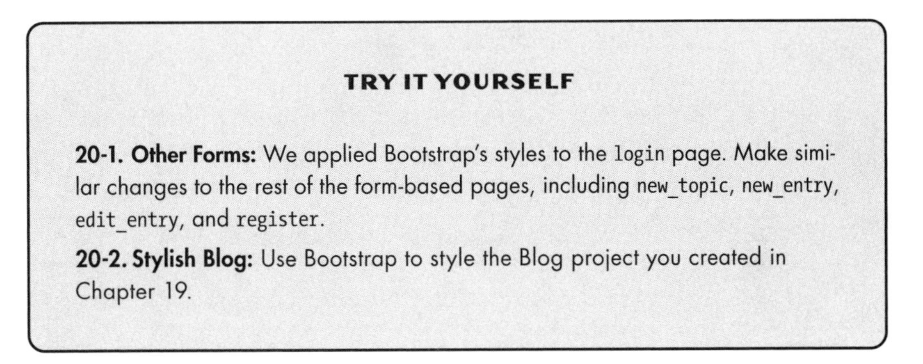

TRY IT YOURSELF

20-1. Other Forms: We applied Bootstrap's styles to the `login` page. Make similar changes to the rest of the form-based pages, including `new_topic`, `new_entry`, `edit_entry`, and `register`.

20-2. Stylish Blog: Use Bootstrap to style the Blog project you created in Chapter 19.

Deploying Learning Log

Now that we have a professional-looking project, let's deploy it to a live server so anyone with an internet connection can use it. We'll use Platform.sh, a web-based platform that allows you to manage the deployment of web applications. We'll get Learning Log up and running on Platform.sh.

Making a Platform.sh Account

To make an account, go to *https://platform.sh* and click the **Free Trial** button. Platform.sh has a free tier that, as of this writing, does not require a credit card. The trial period allows you to deploy an app with minimal resources, which lets you test your project in a live deployment before committing to a paid hosting plan.

NOTE *The specific limits of trial plans tend to change periodically, as hosting platforms fight spam and abuse of resources. You can see the current limits of the free trial at* https://platform.sh/free-trial.

Installing the Platform.sh CLI

To deploy and manage a project on Platform.sh, you'll need the tools available in the Command Line Interface (CLI). To install the latest version of the CLI, visit *https://docs.platform.sh/development/cli.html* and follow the instructions for your operating system.

On most systems, you can install the CLI by running the following command in a terminal:

```
$ curl -fsS https://platform.sh/cli/installer | php
```

After this command has finished running, you will need to open a new terminal window before you can use the CLI.

NOTE *This command will probably not work in a standard terminal on Windows. You can use Windows Subsystem for Linux (WSL) or a Git Bash terminal. If you need to install PHP, you can use the XAMPP installer from* https://apachefriends.org. *If you have any difficulty installing the Platform.sh CLI, see the more detailed installation instructions in Appendix E.*

Installing platformshconfig

You'll also need to install one additional package, `platformshconfig`. This package helps detect whether the project is running on your local system or on a Platform.sh server. In an active virtual environment, issue the following command:

```
(ll_env)learning_log$ pip install platformshconfig
```

We'll use this package to modify the project's settings when it's running on the live server.

Creating a requirements.txt File

The remote server needs to know which packages Learning Log depends on, so we'll use pip to generate a file listing them. Again, from an active virtual environment, issue the following command:

```
(ll_env)learning_log$ pip freeze > requirements.txt
```

The `freeze` command tells pip to write the names of all the packages currently installed in the project into the file *requirements.txt*. Open this file to see the packages and version numbers installed in your project:

requirements.txt

```
asgiref==3.5.2
beautifulsoup4==4.11.1
Django==4.1
```

```
django-bootstrap5==21.3
platformshconfig==2.4.0
soupsieve==2.3.2.post1
sqlparse==0.4.2
```

Learning Log already depends on specific versions of seven different packages, so it requires a matching environment to run properly on a remote server. (We installed three of these packages manually, and four of them were installed automatically as dependencies of these packages.)

When we deploy Learning Log, Platform.sh will install all the packages listed in *requirements.txt*, creating an environment with the same packages we're using locally. Because of this, we can be confident the deployed project will function just like it has on our local system. This approach to managing a project is critical as you start to build and maintain multiple projects on your system.

NOTE *If the version number for a package listed on your system differs from what's shown here, keep the version you have on your system.*

Additional Deployment Requirements

The live server requires two additional packages. These packages are used to serve the project in a production environment, where many users can be making requests at the same time.

In the same directory where *requirements.txt* is saved, make a new file called *requirements_remote.txt*. Add the following two packages to it:

requirements_remote.txt

```
# Requirements for live project.
gunicorn
psycopg2
```

The `gunicorn` package responds to requests as they come in to the remote server; this takes the place of the development server we've been using locally. The `psycopg2` package is required to let Django manage the Postgres database that Platform.sh uses. *Postgres* is an open source database that's extremely well suited to production apps.

Adding Configuration Files

Every hosting platform requires some configuration for a project to run correctly on its servers. In this section, we'll add three configuration files:

.platform.app.yaml This is the main configuration file for the project. This tells Platform.sh what kind of project we're trying to deploy and what kinds of resources our project needs, and it includes commands for building the project on the server.

.platform/routes.yaml This file defines the routes to our project. When a request is received by Platform.sh, this is the configuration that helps direct these requests to our specific project.

.platform/services.yaml This file defines any additional services our project needs.

These are all YAML (YAML Ain't Markup Language) files. *YAML* is a language designed for writing configuration files; it's made to be read easily by both humans and computers. You can write or modify a typical YAML file by hand, but a computer can also read and interpret the file unambiguously.

YAML files are great for deployment configuration, because they give you a good deal of control over what happens during the deployment process.

Making Hidden Files Visible

Most operating systems hide files and folders that begin with a dot, such as *.platform*. When you open a file browser, you won't see these kinds of files and folders by default. But as a programmer, you'll need to see them. Here's how to view hidden files, depending on your operating system:

- On Windows, open Windows Explorer, and then open a folder such as *Desktop*. Click the **View** tab, and make sure **File name extensions** and **Hidden items** are checked.
- On macOS, you can press ⌘-SHIFT-. (dot) in any Finder window to see hidden files and folders.
- On Linux systems such as Ubuntu, you can press CTRL-H in any file browser to display hidden files and folders. To make this setting permanent, open a file browser such as Nautilus and click the options tab (indicated by three lines). Select the **Show Hidden Files** checkbox.

The .platform.app.yaml Configuration File

The first configuration file is the longest, because it controls the overall deployment process. We'll show it in parts; you can either enter it by hand in your text editor or download a copy from the online resources at *https://ehmatthes.github.io/pcc_3e*.

Here's the first part of *.platform.app.yaml*, which should be saved in the same directory as *manage.py*:

.platform.app.yaml

```
❶ name: "ll_project"
  type: "python:3.10"

❷ relationships:
      database: "db:postgresql"

  # The configuration of the app when it's exposed to the web.
❸ web:
      upstream:
          socket_family: unix
      commands:
❹         start: "gunicorn -w 4 -b unix:$SOCKET ll_project.wsgi:application"
❺     locations:
          "/":
              passthru: true
          "/static":
```

```
            root: "static"
            expires: 1h
            allow: true

# The size of the persistent disk of the application (in MB).
❻ disk: 512
```

When you save this file, make sure you include the dot at the beginning of the filename. If you omit the dot, Platform.sh won't find the file and your project will not be deployed.

You don't need to understand everything in *.platform.app.yaml* at this point; I'll highlight the most important parts of the configuration. The file starts off by specifying the `name` of the project, which we're calling `'ll_project'` to be consistent with the name we used when starting the project ❶. We also need to specify the version of Python we're using (3.10 at the time of this writing). You can find a list of supported versions at *https://docs.platform.sh/languages/python.html.*

Next is a section labeled `relationships` that defines other services the project needs ❷. Here the only relationship is to a Postgres database. After that is the `web` section ❸. The `commands:start` section tells Platform.sh what process to use to serve incoming requests. Here we're specifying that `gunicorn` will handle requests ❹. This command takes the place of the `python manage.py runserver` command we've been using locally.

The `locations` section tells Platform.sh where to send incoming requests ❺. Most requests should be passed through to `gunicorn`; our *urls.py* files will tell `gunicorn` exactly how to handle those requests. Requests for static files will be handled separately and will be refreshed once an hour. The last line shows that we're requesting 512MB of disk space on one of Platform.sh's servers ❻.

The rest of *.platform.app.yaml* is as follows:

```
--snip--
disk: 512

# Set a local read/write mount for logs.
❶ mounts:
    "logs":
        source: local
        source_path: logs

# The hooks executed at various points in the lifecycle of the application.
❷ hooks:
    build: |
❸       pip install --upgrade pip
        pip install -r requirements.txt
        pip install -r requirements_remote.txt

        mkdir logs
❹       python manage.py collectstatic
        rm -rf logs
❺   deploy: |
        python manage.py migrate
```

The mounts section ❶ lets us define directories where we can read and write data while the project is running. This section defines a *logs/* directory for the deployed project.

The hooks section ❷ defines actions that are taken at various points during the deployment process. In the build section, we install all the packages that are required to serve the project in the live environment ❸. We also run collectstatic ❹, which collects all the static files needed for the project into one place so they can be served efficiently.

Finally, in the deploy section ❺, we specify that migrations should be run each time the project is deployed. In a simple project, this will have no effect when there have been no changes.

The other two configuration files are much shorter; let's write them now.

The routes.yaml Configuration File

A *route* is the path a request takes as it's processed by the server. When a request is received by Platform.sh, it needs to know where to send the request.

Make a new folder called *.platform*, in the same directory as *manage.py*. Make sure you include the dot at the beginning of the name. Inside that folder, make a file called *routes.yaml* and enter the following:

.platform/routes.yaml

```
# Each route describes how an incoming URL will be processed by Platform.sh.

"https://{default}/":
    type: upstream
    upstream: "ll_project:http"

"https://www.{default}/":
    type: redirect
    to: "https://{default}/"
```

This file makes sure requests like *https://project_url.com* and *www.project_url.com* all get routed to the same place.

The services.yaml Configuration File

This last configuration file specifies services that our project needs in order to run. Save this file in the *.platform/* directory, alongside *routes.yaml*:

.platform/services.yaml

```
# Each service listed will be deployed in its own container as part of your
#   Platform.sh project.

db:
    type: postgresql:12
    disk: 1024
```

This file defines one service, a Postgres database.

Modifying settings.py for Platform.sh

Now we need to add a section at the end of *settings.py* to modify some settings for the Platform.sh environment. Add this code to the very end of *settings.py*:

settings.py

```
--snip--
# Platform.sh settings.
❶ from platformshconfig import Config

config = Config()
❷ if config.is_valid_platform():
❸     ALLOWED_HOSTS.append('.platformsh.site')

❹     if config.appDir:
          STATIC_ROOT = Path(config.appDir) / 'static'
❺     if config.projectEntropy:
          SECRET_KEY = config.projectEntropy

      if not config.in_build():
❻         db_settings = config.credentials('database')
          DATABASES = {
              'default': {
                  'ENGINE': 'django.db.backends.postgresql',
                  'NAME': db_settings['path'],
                  'USER': db_settings['username'],
                  'PASSWORD': db_settings['password'],
                  'HOST': db_settings['host'],
                  'PORT': db_settings['port'],
              },
          }
```

We normally place `import` statements at the beginning of a module, but in this case, it's helpful to keep all the remote-specific settings in one section. Here we import `Config` from `platformshconfig` ❶, which helps determine settings on the remote server. We only modify settings if the method `config.is_valid_platform()` returns `True` ❷, indicating the settings are being used on a Platform.sh server.

We modify `ALLOWED_HOSTS` to allow the project to be served by hosts ending in *.platformsh.site* ❸. All projects deployed to the free tier will be served using this host. If settings are being loaded in the deployed app's directory ❹, we set `STATIC_ROOT` so that static files are served correctly. We also set a more secure `SECRET_KEY` on the remote server ❺.

Finally, we configure the production database ❻. This is only set if the build process has finished running and the project is being served. Everything you see here is necessary to let Django talk to the Postgres server that Platform.sh set up for the project.

Using Git to Track the Project's Files

As discussed in Chapter 17, Git is a version control program that allows you to take a snapshot of the code in your project each time you implement a new feature successfully. If anything goes wrong, you can easily return to the last working snapshot of your project; for example, if you accidentally introduce a bug while working on a new feature. Each snapshot is called a *commit*.

Using Git, you can try implementing new features without worrying about breaking your project. When you're deploying to a live server, you need to make sure you're deploying a working version of your project. To read more about Git and version control, see Appendix D.

Installing Git

Git may already be installed on your system. To find out, open a new terminal window and issue the command **`git --version`**:

```
(ll_env)learning_log$ git --version
git version 2.30.1 (Apple Git-130)
```

If you get a message indicating that Git is not installed, see the installation instructions in Appendix D.

Configuring Git

Git keeps track of who makes changes to a project, even when only one person is working on the project. To do this, Git needs to know your username and email. You must provide a username, but you can make up an email for your practice projects:

```
(ll_env)learning_log$ git config --global user.name "eric"
(ll_env)learning_log$ git config --global user.email "eric@example.com"
```

If you forget this step, Git will prompt you for this information when you make your first commit.

Ignoring Files

We don't need Git to track every file in the project, so we'll tell it to ignore some files. Create a file called *.gitignore* in the folder that contains *manage.py*. Notice that this filename begins with a dot and has no file extension. Here's the code that goes in *.gitignore*:

.gitignore

```
ll_env/
__pycache__/
*.sqlite3
```

We tell Git to ignore the entire *ll_env* directory, because we can re-create it automatically at any time. We also don't track the *__pycache__* directory, which contains the *.pyc* files that are created automatically when the *.py* files are executed. We don't track changes to the local database, because it's a bad habit: if you're ever using SQLite on a server, you might accidentally overwrite the live database with your local test database when you push the project to the server. The asterisk in `*.sqlite3` tells Git to ignore any file that ends with the extension *.sqlite3*.

NOTE *If you're using macOS, add* `.DS_Store` *to your* .gitignore *file. This is a file that stores information about folder settings on macOS, and it has nothing to do with this project.*

Committing the Project

We need to initialize a Git repository for Learning Log, add all the necessary files to the repository, and commit the initial state of the project. Here's how to do that:

```
❶ (ll_env)learning_log$ git init
  Initialized empty Git repository in /Users/eric/.../learning_log/.git/
❷ (ll_env)learning_log$ git add .
❸ (ll_env)learning_log$ git commit -am "Ready for deployment to Platform.sh."
  [main (root-commit) c7ffaad] Ready for deployment to Platform.sh.
   42 files changed, 879 insertions(+)
   create mode 100644 .gitignore
   create mode 100644 .platform.app.yaml
   --snip--
   create mode 100644 requirements_remote.txt
❹ (ll_env)learning_log$ git status
  On branch main
  nothing to commit, working tree clean
  (ll_env)learning_log$
```

We issue the `git init` command to initialize an empty repository in the directory containing Learning Log ❶. We then use the `git add .` command, which adds all the files that aren't being ignored to the repository ❷. (Don't forget the dot.) Next, we issue the command `git commit -am "commit message"`: the `-a` flag tells Git to include all changed files in this commit, and the `-m` flag tells Git to record a log message ❸.

Issuing the `git status` command ❹ indicates that we're on the *main* branch and that our working tree is *clean*. This is the status you'll want to see anytime you push your project to a remote server.

Creating a Project on Platform.sh

At this point, the Learning Log project still runs on our local system and is also configured to run correctly on a remote server. We'll use the Platform.sh CLI to create a new project on the server and then push our project to the remote server.

Make sure you're in a terminal, at the *learning_log/* directory, and issue the following command:

```
  (ll_env)learning_log$ platform login
  Opened URL: http://127.0.0.1:5000
  Please use the browser to log in.
  --snip--
❶ Do you want to create an SSH configuration file automatically? [Y/n] Y
```

This command will open a browser tab where you can log in. Once you're logged in, you can close the browser tab and return to the terminal. If you're prompted about creating an SSH configuration file ❶, enter **Y** so you can connect to the remote server later.

Now we'll create a project. There's a lot of output, so we'll look at the creation process in sections. Start by issuing the **create** command:

```
(ll_env)learning_log$ platform create
* Project title (--title)
Default: Untitled Project
❶ > ll_project

* Region (--region)
The region where the project will be hosted
  --snip--
  [us-3.platform.sh] Moses Lake, United States (AZURE) [514 gC02eq/kWh]
❷ > us-3.platform.sh
* Plan (--plan)
Default: development
Enter a number to choose:
  [0] development
  --snip--
❸ > 0

* Environments (--environments)
The number of environments
Default: 3
❹ > 3

* Storage (--storage)
The amount of storage per environment, in GiB
Default: 5
❺ > 5
```

The first prompt asks for a name for the project ❶, so we use the name **ll_project**. The next prompt asks which region we'd like the server to be in ❷. Choose the server closest to you; for me, that's `us-3.platform.sh`. For the rest of the prompts, you can accept the defaults: a server on the lowest development plan ❸, three environments for the project ❹, and 5GB of storage for the overall project ❺.

There are three more prompts to respond to:

```
Default branch (--default-branch)
The default Git branch name for the project (the production environment)
Default: main
❶ > main

Git repository detected: /Users/eric/.../learning_log
❷ Set the new project ll_project as the remote for this repository? [Y/n] Y

The estimated monthly cost of this project is: $10 USD
❸ Are you sure you want to continue? [Y/n] Y
```

```
The Platform.sh Bot is activating your project
```

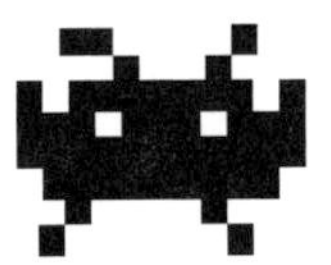

```
The project is now ready!
```

A Git repository can have multiple branches; Platform.sh is asking us if the default branch for the project should be `main` ❶. It then asks if we want to connect the local project's repository to the remote repository ❷. Finally, we're informed that this project will cost about $10 per month if we keep it running beyond the free trial period ❸. If you haven't entered a credit card yet, you shouldn't have to worry about this cost. Platform.sh will simply suspend your project if you exceed the free trial's limits without adding a credit card.

Pushing to Platform.sh

The last step before seeing the live version of the project is to push our code to the remote server. To do that, issue the following command:

```
(ll_env)learning_log$ platform push
❶ Are you sure you want to push to the main (production) branch? [Y/n] Y
--snip--
The authenticity of host 'git.us-3.platform.sh (...)' can't be established.
RSA key fingerprint is SHA256:Tvn...7PM
❷ Are you sure you want to continue connecting (yes/no/[fingerprint])? Y
Pushing HEAD to the existing environment main
  --snip--
  To git.us-3.platform.sh:3pp3mqcexhlvy.git
   * [new branch]      HEAD -> main
```

When you issue the command **`platform push`**, you'll be asked for one more confirmation that you want to push the project ❶. You may also see a message about the authenticity of Platform.sh, if this is your first time connecting to the site ❷. Enter `Y` for each of these prompts, and you'll see a bunch of output scroll by. This output will probably look confusing at first, but if anything goes wrong, it's really useful to have during troubleshooting. If you skim through the output, you can see where Platform.sh installs necessary packages, collects static files, applies migrations, and sets up URLs for the project.

NOTE *You may see an error from something that you can easily diagnose, such as a typo in one of the configuration files. If this happens, fix the error in your text editor, save the file, and reissue the* **`git commit`** *command. Then you can run* **`platform push`** *again.*

Viewing the Live Project

Once the push is complete, you can open the project:

```
(ll_env)learning_log$ platform url
Enter a number to open a URL
  [0] https://main-bvxea6i-wmye2fx7wwqgu.us-3.platformsh.site/
  --snip--
 > 0
```

The `platform url` command lists the URLs associated with a deployed project; you'll be given a choice of several URLs that are all valid for your project. Choose one, and your project should open in a new browser tab! This will look just like the project we've been running locally, but you can share this URL with anyone in the world, and they can access and use your project.

NOTE *When you deploy your project using a trial account, don't be surprised if it sometimes takes longer than usual for a page to load. On most hosting platforms, free resources that are idle are often suspended and only restarted when new requests come in. Most platforms are much more responsive on paid hosting plans.*

Refining the Platform.sh Deployment

Now we'll refine the deployment by creating a superuser, just as we did locally. We'll also make the project more secure by changing the setting `DEBUG` to `False`, so error messages won't show users any extra information that they could use to attack the server.

Creating a Superuser on Platform.sh

The database for the live project has been set up, but it's completely empty. All the users we created earlier only exist in our local version of the project.

To create a superuser on the live version of the project, we'll start an SSH (secure socket shell) session where we can run management commands on the remote server:

```
(ll_env)learning_log$ platform environment:ssh

 ___ _      _    __
| _ \ |__ _| |_ / _|___ _ _ _ __    __| |_
|  _/ / _` |  _|  _/ _ \ '_| '  \ _(_-< ' \
|_| |_\__,_|\__|_| \___/_| |_|_|_(_)__/_||_|

 Welcome to Platform.sh.

❶ web@ll_project.0:~$ ls
accounts  learning_logs  ll_project  logs  manage.py  requirements.txt
    requirements_remote.txt  static
❷ web@ll_project.0:~$ python manage.py createsuperuser
❸ Username (leave blank to use 'web'): ll_admin_live
Email address:
```

```
Password:
Password (again):
Superuser created successfully.
❹ web@ll_project.0:~$ exit
logout
Connection to ssh.us-3.platform.sh closed.
❺ (ll_env)learning_log$
```

When you first run the **`platform environment:ssh`** command, you may get another prompt about the authenticity of this host. If you see this message, enter **`Y`** and you should be logged in to a remote terminal session.

After running the **`ssh`** command, your terminal acts just like a terminal on the remote server. Note that your prompt has changed to indicate that you're in a `web` session associated with the project named `ll_project` ❶. If you issue the **`ls`** command, you'll see the files that have been pushed to the Platform.sh server.

Issue the same **`createsuperuser`** command we used in Chapter 18 ❷. This time, I entered an admin username, **`ll_admin_live`**, that's distinct from the one I used locally ❸. When you're finished working in the remote terminal session, enter the **`exit`** command ❹. Your prompt will indicate that you're working in your local system again ❺.

Now you can add */admin/* to the end of the URL for the live app and log in to the admin site. If others have already started using your project, be aware that you'll have access to all their data! Take this responsibility seriously, and users will continue to trust you with their data.

NOTE *Windows users will use the same commands shown here (such as `ls` instead of `dir`), because you're running a Linux terminal through a remote connection.*

Securing the Live Project

There's one glaring security issue in the way our project is currently deployed: the setting `DEBUG = True` in *settings.py*, which provides debug messages when errors occur. Django's error pages give you vital debugging information when you're developing a project; however, they give way too much information to attackers if you leave them enabled on a live server.

To see how bad this is, go to the home page of your deployed project. Log in to a user's account and add */topics/999/* to the end of the home page URL. Assuming you haven't made thousands of topics, you should see a page with the message *DoesNotExist at /topics/999/*. If you scroll down, you should see a whole bunch of information about the project and the server. You won't want your users to see this, and you certainly wouldn't want this information available to anyone interested in attacking the site.

We can prevent this information from being shown on the live site by setting `DEBUG = False` in the part of *settings.py* that only applies to the deployed version of the project. This way you'll continue to see debugging information locally, where that information is useful, but it won't show up on the live site.

Open *settings.py* in your text editor, and add one line of code to the part that modifies settings for Platform.sh:

settings.py

```
--snip--
if config.is_valid_platform():
    ALLOWED_HOSTS.append('.platformsh.site')
    DEBUG = False
    --snip--
```

All the work to set up configuration for the deployed version of the project has paid off. When we want to adjust the live version of the project, we just change the relevant part of the configuration we set up earlier.

Committing and Pushing Changes

Now we need to commit the changes made to *settings.py* and push the changes to Platform.sh. Here's a terminal session showing the first part of this process:

```
❶ (ll_env)learning_log$ git commit -am "Set DEBUG False on live site."
[main d2ad0f7] Set DEBUG False on live site.
  1 file changed, 1 insertion(+)
❷ (ll_env)learning_log$ git status
On branch main
nothing to commit, working tree clean
(ll_env)learning_log$
```

We issue the `git commit` command with a short but descriptive commit message ❶. Remember the `-am` flag makes sure Git commits all the files that have changed and records the log message. Git recognizes that one file has changed and commits this change to the repository.

Running **`git status`** shows that we're working on the `main` branch of the repository and that there are now no new changes to commit ❷. It's important to check the status before pushing to a remote server. If you don't see a clean status, then some changes haven't been committed and those changes won't be pushed to the server. You can try issuing the `commit` command again; if you're not sure how to resolve the issue, read through Appendix D to better understand how to work with Git.

Now let's push the updated repository to Platform.sh:

```
(ll_env)learning_log$ platform push
Are you sure you want to push to the main (production) branch? [Y/n] Y
Pushing HEAD to the existing environment main
--snip--
  To git.us-3.platform.sh:wmye2fx7wwqgu.git
     fce0206..d2ad0f7  HEAD -> main
(ll_env)learning_log$
```

Platform.sh recognizes that the repository has been updated, and it rebuilds the project to make sure all the changes have been taken into account. It doesn't rebuild the database, so we haven't lost any data.

To make sure this change took effect, visit the */topics/999/* URL again. You should see just the message *Server Error (500)*, with no sensitive information about the project at all.

Creating Custom Error Pages

In Chapter 19, we configured Learning Log to return a 404 error if the user requests a topic or entry that doesn't belong to them. Now you've seen a 500 server error as well. A 404 error usually means your Django code is correct, but the object being requested doesn't exist. A 500 error usually means there's an error in the code you've written, such as an error in a function in *views.py*. Django currently returns the same generic error page in both situations, but we can write our own 404 and 500 error page templates that match Learning Log's overall appearance. These templates belong in the root template directory.

Making Custom Templates

In the *learning_log* folder, make a new folder called *templates*. Then make a new file called *404.html*; the path to this file should be *learning_log/templates/404.html*. Here's the code for this file:

404.html

```
{% extends "learning_logs/base.html" %}

{% block page_header %}
  <h2>The item you requested is not available. (404)</h2>
{% endblock page_header %}
```

This simple template provides the generic 404 error page information but is styled to match the rest of the site.

Make another file called *500.html* using the following code:

500.html

```
{% extends "learning_logs/base.html" %}

{% block page_header %}
  <h2>There has been an internal error. (500)</h2>
{% endblock page_header %}
```

These new files require a slight change to *settings.py*.

settings.py

```
--snip--
TEMPLATES = [
    {
        'BACKEND': 'django.template.backends.django.DjangoTemplates',
        'DIRS': [BASE_DIR / 'templates'],
        'APP_DIRS': True,
        --snip--
    },
]
--snip--
```

This change tells Django to look in the root template directory for the error page templates and any other templates that aren't associated with a particular app.

Pushing the Changes to Platform.sh

Now we need to commit the changes we just made and push them to Platform.sh:

```
❶ (ll_env)learning_log$ git add .
❷ (ll_env)learning_log$ git commit -am "Added custom 404 and 500 error pages."
   3 files changed, 11 insertions(+), 1 deletion(-)
   create mode 100644 templates/404.html
   create mode 100644 templates/500.html
❸ (ll_env)learning_log$ platform push
  --snip--
    To git.us-3.platform.sh:wmye2fx7wwqgu.git
       d2ad0f7..9f042ef  HEAD -> main
  (ll_env)learning_log$
```

We issue the `git add .` command ❶ because we created some new files in the project. Then we commit the changes ❷ and push the updated project to Platform.sh ❸.

Now when an error page appears, it should have the same styling as the rest of the site, making for a smoother user experience when errors arise.

Ongoing Development

You might want to further develop Learning Log after your initial push to a live server, or you might want to develop your own projects to deploy. When doing so, there's a fairly consistent process for updating your projects.

First, you'll make the necessary changes to your local project. If your changes result in any new files, add those files to the Git repository using the command `git add .` (making sure to include the dot at the end of the command). Any change that requires a database migration will need this command, because each migration generates a new migration file.

Second, commit the changes to your repository using `git commit -am "commit message"`. Then push your changes to Platform.sh, using the command `platform push`. Visit your live project and make sure the changes you expect to see have taken effect.

It's easy to make mistakes during this process, so don't be surprised when something goes wrong. If the code doesn't work, review what you've done and try to spot the mistake. If you can't find the mistake or you can't figure out how to undo it, refer to the suggestions for getting help in Appendix C. Don't be shy about asking for help: everyone else learned to build projects by asking the same questions you're likely to ask, so someone will be happy to help you. Solving each problem that arises helps you steadily develop your skills until you're building meaningful, reliable projects and answering other people's questions as well.

Deleting a Project on Platform.sh

It's great practice to run through the deployment process a number of times with the same project or with a series of small projects, to get the hang of deployment. But you'll need to know how to delete a project that's been deployed. Platform.sh also limits the number of projects you can host for free, and you don't want to clutter your account with practice projects.

You can delete a project using the CLI:

```
(ll_env)learning_log$ platform project:delete
```

You'll be asked to confirm that you want to take this destructive action. Respond to the prompts, and your project will be deleted.

The command `platform create` also gave the local Git repository a reference to the remote repository on Platform.sh's servers. You can remove this remote from the command line as well:

```
(ll_env)learning_log$ git remote
platform
(ll_env)learning_log$ git remote remove platform
```

The command `git remote` lists the names of all remote URLs associated with the current repository. The command `git remote remove remote_name` deletes these remote URLs from the local repository.

You can also delete a project's resources by logging in to the Platform.sh website and visiting your dashboard at *https://console.platform.sh*. This page lists all your active projects. Click the three dots in a project's box, and click **Edit Plan**. This is a pricing page for the project; click the **Delete Project** button at the bottom of the page, and you'll be shown a confirmation page where you can follow through with the deletion. Even if you deleted your project using the CLI, it's a good idea to familiarize yourself with the dashboard of any hosting provider you deploy to.

NOTE *Deleting a project on Platform.sh does nothing to your local version of the project. If no one has used your deployed project and you're just practicing the deployment process, it's perfectly reasonable to delete your project on Platform.sh and redeploy it. Just be aware that if things stop working, you may have run into the host's free-tier limitations.*

TRY IT YOURSELF

20-3. Live Blog: Deploy the Blog project you've been working on to Platform.sh. Make sure you set `DEBUG` to `False`, so users don't see the full Django error pages when something goes wrong.

(continued)

20-4. Extended Learning Log: Add one feature to Learning Log, and push the change to your live deployment. Try a simple change, such as writing more about the project on the home page. Then try adding a more advanced feature, such as giving users the option of making a topic public. This would require an attribute called `public` as part of the `Topic` model (this should be set to `False` by default) and a form element on the new_topic page that allows the user to change a topic from private to public. You'd then need to migrate the project and revise *views.py* so any topic that's public is visible to unauthenticated users as well.

Summary

In this chapter, you learned to give your projects a simple but professional appearance using the Bootstrap library and the django-bootstrap5 app. With Bootstrap, the styles you choose will work consistently on almost any device people use to access your project.

You learned about Bootstrap's templates and used the *Navbar static* template to create a simple look and feel for Learning Log. You used a jumbotron to make a home page's message stand out, and learned to style all the pages in a site consistently.

In the final part of the project, you learned how to deploy a project to a remote server so anyone can access it. You made a Platform.sh account and installed some tools that help manage the deployment process. You used Git to commit the working project to a repository, and then pushed the repository to a remote server on Platform.sh. Finally, you learned to begin securing your app by setting `DEBUG = False` on the live server. You also made custom error pages, so the inevitable errors that come up will look well-handled.

Now that you've finished Learning Log, you can start building your own projects. Start simple, and make sure the project works before adding complexity. Enjoy your continued learning, and good luck with your projects!

A

INSTALLATION AND TROUBLESHOOTING

There are many versions of Python available and numerous ways to set it up on each operating system. If the approach in Chapter 1 didn't work, or if you want to install a different version of Python than the one currently installed, the instructions in this appendix can help.

Python on Windows

The instructions in Chapter 1 show you how to install Python using the official installer at *https://python.org*. If you couldn't get Python to run after using the installer, the troubleshooting instructions in this section should help you get Python up and running.

Using py Instead of python

If you run a recent Python installer and then issue the command **python** in a terminal, you should see the Python prompt for a terminal session (>>>).

When Windows doesn't recognize the `python` command, it will either open the Microsoft Store because it thinks Python isn't installed, or you'll get a message such as "Python was not found." If the Microsoft Store opens, close it; it's better to use the official Python installer from *https://python.org* than the one that Microsoft maintains.

The simplest solution, without making any changes to your system, is to try the **`py`** command. This is a Windows utility that finds the latest version of Python installed on your system and runs that interpreter. If this command works and you want to use it, simply use **`py`** anywhere you see the `python` or `python3` command in this book.

Rerunning the Installer

The most common reason `python` doesn't work is that people forget to select the Add Python to PATH option when running the installer; this is an easy mistake to make. The `PATH` variable is a system setting that tells Python where to look for commonly used programs. In this case, Windows doesn't know how to find the Python interpreter.

The simplest fix in this situation is to run the installer again. If there's a newer installer available from *https://python.org*, download the new installer and run it, making sure to check the **Add Python to PATH** box.

If you already have the latest installer, run it again and select the **Modify** option. You'll see a list of optional features; keep the default options selected on this screen. Then click **Next** and check the **Add Python to Environment Variables** box. Finally, click **Install**. The installer will recognize that Python is already installed, and it will add the location of the Python interpreter to the `PATH` variable. Make sure you close any open terminals, because they'll still be using the old `PATH` variable. Open a new terminal window and issue the command **`python`** again; you should see a Python prompt (`>>>`).

Python on macOS

The installation instructions in Chapter 1 use the official Python installer at *https://python.org*. The official installer has been working well for years now, but there are a few things that can get you off track. This section will help if anything isn't working in a straightforward manner.

Accidentally Installing Apple's Version of Python

If you run the `python3` command and Python is not yet installed on your system, you'll most likely see a message that the *command line developer tools* need to be installed. The best approach at this point is to close the pop-up showing this message, download the Python installer from *https://python.org*, and run the installer.

If you choose to install the command line developer tools at this point, macOS will install Apple's version of Python along with the developer tools. The only issue with this is that Apple's version of Python is usually somewhat behind the latest official version of Python. However, you can still download and run the official installer from *https://python.org*, and `python3` will then point to the newer version. Don't worry about having the developer tools installed; there are some useful tools in there, including the Git version control system discussed in Appendix D.

Python 2 on Older Versions of macOS

On older versions of macOS, before Monterey (macOS 12), an outdated version of Python 2 was installed by default. On these systems, the command `python` points to the outdated system interpreter. If you're using a version of macOS with Python 2 installed, make sure you use the **`python3`** command, and you'll always be using the version of Python you installed.

Python on Linux

Python is included by default on almost every Linux system. However, if the default version on your system is earlier than Python 3.9, you should install the latest version. You can also install the latest version if you want the most recent features, like Python's improved error messages. The following instructions should work for most apt-based systems.

Using the Default Python Installation

If you want to use the version of Python that `python3` points to, make sure you have these three additional packages installed:

```
$ sudo apt install python3-dev python3-pip python3-venv
```

These packages include tools that are useful for developers and tools that let you install third-party packages, like the ones used in the projects section of this book.

Installing the Latest Version of Python

We'll use a package called `deadsnakes`, which makes it easy to install multiple versions of Python. Enter the following commands:

```
$ sudo add-apt-repository ppa:deadsnakes/ppa
$ sudo apt update
$ sudo apt install python3.11
```

These commands will install Python 3.11 onto your system.

Enter the following command to start a terminal session that runs Python 3.11:

```
$ python3.11
>>>
```

Anywhere you see the command `python` in this book, use **`python3.11`** instead. You'll also want to use this command when you run programs from the terminal.

You'll need to install two more packages to make the most of your Python installation:

```
$ sudo apt install python3.11-dev python3.11-venv
```

These packages include modules you'll need when installing and running third-party packages, like the ones used in the projects in the second half of the book.

NOTE *The* deadsnakes *package has been actively maintained for a long time. When newer versions of Python come out, you can use these same commands, replacing* python3.11 *with the latest version currently available.*

Checking Which Version of Python You're Using

If you're having any issues running Python or installing additional packages, it can be helpful to know exactly which version of Python you're using. You may have multiple versions of Python installed and not be clear about which version is currently being used.

Issue the following command in a terminal:

```
$ python --version
Python 3.11.0
```

This tells you exactly which version the command python is currently pointing to. The shorter command python -V will give the same output.

Python Keywords and Built-in Functions

Python comes with its own set of keywords and built-in functions. It's important to be aware of these when you're naming things in Python: your names cannot be the same as these keywords and shouldn't be the same as the function names, or you'll overwrite the functions.

In this section, we'll list Python's keywords and built-in function names, so you'll know which names to avoid.

Python Keywords

Each of the following keywords has a specific meaning, and you'll see an error if you try to use any of them as a variable name.

```
False      await      else       import     pass
None       break      except     in         raise
True       class      finally    is         return
and        continue   for        lambda     try
as         def        from       nonlocal   while
assert     del        global     not        with
async      elif       if         or         yield
```

Python Built-in Functions

You won't get an error if you use one of the following readily available built-in functions as a variable name, but you'll override the behavior of that function:

abs()	complex()	hash()	min()	slice()
aiter()	delattr()	help()	next()	sorted()
all()	dict()	hex()	object()	staticmethod()
any()	dir()	id()	oct()	str()
anext()	divmod()	input()	open()	sum()
ascii()	enumerate()	int()	ord()	super()
bin()	eval()	isinstance()	pow()	tuple()
bool()	exec()	issubclass()	print()	type()
breakpoint()	filter()	iter()	property()	vars()
bytearray()	float()	len()	range()	zip()
bytes()	format()	list()	repr()	__import__()
callable()	frozenset()	locals()	reversed()	
chr()	getattr()	map()	round()	
classmethod()	globals()	max()	set()	
compile()	hasattr()	memoryview()	setattr()	

B

TEXT EDITORS AND IDES

Programmers spend a lot of time writing, reading, and editing code, and using a text editor or an IDE (integrated development environment) to make this work as efficient as possible is essential. A good editor will do simple tasks, like highlighting your code's structure so you can catch common bugs as you're working. But it won't do so much that it distracts you from your thinking. Editors also have useful features like automatic indenting, markers to show appropriate line length, and keyboard shortcuts for common operations.

An *IDE* is a text editor with a number of other tools included, like interactive debuggers and code introspection. An IDE examines your code as you enter it and tries to learn about the project you're building. For example, when you start typing the name of a function, an IDE might show you all the arguments that function accepts. This behavior can be very helpful when everything works and you understand what you're seeing. But it can also be overwhelming as a beginner and difficult to troubleshoot when you aren't sure why your code isn't working in the IDE.

These days, the lines have blurred between text editors and IDEs. Most popular editors have some features that used to be exclusive to IDEs. Likewise, most IDEs can be configured to run in a lighter mode that's less distracting as you work, but lets you use the more advanced features when you need them.

If you already have an editor or IDE installed that you like, and if it's already configured to work with a recent version of Python that's installed on your system, then I encourage you to stick with what you already know. Exploring different editors can be fun, but it's also a way to avoid the work of learning a new language.

If you don't already have an editor or IDE installed, I recommend VS Code for a number of reasons:

- It's free, and it's released under an open source license.
- It can be installed on all major operating systems.
- It's beginner-friendly but also powerful enough that many professional programmers use it as their main editor.
- It finds the versions of Python you have installed, and it typically does not require any configuration to run your first programs.
- It has an integrated terminal, so your output appears in the same window as your code.
- A Python extension is available that makes the editor highly efficient for writing and maintaining Python code.
- It's highly customizable, so you can tune it to match the way you work with code.

In this appendix, you'll learn how to start configuring VS Code so that it works well for you. You'll also learn some shortcuts that let you work more efficiently. Being a fast typist is not as important as many people think in programming, but understanding your editor and knowing how to use it efficiently is quite helpful.

With all that said, VS Code doesn't work for everyone. If it doesn't work well on your system for some reason, or if it's distracting you as you work, there are a number of other editors that you might find more appealing. This appendix includes a brief description of some of the other editors and IDEs you should consider.

Working Efficiently with VS Code

In Chapter 1, you installed VS Code and added the Python extension as well. This section will show you some further configurations you can make, plus shortcuts for working efficiently with your code.

Configuring VS Code

There are a few ways to change the default configuration settings for VS Code. Some changes can be made through the interface, and some will require changes in configuration files. These changes will sometimes take effect for everything you do in VS Code, while others will affect only the files within the folder that contains the configuration file.

For example, if you have a configuration file in your *python_work* folder, those settings will only affect the files in that folder (and its subfolders). This is a good feature, because it means you can have project-specific settings that override your global settings.

Using Tabs and Spaces

If you use a mix of tabs and spaces in your code, it can cause problems in your programs that are difficult to diagnose. When working in a *.py* file with the Python extension installed, VS Code is configured to insert four spaces whenever you press the TAB key. If you're writing only your own code and you have the Python extension installed, you'll likely never have an issue with tabs and spaces.

However, your installation of VS Code may not be configured correctly. Also, at some point, you may end up working on a file that has only tabs or a mix of tabs and spaces. If you suspect any issue with tabs and spaces, look at the status bar at the bottom of the VS Code window and click either **Spaces** or **Tab Size**. A drop-down menu will appear that lets you switch between using tabs and using spaces. You can also change the default indentation level and convert all indentation in the file to either tabs or spaces.

If you're looking at some code and you're not sure whether the indentation consists of tabs or spaces, highlight several lines of code. This will make the invisible whitespace characters visible. Each space will show up as a dot, and each tab will show up as an arrow.

NOTE *In programming, spaces are preferred over tabs because spaces can be interpreted unambiguously by all tools that work with a code file. The width of tabs can be interpreted differently by different tools, which leads to errors that can be extremely difficult to diagnose.*

Changing the Color Theme

VS Code uses a dark theme by default. If you want to change this, click **File** (**Code** in the menu bar on macOS), then click **Preferences** and choose **Color Theme**. A drop-down list will appear, and it will let you choose a theme that works well for you.

Setting the Line Length Indicator

Most editors allow you to set up a visual cue, usually a vertical line, to show where your lines should end. In the Python community, the convention is to limit lines to 79 characters or less.

To set this feature, click **Code** and then **Preferences**, and then choose **Settings**. In the dialog that appears, enter `rulers`. You'll see a setting for

Editor: Rulers; click the link labeled *Edit in settings.json*. In the file that appears, add the following to the `editor.rulers` setting:

settings.json

```
    "editor.rulers": [
    80,
    ]
```

This will add a vertical line in the editing window at the 80-character position. You can have more than one vertical line; for example, if you want an additional line at 120 characters, the value for your setting would be `[80, 120]`. If you don't see the vertical lines, make sure you saved the settings file; you may also need to quit and reopen VS Code for the changes to take effect on some systems.

Simplifying the Output

By default, VS Code shows the output of your programs in an embedded terminal window. This output includes the commands that are being used to run the file. For many situations, this is ideal, but it might be more distracting than you want when you're first learning Python.

To simplify the output, close all the tabs that are open in VS Code and then quit VS Code. Launch VS Code again and open the folder that contains the Python files you're working on; this could just be the *python_work* folder where *hello_world.py* is saved.

Click the Run/Debug icon (which looks like a triangle with a small bug), and then click **Create a *launch.json* File**. Select the Python options in the prompts that appear. In the *launch.json* file that opens, make the following change:

launch.json

```
{
    --snip--
    "configurations": [
        {
            --snip--
            "console": "internalConsole",
            "justMyCode": true
        }
    ]
}
```

Here, we're changing the `console` setting from `integratedTerminal` to `internalConsole`. After saving the settings file, open a *.py* file such as *hello_world.py*, and run it by pressing CTRL-F5. In the output pane of VS Code, click **Debug Console** if it's not already selected. You should see only your program's output, and the output should be refreshed every time you run a program.

NOTE *The Debug Console is read-only. It won't work for files that use the `input()` function, which you'll start using in Chapter 7. When you need to run these programs, you can either change the `console` setting back to the default `integratedTerminal`, or you can run these programs in a separate terminal window as described in "Running Python Programs from a Terminal" on page 11.*

Exploring Further Customizations

You can customize VS Code in many ways to help you work more efficiently. To start exploring the customizations available, click **Code** and then **Preferences**, and then choose **Settings**. You'll see a list titled Commonly Used; click any of the subheadings to see some common ways you can modify your installation of VS Code. Take some time to see if there are any that make VS Code work better for you, but don't get so lost in configuring your editor that you put off learning how to use Python!

VS Code Shortcuts

All editors and IDEs offer efficient ways to do common tasks that everyone needs to do when writing and maintaining code. For example, you can easily indent a single line of code or an entire block of code; you can just as easily move a block of lines up or down in a file.

There are too many shortcuts to describe fully here. This section will share just a few that you'll likely find helpful as you're writing your first Python files. If you end up using a different editor than VS Code, make sure you learn how to do these same tasks efficiently in the editor you've chosen.

Indenting and Unindenting Code Blocks

To indent an entire block of code, highlight it and press CTRL-], or ⌘-] on macOS. To unindent a block of code, highlight it and press CTRL-[, or ⌘-[on macOS.

Commenting Out Blocks of Code

To temporarily disable a block of code, you can highlight the block and comment it so Python will ignore it. Highlight the section of code you want to ignore and press CTRL-/, or ⌘-/ on macOS. The selected lines will be commented out with a hash mark (#) indented at the same level as the line of code, to indicate these are not regular comments. When you want to uncomment the block of code, highlight the block and reissue the same command.

Moving Lines Up or Down

As your programs grow more complex, you may find that you want to move a block of code up or down within a file. To do so, highlight the code you want to move and press ALT-up arrow, or Option-up arrow on macOS. The same key combination with the down arrow will move the block down in the file.

If you're moving a single line up or down, you can click anywhere in that line; you don't need to highlight the whole line to move it.

Hiding the File Explorer

The integrated file explorer in VS Code is really convenient. However, it can be distracting when you're writing code and can take up valuable space on a smaller screen. The command CTRL-B, or ⌘-B on macOS, toggles the visibility of the file explorer pane.

Finding Additional Shortcuts

Working efficiently in an editing environment takes practice, but it also takes intention. When you're learning to work with code, try to notice the things you do repeatedly. Any action you take in your editor likely has a shortcut; if you're clicking menu items to carry out editing tasks, look for the shortcuts for those actions. If you're switching between your keyboard and mouse frequently, look for the navigation shortcuts that keep you from reaching for your mouse so often.

You can see all the keyboard shortcuts in VS Code by clicking **Code** and then **Preferences**, and then choosing **Keyboard Shortcuts**. You can use the search bar to find a particular shortcut, or you can scroll through the list to find shortcuts that might help you work more efficiently.

Remember, it's better to focus on the code that you're working on, and avoid spending too much time on the tools you're using.

Other Text Editors and IDEs

You'll hear about and see people using a number of other text editors. Most of them can be configured to help you in the same way you've customized VS Code. Here's a small selection of text editors you might hear about.

IDLE

IDLE is a text editor that's included with Python. It's a little less intuitive to work with than other, more modern editors. However, you'll see references to it in other tutorials aimed at beginners, so you might want to give it a try.

Geany

Geany is a simple text editor that displays all of your output in a separate terminal window, which helps you become comfortable using terminals. Geany has a very minimalist interface, but it's powerful enough that a significant number of experienced programmers still use it.

If you find VS Code too distracting and full of too many features, consider using Geany instead.

Sublime Text

Sublime Text is another minimalist editor that you should consider using if you find VS Code too busy. Sublime Text has a really clean interface and is known for working well even on very large files. It's an editor that will get out of your way and let you focus on the code you're writing.

Sublime Text has an unlimited free trial, but it's not free or open source. If you decide you like it and can afford to purchase a full license, you should do so. The purchase is a one-time fee; it's not a software subscription.

Emacs and Vim

Emacs and *Vim* are two popular editors favored by many experienced programmers, because they're designed so you can use them without your hands ever having to leave the keyboard. This makes writing, reading, and modifying code very efficient, once you learn how the editor works. It also means both editors have a fairly steep learning curve. Vim is included on most Linux and macOS machines, and both Emacs and Vim can be run entirely inside a terminal. For this reason, they're often used to write code on servers through remote terminal sessions.

Programmers will often recommend that you give them a try, but many proficient programmers forget how much new programmers are already trying to learn. It's good to be aware of these editors, but you should hold off on using them until you're comfortable working with code in a more user-friendly editor that lets you focus on learning to program, rather than learning to use an editor.

PyCharm

PyCharm is a popular IDE among Python programmers because it was built to work specifically with Python. The full version requires a paid subscription, but a free version called the PyCharm Community Edition is also available, and many developers find it useful.

If you try PyCharm, be aware that, by default, it sets up an isolated environment for each of your projects. This is usually a good thing, but it can lead to unexpected behavior if you don't understand what it's doing for you.

Jupyter Notebooks

Jupyter Notebook is a different kind of tool than traditional text editors or IDEs, in that it's a web app primarily built of blocks; each block is either a code block or a text block. The text blocks are rendered in Markdown, so you can include simple formatting in your text blocks.

Jupyter Notebooks were developed to support the use of Python in scientific applications, but they have since expanded to become useful in a wide variety of situations. Rather than just writing comments inside a *.py* file, you can write clear text with simple formatting, such as headers, bulleted lists, and hyperlinks in between sections of code. Every code block can be run independently, allowing you to test small pieces of your program, or you can run all the code blocks at once. Each code block has its own output area, and you can toggle the output areas on or off as needed.

Jupyter Notebooks can be confusing at times because of the interactions between different cells. If you define a function in one cell, that function is available to other cells as well. This is beneficial most of the time, but it can be confusing in longer notebooks and if you don't fully understand how the Notebook environment works.

If you're doing any scientific or data-focused work in Python, you'll almost certainly see Jupyter Notebooks at some point.

C

GETTING HELP

Everyone gets stuck at some point when they're learning to program. So, one of the most important skills to learn as a programmer is how to get unstuck efficiently. This appendix outlines several ways to help you get going again when programming gets confusing.

First Steps

When you're stuck, your first step should be to assess your situation. Before you ask for help from anyone else, answer the following three questions clearly:

- What are you trying to do?
- What have you tried so far?
- What results have you been getting?

Make your answers as specific as possible. For the first question, explicit statements like "I'm trying to install the latest version of Python on my new Windows laptop" are detailed enough for others in the Python community to help you. Statements like "I'm trying to install Python" don't provide enough information for others to offer much help.

Your answer to the second question should provide enough detail so you won't be advised to repeat what you've already tried: "I went to *https://python.org/downloads* and clicked the Download button for my system. Then I ran the installer" is more helpful than "I went to the Python website and downloaded something."

For the third question, it's helpful to know the exact error messages you received, so you can use them to search online for a solution or provide them when asking for help.

Sometimes, just answering these three questions before you ask for help from others allows you to see something you're missing, and helps get you unstuck without having to go any further. Programmers even have a name for this: *rubber duck debugging*. The idea is that if you clearly explain your situation to a rubber duck (or any inanimate object) and ask it a specific question, you'll often be able to answer your own question. Some programming teams even keep a real rubber duck around to encourage people to "talk to the duck."

Try It Again

Just going back to the start and trying again can be enough to solve many problems. Say you're trying to write a `for` loop based on an example in this book. You might have only missed something simple, like a colon at the end of the `for` line. Going through the steps again might help you avoid repeating the same mistake.

Take a Break

If you've been working on the same problem for a while, taking a break is one of the best tactics you can try. When we work on the same task for long periods of time, our brains start to zero in on only one solution. We lose sight of the assumptions we've made, and taking a break helps us get a fresh perspective on the problem. It doesn't need to be a long break, just something that gets you out of your current mindset. If you've been sitting for a long time, do something physical: take a short walk, go outside for a bit, or perhaps drink a glass of water or eat a light snack.

If you're getting frustrated, it might be worth putting your work away for the day. A good night's sleep almost always makes a problem more approachable.

Refer to This Book's Resources

The online resources for this book, available at *https://ehmatthes.github.io/pcc_3e*, include a number of helpful sections about setting up your system and working through each chapter. If you haven't done so already, take a look at these resources and see if there's anything that helps your situation.

Searching Online

Chances are good that someone else has had the same problem you're having and has written about it online. Good searching skills and specific inquiries will help you find existing resources to solve the issue you're facing. For example, if you're struggling to install the latest version of Python on a new Windows system, searching for *install python windows* and limiting the results to resources from the last year might direct you to a clear answer.

Searching the exact error message can be extremely helpful too. For example, say you get the following error when you try to run a Python program from a terminal on a new Windows system:

```
> python hello_world.py
Python was not found; run without arguments to install from the Microsoft
  Store...
```

Searching for the full phrase, "Python was not found; run without arguments to install from the Microsoft Store," will probably yield some good advice.

When you start searching for programming-related topics, a few sites will appear repeatedly. I'll describe some of these sites briefly, so you'll know how helpful they're likely to be.

Stack Overflow

Stack Overflow (*https://stackoverflow.com*) is one of the most popular question-and-answer sites for programmers, and it will often appear in the first page of results on Python-related searches. Members post questions when they're stuck, and other members try to give helpful responses. Users can vote for the responses they find most helpful, so the best answers are usually the first ones you'll find.

Many basic Python questions have very clear answers on Stack Overflow, because the community has refined them over time. Users are encouraged to post updates, too, so responses tend to stay relatively current. At the time of this writing, almost two million Python-related questions have been answered on Stack Overflow.

There's one expectation you should be aware of before posting on Stack Overflow. Questions are meant to be the shortest example of the kind of issue you're facing. If you post 5–20 lines of code that generate the error you're facing, and if you address the questions mentioned in "First Steps" on page 477 earlier in this appendix, someone will probably help you. If you share a link to a project with multiple large files, people will be very unlikely to help. There's a great guide to writing up a good question at *https://stackoverflow.com/help/how-to-ask*. The suggestions in this guide are applicable to getting help in any community of programmers.

The Official Python Documentation

The official Python documentation (*https://docs.python.org*) is a bit more hit-or-miss for beginners, because its purpose is more to document the language than to provide explanations. The examples in the official documentation should work, but you might not understand everything shown. Still, it's a good resource to check when it comes up in your searches, and it will become more useful to you as you continue building your understanding of Python.

Official Library Documentation

If you're using a specific library, such as Pygame, Matplotlib, or Django, links to the official documentation for it will often appear in searches. For example, *https://docs.djangoproject.com* is very helpful when working with Django. If you're planning to work with any of these libraries, it's a good idea to become familiar with their official documentation.

r/learnpython

Reddit is made up of a number of subforums called *subreddits*. The *r/learnpython* subreddit (*https://reddit.com/r/learnpython*) is very active and supportive. You can read others' questions and post your own as well. You will often get multiple perspectives about the questions you raise, which can be really helpful in gaining a deeper understanding of the topic you're working on.

Blog Posts

Many programmers maintain blogs and share posts about the parts of the language they're working with. You should look for a date on the blog posts you find, to see how applicable the information is likely to be for the version of Python you're using.

Discord

Discord is an online chat environment with a Python community where you can ask for help and follow Python-related discussions.

To check it out, head to *https://pythondiscord.com* and click the **Discord** link at the upper right. If you already have a Discord account, you can log in with your existing account. If you don't have an account, enter a username and follow the prompts to complete your Discord registration.

If this is your first time visiting the Python Discord, you'll need to accept the rules for the community before participating fully. Once you've done that, you can join any of the channels that interest you. If you're looking for help, be sure to post in one of the Python Help channels.

Slack

Slack is another online chat environment. It is often used for internal company communications, but there are also many public groups you can join. If you want to check out Python Slack groups, start with *https://pyslackers.com*. Click the **Slack** link at the top of the page, then enter your email address to get an invitation.

Once you're in the Python Developers workspace, you'll see a list of channels. Click **Channels** and then choose the topics that interest you. You might want to start with the *#help* and *#django* channels.

D

USING GIT FOR VERSION CONTROL

Version control software allows you to take snapshots of a project whenever it's in a working state. When you make changes to a project—for example, when you implement a new feature—you can go back to a previous working state if the project's current state isn't functioning well.

Using version control software gives you the freedom to work on improvements and make mistakes without worrying about ruining your project. This is especially critical in large projects, but can also be helpful in smaller projects, even when you're working on programs contained in a single file.

In this appendix, you'll learn to install Git and use it for version control in the programs you're working on now. *Git* is the most popular version control software in use today. Many of its advanced tools help teams collaborate on large projects, but its most basic features also work well for solo developers. Git implements version control by tracking the changes made to every file in a project; if you make a mistake, you can just return to a previously saved state.

Installing Git

Git runs on all operating systems, but there are different approaches to installing it on each system. The following sections provide specific instructions for each operating system.

Git is included on some systems by default, and is often bundled with other packages that you might have already installed. Before trying to install Git, see if it's already on your system. Open a new terminal window and issue the command **git --version**. If you see output listing a specific version number, Git is installed on your system. If you see a message prompting you to install or update Git, follow the onscreen instructions.

If you don't see any onscreen instructions and you're using Windows or macOS, you can download an installer from *https://git-scm.com*. If you're a Linux user with an apt-compatible system, you can install Git with the command **sudo apt install git**.

Configuring Git

Git keeps track of who makes changes to a project, even when only one person is working on the project. To do this, Git needs to know your username and email. You must provide a username, but you can make up a fake email address:

```
$ git config --global user.name "username"
$ git config --global user.email "username@example.com"
```

If you forget this step, Git will prompt you for this information when you make your first commit.

It's also best to set the default name for the main branch in each project. A good name for this branch is `main`:

```
$ git config --global init.defaultBranch main
```

This configuration means that each new project you use Git to manage will start out with a single branch of commits called *main*.

Making a Project

Let's make a project to work with. Create a folder somewhere on your system called *git_practice*. Inside the folder, make a simple Python program:

hello_git.py

```
print("Hello Git world!")
```

We'll use this program to explore Git's basic functionality.

Ignoring Files

Files with the extension *.pyc* are automatically generated from *.py* files, so we don't need Git to keep track of them. These files are stored in a directory called *__pycache__*. To tell Git to ignore this directory, make a special file called *.gitignore*—with a dot at the beginning of the filename and no file extension—and add the following line to it:

.gitignore

```
__pycache__/
```

This file tells Git to ignore any file in the *__pycache__* directory. Using a *.gitignore* file will keep your project clutter-free and easier to work with.

You might need to modify your file browser's settings so hidden files (files whose names begin with a dot) will be shown. In Windows Explorer, check the box in the View menu labeled **Hidden Items**. On macOS, press ⌘-SHIFT-. (dot). On Linux, look for a setting labeled Show Hidden Files.

NOTE *If you're on macOS, add one more line to* .gitignore. *Add the name* `.DS_Store`*; these are hidden files that contain information about each directory on macOS, and they will clutter up your project if you don't add them to* .gitignore.

Initializing a Repository

Now that you have a directory containing a Python file and a *.gitignore* file, you can initialize a Git repository. Open a terminal, navigate to the *git_practice* folder, and run the following command:

```
git_practice$ git init
Initialized empty Git repository in git_practice/.git/
git_practice$
```

The output shows that Git has initialized an empty repository in *git_practice*. A *repository* is the set of files in a program that Git is actively tracking. All the files Git uses to manage the repository are located in the hidden directory *.git*, which you won't need to work with at all. Just don't delete that directory, or you'll lose your project's history.

Checking the Status

Before doing anything else, let's look at the project's status:

```
  git_practice$ git status
❶ On branch main
  No commits yet

❷ Untracked files:
    (use "git add <file>..." to include in what will be committed)
        .gitignore
        hello_git.py

❸ nothing added to commit but untracked files present (use "git add" to track)
  git_practice$
```

In Git, a *branch* is a version of the project you're working on; here you can see that we're on a branch named main ❶. Each time you check your project's status, it should show that you're on the branch main. You then see that we're about to make the initial commit. A *commit* is a snapshot of the project at a particular point in time.

Git informs us that untracked files are in the project ❷, because we haven't told it which files to track yet. Then we're told that there's nothing added to the current commit, but untracked files are present that we might want to add to the repository ❸.

Adding Files to the Repository

Let's add the two files to the repository and check the status again:

```
❶ git_practice$ git add .
❷ git_practice$ git status
  On branch main
  No commits yet

  Changes to be committed:
    (use "git rm --cached <file>..." to unstage)
❸       new file:   .gitignore
        new file:   hello_git.py

  git_practice$
```

The command git add . adds to the repository all files within a project that aren't already being tracked ❶, as long as they're not listed in *.gitignore*. It doesn't commit the files; it just tells Git to start paying attention to them. When we check the status of the project now, we can see that Git recognizes some changes that need to be committed ❷. The label *new file* means these files were newly added to the repository ❸.

Making a Commit

Let's make the first commit:

```
❶ git_practice$ git commit -m "Started project."
❷ [main (root-commit) cea13dd] Started project.
❸ 2 files changed, 5 insertions(+)
   create mode 100644 .gitignore
   create mode 100644 hello_git.py
❹ git_practice$ git status
  On branch main
  nothing to commit, working tree clean
  git_practice$
```

We issue the command git commit -m "*message*" ❶ to make a snapshot of the project. The -m flag tells Git to record the message that follows (Started project.) in the project's log. The output shows that we're on the main branch ❷ and that two files have changed ❸.

When we check the status now, we can see that we're on the main branch, and we have a clean working tree ❹. This is the message you should see each time you commit a working state of your project. If you get a different message, read it carefully; it's likely you forgot to add a file before making a commit.

Checking the Log

Git keeps a log of all commits made to the project. Let's check the log:

```
git_practice$ git log
commit cea13ddc51b885d05a410201a54faf20e0d2e246 (HEAD -> main)
Author: eric <eric@example.com>
Date:   Mon Jun 6 19:37:26 2022 -0800

    Started project.
git_practice$
```

Each time you make a commit, Git generates a unique, 40-character reference ID. It records who made the commit, when it was made, and the message recorded. You won't always need all of this information, so Git provides an option to print a simpler version of the log entries:

```
git_practice$ git log --pretty=oneline
cea13ddc51b885d05a410201a54faf20e0d2e246 (HEAD -> main) Started project.
git_practice$
```

The --pretty=oneline flag provides the two most important pieces of information: the reference ID of the commit and the message recorded for the commit.

The Second Commit

To see the real power of version control, we need to make a change to the project and commit that change. Here we'll just add another line to *hello_git.py*:

hello_git.py

```
print("Hello Git world!")
print("Hello everyone.")
```

When we check the status of the project, we'll see that Git has noticed the file that changed:

```
git_practice$ git status
❶ On branch main
Changes not staged for commit:
  (use "git add <file>..." to update what will be committed)
  (use "git restore <file>..." to discard changes in working directory)
```

```
❷ modified:   hello_git.py

❸ no changes added to commit (use "git add" and/or "git commit -a")
git_practice$
```

We see the branch we're working on ❶, the name of the file that was modified ❷, and that no changes have been committed ❸. Let's commit the change and check the status again:

```
❶ git_practice$ git commit -am "Extended greeting."
[main 945fa13] Extended greeting.
 1 file changed, 1 insertion(+), 1 deletion(-)
❷ git_practice$ git status
On branch main
nothing to commit, working tree clean
❸ git_practice$ git log --pretty=oneline
945fa13af128a266d0114eebb7a3276f7d58ecd2 (HEAD -> main) Extended greeting.
cea13ddc51b885d05a410201a54faf20e0d2e246 Started project.
git_practice$
```

We make a new commit, passing the `-am` flags when we use the command `git commit` ❶. The `-a` flag tells Git to add all modified files in the repository to the current commit. (If you create any new files between commits, reissue the `git add .` command to include the new files in the repository.) The `-m` flag tells Git to record a message in the log for this commit.

When we check the project's status, we see that we once again have a clean working tree ❷. Finally, we see the two commits in the log ❸.

Abandoning Changes

Now let's look at how to abandon a change and go back to the previous working state. First, add a new line to *hello_git.py*:

hello_git.py

```
print("Hello Git world!")
print("Hello everyone.")

print("Oh no, I broke the project!")
```

Save and run this file.

We check the status and see that Git notices this change:

```
git_practice$ git status
On branch main
Changes not staged for commit:
  (use "git add <file>..." to update what will be committed)
  (use "git restore <file>..." to discard changes in working directory)

❶     modified:   hello_git.py

no changes added to commit (use "git add" and/or "git commit -a")
git_practice$
```

Git sees that we modified *hello_git.py* ❶, and we can commit the change if we want to. But this time, instead of committing the change, we'll go back to the last commit when we knew our project was working. We won't do anything to *hello_git.py*: we won't delete the line or use the Undo feature in the text editor. Instead, enter the following commands in your terminal session:

```
git_practice$ git restore .
git_practice$ git status
On branch main
nothing to commit, working tree clean
git_practice$
```

The command `git restore` *filename* allows you to abandon all changes since the last commit in a specific file. The command `git restore .` abandons all changes made in all files since the last commit; this action restores the project to the last committed state.

When you return to your text editor, you'll see that *hello_git.py* has changed back to this:

```
print("Hello Git world!")
print("Hello everyone.")
```

Although going back to a previous state might seem trivial in this simple project, if we were working on a large project with dozens of modified files, all the files that had changed since the last commit would be restored. This feature is incredibly useful: you can make as many changes as you want when implementing a new feature, and if they don't work, you can discard them without affecting the project. You don't have to remember those changes and manually undo them. Git does all of that for you.

NOTE *You might have to refresh the file in your editor to see the restored version.*

Checking Out Previous Commits

You can revisit any commit in your log, using the `checkout` command, by using the first six characters of a reference ID. After checking out and reviewing an earlier commit, you can return to the latest commit or abandon your recent work and pick up development from the earlier commit:

```
git_practice$ git log --pretty=oneline
945fa13af128a266d0114eebb7a3276f7d58ecd2 (HEAD -> main) Extended greeting.
cea13ddc51b885d05a410201a54faf20e0d2e246 Started project.
git_practice$ git checkout cea13d
Note: switching to 'cea13d'.

❶ You are in 'detached HEAD' state. You can look around, make experimental
changes and commit them, and you can discard any commits you make in this
state without impacting any branches by switching back to a branch.
```

```
If you want to create a new branch to retain commits you create, you may
do so (now or later) by using -c with the switch command. Example:

  git switch -c <new-branch-name>

❷ Or undo this operation with:

  git switch -

Turn off this advice by setting config variable advice.detachedHead to false

HEAD is now at cea13d Started project.
git_practice$
```

When you check out a previous commit, you leave the main branch and enter what Git refers to as a *detached HEAD* state ❶. *HEAD* is the current committed state of the project; you're *detached* because you've left a named branch (main, in this case).

To get back to the main branch, you follow the suggestion ❷ to undo the previous operation:

```
git_practice$ git switch -
Previous HEAD position was cea13d Started project.
Switched to branch 'main'
git_practice$
```

This command brings you back to the main branch. Unless you want to work with some more advanced features of Git, it's best not to make any changes to your project when you've checked out a previous commit. However, if you're the only one working on a project and you want to discard all of the more recent commits and go back to a previous state, you can reset the project to a previous commit. Working from the main branch, enter the following:

```
❶ git_practice$ git status
  On branch main
  nothing to commit, working directory clean
❷ git_practice$ git log --pretty=oneline
  945fa13af128a266d0114eebb7a3276f7d58ecd2 (HEAD -> main) Extended greeting.
  cea13ddc51b885d05a410201a54faf20e0d2e246 Started project.
❸ git_practice$ git reset --hard cea13d
  HEAD is now at cea13dd Started project.
❹ git_practice$ git status
  On branch main
  nothing to commit, working directory clean
❺ git_practice$ git log --pretty=oneline
  cea13ddc51b885d05a410201a54faf20e0d2e246 (HEAD -> main) Started project.
  git_practice$
```

We first check the status to make sure we're on the main branch ❶. When we look at the log, we see both commits ❷. We then issue the git reset --hard command with the first six characters of the reference ID of the commit we want to go back to permanently ❸. We check the status again and see we're on the `main` branch with nothing to commit ❹. When we look at the log again, we see that we're at the commit we wanted to start over from ❺.

Deleting the Repository

Sometimes you'll mess up your repository's history and won't know how to recover it. If this happens, first consider asking for help using the approaches discussed in Appendix C. If you can't fix it and you're working on a solo project, you can continue working with the files but get rid of the project's history by deleting the *.git* directory. This won't affect the current state of any of the files, but it will delete all commits, so you won't be able to check out any other states of the project.

To do this, either open a file browser and delete the *.git* repository or delete it from the command line. Afterward, you'll need to start over with a fresh repository to start tracking your changes again. Here's what this entire process looks like in a terminal session:

```
❶ git_practice$ git status
  On branch main
  nothing to commit, working directory clean
❷ git_practice$ rm -rf .git/
❸ git_practice$ git status
  fatal: Not a git repository (or any of the parent directories): .git
❹ git_practice$ git init
  Initialized empty Git repository in git_practice/.git/
❺ git_practice$ git status
  On branch main
  No commits yet

  Untracked files:
    (use "git add <file>..." to include in what will be committed)
        .gitignore
        hello_git.py

  nothing added to commit but untracked files present (use "git add" to track)
❻ git_practice$ git add .
  git_practice$ git commit -m "Starting over."
  [main (root-commit) 14ed9db] Starting over.
   2 files changed, 5 insertions(+)
   create mode 100644 .gitignore
   create mode 100644 hello_git.py
❼ git_practice$ git status
  On branch main
  nothing to commit, working tree clean
  git_practice$
```

We first check the status and see that we have a clean working directory ❶. Then we use the command `rm -rf .git/` to delete the *.git* directory (`del .git` on Windows) ❷. When we check the status after deleting the *.git* folder, we're told that this is not a Git repository ❸. All the information Git uses to track a repository is stored in the *.git* folder, so removing it deletes the entire repository.

We're then free to use `git init` to start a fresh repository ❹. Checking the status shows that we're back at the initial stage, awaiting the first commit ❺. We add the files and make the first commit ❻. Checking the status now shows us that we're on the new `main` branch with nothing to commit ❼.

Using version control takes a bit of practice, but once you start using it, you'll never want to work without it again.

E

TROUBLESHOOTING DEPLOYMENTS

Deploying an app is tremendously satisfying when it works, especially if you've never done it before. However, there are many obstacles that can arise in the deployment process, and unfortunately, some of these issues can be difficult to identify and address. This appendix will help you understand modern approaches to deployment and give you specific ways to troubleshoot the deployment process when things aren't working.

If the additional information in this appendix isn't enough to help you get through the deployment process successfully, see the online resources at *https://ehmatthes.github.io/pcc_3e*; the updates there will almost certainly help you carry out a successful deployment.

Understanding Deployments

When you're trying to troubleshoot a particular deployment attempt, it's helpful to have a clear understanding of how a typical deployment works. *Deployment* refers to the process of taking a project that works on your local system, and copying that project to a remote server in a way that allows it to respond to requests from any user on the internet. The remote environment differs from a typical local system in a number of important ways: it's probably not the same operating system (OS) as the one you're using, and it's most likely one of many virtual servers on a single physical server.

When you deploy a project, or *push* it to the remote server, the following steps need to be taken:

- Create a virtual server on a physical machine at a datacenter.
- Establish a connection between the local system and the remote server.
- Copy the project's code to the remote server.
- Identify all of the project's dependencies and install them on the remote server.
- Set up a database and run any existing migrations.
- Copy static files (CSS, JavaScript files, and media files) to a place where they can be served efficiently.
- Start a server to handle incoming requests.
- Start routing incoming requests to the project, once it's ready to handle requests.

When you consider all that goes into a deployment, it's no wonder deployments often fail. Fortunately, once you gain an understanding of what should be happening, you'll stand a better chance of identifying what went wrong. If you can identify what went wrong, you might be able to identify a fix that will make the next deployment attempt successful.

You can develop locally on one kind of OS and push to a server running a different OS. It's important to know what kind of system you're pushing to, because that can inform some of your troubleshooting work. At the time of this writing, a basic remote server on Platform.sh runs Debian Linux; most remote servers are Linux-based systems.

Basic Troubleshooting

Some troubleshooting steps are specific to each OS, but we'll get to that in a moment. First, let's consider the steps everyone should try when troubleshooting a deployment.

Your best resource is the output generated during the attempted push. This output can look intimidating; if you're new to deploying apps, it can look highly technical, and there's usually a lot of it. The good news is you don't need to understand everything in the output. You should have two goals when skimming log output: identify any deployment steps that worked, and identify any steps that didn't. If you can do this, you might be able to figure out what to change in your project, or in your deployment process, to make your next push successful.

Follow Onscreen Suggestions

Sometimes, the platform you're pushing to will generate a message that has a clear suggestion for how to address the issue. For example, here's the message you'll see if you create a Platform.sh project before initializing a Git repository, and then try to push the project:

```
$ platform push
❶ Enter a number to choose a project:
    [0] ll_project (votohz445ljyg)
   > 0

❷   [RootNotFoundException]
    Project root not found. This can only be run from inside a project
      directory.

❸ To set the project for this Git repository, run:
      platform project:set-remote [id]
```

We're trying to push a project, but the local project hasn't been associated with a remote project yet. So, the Platform.sh CLI asks which remote project we want to push to ❶. We enter `0`, to select the only project listed. But next, we see a `RootNotFoundException` ❷. This happens because Platform.sh looks for a *.git* directory when it inspects the local project, to figure out how to connect the local project with the remote project. In this case, since there was no *.git* directory when the remote project was created, that connection was never established. The CLI suggests a fix ❸; it's telling us that we can specify the remote project that should be associated with this local project, using the `project:set-remote` command.

Let's try this suggestion:

```
$ platform project:set-remote votohz445ljyg
Setting the remote project for this repository to: ll_project (votohz445ljyg)

The remote project for this repository is
    now set to: ll_project (votohz445ljyg)
```

In the previous output, the CLI showed the ID of this remote project, `votohz4451jyg`. So we run the command that's suggested, using this ID, and the CLI is able to make the connection between the local project and the remote project.

Now let's try to push the project again:

```
$ platform push
Are you sure you want to push to the main (production) branch? [Y/n] y
Pushing HEAD to the existing environment main
--snip--
```

This was a successful push; following the onscreen suggestion worked.

You should be careful about running commands that you don't fully understand. However, if you have good reason to believe that a command can do little harm, and if you trust the source of the recommendation, it might be reasonable to try the suggestions offered by the tools you're using.

NOTE *Keep in mind there are individuals who will tell you to run commands that will wipe your system or expose your system to remote exploitation. Following the suggestions of a tool provided by a company or organization you trust is different from following the suggestions of random people online. Anytime you're dealing with remote connections, proceed with an abundance of caution.*

Read the Log Output

As mentioned earlier, the log output that you see when you run a command like `platform push` can be both informative and intimidating. Read through the following snippet of log output, taken from a different attempt at using `platform push`, and see if you can spot the issue:

```
--snip--
Collecting soupsieve==2.3.2.post1
  Using cached soupsieve-2.3.2.post1-py3-none-any.whl (37 kB)
Collecting sqlparse==0.4.2
  Using cached sqlparse-0.4.2-py3-none-any.whl (42 kB)
Installing collected packages: platformshconfig, sqlparse,...
Successfully installed Django-4.1 asgiref-3.5.2 beautifulsoup4-4.11.1...
W: ERROR: Could not find a version that satisfies the requirement gunicorrn
W: ERROR: No matching distribution found for gunicorrn

130 static files copied to '/app/static'.

Executing pre-flight checks...
--snip--
```

When a deployment attempt fails, a good strategy is to look through the log output and see if you can spot anything that looks like warnings or errors. Warnings are fairly common; they're often messages about upcoming changes in a project's dependencies, to help developers address issues before they cause actual failures.

A successful push may have warnings, but it shouldn't have any errors. In this case, Platform.sh couldn't find a way to install the requirement `gunicorrn`. This is a typo in the *requirements_remote.txt* file, which was supposed to include `gunicorn` (with one *r*). It's not always easy to spot the root issue in log output, especially when the problem causes a bunch of cascading errors and warnings. Just like when reading a traceback on your local system, it's a good idea to look closely at the first few errors that are listed, and also the last few errors. Most of the errors in between tend to be internal packages complaining that something went wrong, and passing messages about the error to other internal packages. The actual error we can fix is usually one of the first or last errors listed.

Sometimes, you'll be able to spot the error, and other times, you'll have no idea what the output means. It's certainly worth a try, and using log output to successfully diagnose an error is a tremendously satisfying feeling. As you spend more time looking through log output, you'll get better at identifying the information that's most meaningful to you.

OS-Specific Troubleshooting

You can develop on any operating system you like and push to any host you like. The tools for pushing projects have developed enough that they'll modify your project as needed to run correctly on the remote system. However, there are some OS-specific issues that can arise.

In the Platform.sh deployment process, one of the most likely sources of difficulties is installing the CLI. Here's the command to do so:

```
$ curl -fsS https://platform.sh/cli/installer | php
```

The command starts with `curl`, a tool that lets you request remote resources, accessed through a URL, within a terminal. Here, it's being used to download the CLI installer from a Platform.sh server. The `-fsS` section of the command is a set of flags that modify how curl runs. The `f` flag tells curl to suppress most error messages, so the CLI installer can handle them instead of reporting them all to you. The `s` flag tells curl to run silently; it lets the CLI installer decide what information to show in the terminal. The `S` flag tells curl to show an error message if the overall command fails. The `| php` at the end of the command tells your system to run the downloaded installer file using a PHP interpreter, because the Platform.sh CLI is written in PHP.

This means your system needs curl and PHP in order to install the Platform.sh CLI. To use the CLI, you'll also need Git, and a terminal that can run Bash commands. *Bash* is a language that's available in most server environments. Most modern systems have plenty of room for multiple tools like this to be installed.

The following sections will help you address these requirements for your OS. If you don't already have Git installed, see the instructions for installing Git on page 484 in Appendix D and then go to the section here that's applicable to your OS.

NOTE *An excellent tool for understanding terminal commands like the one shown here is* https://explainshell.com. *Enter the command you're trying to understand, and the site will show you the documentation for all the parts of your command. Try it out with the command used to install the Platform.sh CLI.*

Deploying from Windows

Windows has seen a resurgence in popularity with programmers in recent years. Windows has integrated many different elements of other operating systems, providing users with a number of options for how to do local development work and interact with remote systems.

One of the most significant difficulties in deploying from Windows is that the core Windows operating system is not the same as what a Linux-based remote server uses. A base Windows system has a different set of tools and languages than a base Linux system, so to carry out deployment work from Windows, you'll need to choose how to integrate Linux-based tool sets into your local environment.

Windows Subsystem for Linux

One popular approach is to use *Windows Subsystem for Linux* (*WSL*), an environment that allows Linux to run directly on Windows. If you have WSL set up, using the Platform.sh CLI on Windows becomes as easy as using it on Linux. The CLI won't know it's running on Windows; it will just see the Linux environment you're using it in.

Setting up WSL is a two-step process: you first install WSL, and then choose a Linux distribution to install into the WSL environment. Setting up a WSL environment is more than can be described here; if you're interested in this approach and don't already have it set up, see the documentation at *https://docs.microsoft.com/en-us/windows/wsl/about*. Once you have WSL set up, you can follow the instructions in the Linux section of this appendix to continue your deployment work.

Git Bash

Another approach to building a local environment that you can deploy from uses *Git Bash*, a terminal environment that's compatible with Bash but runs on Windows. Git Bash is installed along with Git when you use the installer from *https://git-scm.com*. This approach can work, but it isn't as streamlined as WSL. In this approach, you'll have to use a Windows terminal for some steps and a Git Bash terminal for others.

First you'll need to install PHP. You can do this with *XAMPP*, a package that bundles PHP with a few other developer-focused tools. Go to *https://apachefriends.org* and click the button to download XAMPP for Windows. Open the installer and run it; if you see a warning about User Account Control (UAC) restrictions, click **OK**. Accept all of the installer's defaults.

When the installer finishes running, you'll need to add PHP to your system's path; this will tell Windows where to look when you want to run PHP. In the Start menu, enter **`path`** and click **Edit the System Environment Variables**; click the button labeled **Environment Variables**. You should see the variable `Path` highlighted; click **Edit** under this pane. Click **New** to add a new path to the current list of paths. Assuming you kept the default settings when running the XAMPP installer, add **`C:\xampp\php`** in the box that appears, then click **OK**. When you're finished, close all of the system dialogs that are still open.

With these requirements taken care of, you can install the Platform.sh CLI. You'll need to use a Windows terminal with administrator privileges; enter **`command`** into the Start menu, and under the Command Prompt app, click **Run as administrator**. In the terminal that appears, enter the following command:

```
> curl -fsS https://platform.sh/cli/installer | php
```

This will install the Platform.sh CLI, as described earlier.

Finally, you'll work in Git Bash. To open a Git Bash terminal, go to the Start menu and search for **`git bash`**. Click the **Git Bash app** that appears; you should see a terminal window open. You can use traditional Linux-based commands like `ls` in this terminal, as well as Windows-based commands like `dir`. To make sure the installation was successful, issue the **`platform list`** command. You should see a list of all the commands in the Platform.sh CLI. From this point forward, carry out all of your deployment work using the Platform.sh CLI inside a Git Bash terminal window.

Deploying from macOS

The macOS operating system is not based on Linux, but they were both developed on similar principles. What this means, practically, is that a lot of the commands and workflows that you use on macOS will work in a remote server environment as well. You might need to install some developer-focused resources in order to have all of these tools available in your local macOS environment. If you get a prompt to install the *command line developer tools* at any point in your work, click **Install** to approve the installation.

The most likely difficulty when installing the Platform.sh CLI is making sure PHP is installed. If you see a message that the `php` command is not found, you'll need to install PHP. One of the easiest ways to install PHP is by using the *Homebrew* package manager, which facilitates the installation of a wide variety of packages that programmers depend on. If you don't already have Homebrew installed, visit *https://brew.sh* and follow the instructions to install it.

Once Homebrew is installed, use the following command to install PHP:

```
$ brew install php
```

This will take a while to run, but once it has completed, you should be able to successfully install the Platform.sh CLI.

Deploying from Linux

Because most server environments are Linux-based, you should have very little difficulty installing and using the Platform.sh CLI. If you try to install the CLI on a system with a fresh installation of Ubuntu, it will tell you exactly which packages you need:

```
$ curl -fsS https://platform.sh/cli/installer | php
Command 'curl' not found, but can be installed with:
sudo apt  install curl
Command 'php' not found, but can be installed with:
sudo apt install php-cli
```

The actual output will have more information about a few other packages that would work, plus some version information. The following command will install curl and PHP:

```
$ sudo apt install curl php-cli
```

After running this command, the Platform.sh CLI installation command should run successfully. Since your local environment is quite similar to most Linux-based hosting environments, much of what you learn about working in your terminal will carry over to working in a remote environment as well.

Other Deployment Approaches

If Platform.sh doesn't work for you, or if you want to try a different approach, there are many hosting platforms to choose from. Some work similarly to the process described in Chapter 20, and some have a much different approach to carrying out the steps described at the beginning of this appendix:

- Platform.sh allows you to use a browser to carry out the steps we used the CLI for. If you like browser-based interfaces better than terminal-based workflows, you may prefer this approach.
- There are a number of other hosting providers that offer both CLI- and browser-based approaches. Some of these providers offer terminals within their browser, so you don't have to install anything on your system.
- Some providers allow you to push your project to a remote code hosting site like GitHub, and then connect your GitHub repository to the hosting site. The host then pulls your code from GitHub, instead of requiring you to push your code from your local system directly to the host. Platform.sh supports this kind of workflow as well.
- Some providers offer an array of services that you select from, in order to put together an infrastructure that works for your project. This typically requires you to have a deeper understanding of the deployment process, and what a remote server needs in order to serve a project. These hosts include Amazon Web Services (AWS) and Microsoft's Azure platform. It can be much harder to track your costs in these kinds of platforms, because each service can accrue charges independently.
- Many people host their projects on a virtual private server (VPS). In this approach, you rent a virtual server that acts just like a remote computer, log in to the server, install the software needed to run your project, copy your code over, set the right connections, and allow your server to start accepting requests.

New hosting platforms and approaches appear on a regular basis; find one that looks appealing to you, and invest the time to learn that provider's deployment process. Maintain your project long enough so that you get to know what works well with your provider's approach and what doesn't. No hosting platform is going to be perfect; you'll need to make an ongoing judgement call about whether the provider you're currently using is good enough for your use case.

I'll offer one last word of caution about choosing a deployment platform and an overall approach to deployment. Some people will enthusiastically steer you toward overly complex deployment approaches and services that are meant to make your project highly reliable and capable of serving millions of users simultaneously. Many programmers spend lots of time, money, and energy building out a complex deployment strategy, only to find that hardly anyone is using their project. Most Django projects can be set up on a small hosting plan and tuned to serve thousands of requests per minute. If your project is getting anything less than this level of traffic, take the time to configure your deployment to work well on a minimal platform before investing in infrastructure that's meant for some of the largest sites in the world.

Deployment is incredibly challenging at times, but just as satisfying when your live project works well. Enjoy the challenge, and ask for help when you need it.

INDEX

Symbols

- + (addition), 26
- += (addition in place), 122
- * (arbitrary arguments), 146
- ** (arbitrary keyword arguments), 148
- {} (braces)
  - dictionaries, 92
  - sets, 104
- @ (decorator), 221, 424
- / (division), 26
- == (equality), 72, 74
- ** (exponent), 26
- > (greater than), 75
- >= (greater than or equal to), 75
- # (hash mark), for comments, 29
- != (inequality), 74
- < (less than), 75
- <= (less than or equal to), 75
- [] (list), 34
- % (modulo), 116
- += (multiline strings), 115
- * (multiplication), 26
- \n (newline), 22
- >>> (Python prompt), 4
- - (subtraction), 26
- \t (tab), 22
- _ (underscore)
  - in file and folder names, 10
  - in numbers, 28
  - in variable names, 17

A

- aliases, 151–152, 178–179
- *alice.py*, 195–197
- *Alien Invasion. See also* Pygame
  - aliens, 256–274
    - building fleet, 259–262
    - checking edges, 265
    - collisions, with bullets, 267
    - collisions, with ship, 270–273
    - controlling fleet direction, 264–266
    - creating an alien, 256–258
    - dropping fleet, 265–266
    - reaching bottom of screen, 273–274
    - rebuilding fleet, 268–269
  - bullets, 247–253, 266–270
    - collisions, with aliens, 267
    - deleting old, 250–251
    - firing, 249–250
    - larger, 268
    - limiting number of, 251–252
    - settings, 247
    - speeding up, 269
  - classes
    - `Alien`, 257
    - `AlienInvasion`, 229
    - `Bullet`, 247–248
    - `Button`, 278–279
    - `GameStats`, 271
    - `Scoreboard`, 286–287
    - `Settings`, 232
    - `Ship`, 234–235
  - ending the game, 274–275
  - initializing dynamic settings, 283–285
  - levels
    - modifying speed settings, 283–285
    - resetting the speed, 285
    - displaying, 294–296
  - moving fleet, 263–266
  - planning, 228
  - Play button, 278–283
    - `Button` class, 278–279
    - deactivating, 282

Alien Invasion (continued)
Play button *(continued)*
drawing, 279–280
hiding mouse cursor, 282–283
resetting game, 281–282
starting game, 281
reviewing the project, 256
scoring, 286–298
all hits, 290
high score, 292–294
increasing point values, 290–291
level, 294–296
number of ships, 296–299
resetting, 289–290
rounding and formatting, 291–292
score attribute, 286
updating, 289
settings, storing, 232–233
ship, 233–244
adjusting speed, 242–243
continuous movement, 239–242
finding an image, 233–234
limiting range, 243–244
amusement_park.py, 80–82
`and` keyword, 75
antialiasing, 279
API. *See* application programming interface
apostrophe.py, 24–25
`append()` method, 37–38
application programming interface (API), 355
API call, 355–357
GitHub API, 368
Hacker News API, 368–371
processing an API response, 357–362
rate limits, 362
requesting data, 356–357
visualizing results, 362–368
arguments, 131. *See also under* functions
`as` keyword, 151–152
assertions, 213, 217–218
attributes, 159. *See also under* classes

B

banned_users.py, 76–77
bicycles.py, 34–35
Boolean values, 77
Bootstrap, 433. *See also unxder* Django
braces ({})
dictionaries, 92
sets, 104
`break` statement, 121
built-in functions, 467

C

calls (functions), 130, 132–135
car.py, 162–178
cars.py, 43–45, 72
cities.py, 121
classes
attributes, 159
accessing, 160
default values, 163–164
modifying, 164–166
creating, 158–161
importing, 173–179
multiple classes, 175–176
single classes, 174–175
inheritance, 167–172
attributes and methods, 169
child classes, 167–170
composition, 170
`__init__()` method, 167–169
instances as attributes, 170–172
overriding methods, 170
parent classes, 170
subclasses, 168
`super()` function, 168
superclasses, 168
instances, 157
methods, 159
calling, 160
chaining, 185
`__init()__` method, 159
modeling real-world objects, 172–173
multiple instances, 161
naming conventions, 158
objects, 157
style guidelines, 181

comma-separated value files. *See* CSV files
comment.py, 29
comments, 29–30
conditional tests, 72–77. *See also* if statements
confirmed_users.py, 124–125
constants, 28
`continue` statement, 122
counting.py, 117–118, 122–123
CSV files, 330–341
    `csv.reader()` function, 330–333
    error checking, 338–341
    file headers, 330–332

D

data analysis, 301
databases. *See under* Django
data visualization, 301. *See also* Matplotlib; Plotly
`datetime` module, 333–335
death_valley_highs_lows.py, 339–341
decorators, 221–223, 423–425
default values
    class attributes, 163–164
    function parameters, 134–135
definition (functions), 130
`def` keyword, 130
`del` statement
    with dictionaries, 96
    with lists, 38–40
dice_visual_d6d10.py, 326–327
dice_visual.py, 324–326
dictionaries
    defining, 92
    empty, 94
    formatting larger, 96–97
    `KeyError`, 98
    key-value pairs, 92
        adding, 93–94
        removing, 96
    looping through
        keys, 101–102
        keys in order, 102–103
        key-value pairs, 99–101
        values, 103–104
    methods
        `get()`, 97–98
        `items()`, 99–101
        `keys()`, 101–103
        `values()`, 103–104
    nesting
        dictionaries in dictionaries, 110–111
        dictionaries in lists, 105–108
        lists in dictionaries, 108–109
    ordering in, 94, 102–103
    sorting a list of, 370
    values
        accessing, 92–93, 97–98
        modifying, 94–96
die.py, 320
die_visual.py, 320–321
dimensions.py, 66–67
div (HTML), 437
division_calculator.py, 192–195
Django. *See also* Git; Learning Log project
    accounts app, 415–423
        creating app, 415–416
        logging out, 419–420
        login page, 416–419
        registration page, 420–423
    admin site, 381–386
    associating data with a user, 425–430
    Bootstrap, 434–445
        card, 443
        collapsible navigation, 437
        container, 440
        django-boostrap5 app, 434
        documentation, 444
        HTML headers, 435–436
        jumbotron, 440–441
        list groups, 443
        navigation bar, 436–439
        styling forms, 441–442
    commands
        `createsuperuser`, 382
        `flush`, 427
        `makemigrations`, 381, 385, 426
        `migrate`, 377
        `runserver`, 377–378, 383, 392
        `shell`, 386
        `startapp`, 379, 415
        `startproject`, 376
    creating new projects, 376

Django *(continued)*
databases
cascading delete, 384
creating, 376
foreign keys, 384, 425
many-to-one relationships, 384
migrating, 377, 381, 385, 426
non-nullable field, 427
Postgres, 447
queries, 398, 428
querysets, 386–387, 395, 398, 426–428
resetting, 427
SQLite, 377
deployment, 445–461, 493–501
committing the project, 453
configuration files, 447–450
creating Platform.sh project, 453–455
creating superuser, 456–457
custom error pages, 459–460
deleting projects, 461
free trial limits, 446
`gunicorn`, 447
ignoring files, 452–453
installing Platform.sh CLI, 446, 497–500
installing `platformshconfig`, 446
other deployment approaches, 500
Platform.sh, 445
Postgres database, 447, 450–451
`psycopg2`, 447
pushing a project, 455
pushing changes, 458, 460
requirements.txt, 446
securing project, 457–460
settings, 451
SSH sessions, 456–457
troubleshooting, 494–501
using Git, 451
viewing project, 456
development server, 377–378, 383, 392
documentation
model fields, 380
queries, 388
templates, 400
forms, 404–423, 429–430
`csrf_token`, 407
GET and POST requests, 406
`ModelForm`, 404, 408
processing forms, 405–406, 409–410, 412–413, 421–422, 429–430
`save()` method, 405–406, 409–410, 430
templates, 407, 410–411, 413, 417, 419, 422
validation, 404–406
widgets, 408
HTML
anchor tag (`<a>`), 393
`<body>` element, 437
comments, 437
`<div>` elements, 437
`<main>` element, 440
margins, 440
padding, 440
`<p>` elements, 391
`<span>` elements, 438
HTTP 404 error, 428–429, 459–460
`INSTALLED_APPS`, 380
installing, 375–376
localhost, 378
logging out, 419–420
`@login_required` decorator, 423–424
login template, 417
mapping URLs, 388–390, 397–398
migrating the database, 426–427
models, 379
activating, 380–381
defining, 379, 384
foreign keys, 384, 425
registering with admin, 382–383, 385–386
`__str__()` method, 380, 384
projects (vs. apps), 379
`redirect()` function, 405–406
release cycle, 376
restricting access to data, 427–430

settings
ALLOWED_HOSTS, 451
DEBUG, 457–458
INSTALLED_APPS, 380–381, 415–416, 434
LOGIN_REDIRECT_URL, 417–418
LOGIN_URL, 424
LOGOUT_REDIRECT_URL, 420
SECRET_KEY, 451
shell, 386–387, 426–427
starting an app, 379
styling. *See* Django: Bootstrap
superusers, 382, 456–457
templates
block tags, 393
child template, 393–394
context dictionary, 395
filters, 399
forms in, 407
indentation in, 393
inheritance, 392–394
linebreaks, 399
links in, 392–393, 399
loops in, 395–397
parent template, 392–393
template tags, 393
timestamps in, 398–399
user object, 418
writing, 390–392
URLs. *See* Django: mapping URLs
UserCreationForm, 421–422
user ID values, 426
versions, 376
view functions, 388, 390
virtual environments, 374–375
docstrings, 130, 153, 181
dog.py, 158–162
dot notation, 150, 160

E

earthquakes. *See* mapping earthquakes
electric_car.py, 167–173
module, 177–179
encoding argument, 195–196
enumerate() function, 331
eq_explore_data.py, 343–347
equality operator (==), 72, 74
eq_world_map.py, 347–352
even_numbers.py, 58
even_or_odd.py, 117
exceptions, 183, 192–199
deciding which errors to report, 199
else block, 194–195
failing silently, 198–199
FileNotFound error, 195–196
handling exceptions, 192–196
preventing crashes, 193–195
try-except blocks, 193
ZeroDivisionError, 192–195
exponents (**), 26

F

favorite_languages.py, 96–97, 100–104, 109
file_reader.py, 184–187
files
encoding argument, 195–196
FileNotFound error, 195–196
file paths, 186
absolute, 186
exists() method, 203–204
pathlib module, 184
Path objects, 184–186, 330
relative, 186
from strings, 198
on Windows, 186
read_text() method, 185, 195–196
splitlines() method, 186–187
write_text() method, 190–191
first_numbers.py, 57
fixtures, 221–223
flags, 120–121
floats, 26–28
foods.py, 63–64
for loops, 49–56, 99–104. *See also* dictionaries; lists
formatted_name.py, 137–139
f-strings
format specifiers, 291–292
using variables in, 20–21
full_name.py, 21
functions, 129–155
arguments
arbitrary, 146–149
default values, 134–135
errors, 136
keyword, 133–134

functions *(continued)*
arguments *(continued)*
lists as, 142–145
optional, 138–139
positional, 131–133
body, 130
built-in, 467
calling functions, 130, 132–135
defining, 130
importing, 149–153
aliases, 151–152
entire modules, 150–151
specific functions, 151
modifying a list in a function, 142–145
modules, 149–153
parameters, 131
return values, 137–141
style guidelines, 153

G

GeoJSON files, 342–347, 350–351
GET requests, 406. *See* Django: forms
getting help
Discord, 480
official Python documentation, 479–480
online resources, xxxv, 478
r/learnpython, 480
rubber duck debugging, 478
searching online, 479
Slack, 481
Stack Overflow, 479
three main questions, 477–478
Git, 356, 451–453, 483–492. *See also* Django: deployment
abandoning changes, 488–489
adding files, 486
branches, 486
checking out previous commits, 489–491
commits, 486–488
configuring, 452, 484
deleting a repository, 491–492
.gitignore, 484
HEAD, 490
ignoring files, 484
initializing a repository, 485
installing, 484
log, 487
repositories, 356
status, 485–486
GitHub, 356
greeter.py, 114–115, 130–131
greet_users.py, 142

H

Hacker News API, 368–371
hash mark (#), for comments, 29
hello_git.py, 484–491
hello_world.py, 10–12, 15–19
hidden files, 448, 485
hn_article.py, 368–369
hn_submissions.py, 369–371

I

IDE (integrated development environment), 469–470
`if` statements
`and` keyword, 75
Boolean expressions, 77
checking for
equality (`==`), 72
inequality (`!=`), 74
item in list, 76
item not in list, 76
list not empty, 86–87
`elif` statement, 80–83
`else` statement, 79–80
`if` statements and lists, 85–88
ignoring case, 73–74
numerical comparisons, 74–76
`or` keyword, 76
simple, 78
style guidelines, 89
testing multiple conditions, 82–83
immutable, 65
`import *`, 152, 177
`import this`, 30–31
indentation errors, 53–56
index errors, 46–47
inheritance, 167–173. *See also* *under* classes
`input()` function, 114–116
numerical input, 115–116
writing prompts, 114–115

insert() method, 38
itemgetter() function, 370
items() method, 99–101

J

JSON files
    GeoJSON files, 342–347, 350–351
    JSON data format, 201
    json.dumps() function, 201–204, 343–344, 368
    json.loads() function, 201–204, 343–344

K

keys() method, 101–103
key-value pairs, 92. *See also* dictionaries
keyword arguments, 133–134
keywords, 466

L

language_survey.py, 219
Learning Log project, 373
    files, 392
        404.html, 459
        500.html, 459
        accounts/urls.py, 416, 420
        accounts/views.py, 421–422
        admin.py, 382–383
        base.html, 392–393, 396, 418–419, 422, 435–440
        edit_entry.html, 413
        forms.py, 404, 408–409
        .gitignore, 452–453
        index.html, 390–394, 440–441
        learning_logs/urls.py, 389–390, 394–395, 397–398, 405, 409, 412
        learning_logs/views.py, 390, 395, 398, 405–406, 409–410, 412–413, 423–425, 428–430
        ll_project/urls.py, 388–389, 416
        login.html, 417, 441–442
        models.py, 379–380, 384
        new_entry.html, 410
        new_topic.html, 407
        .platform.app.yaml, 448–450
        register.html, 422
        requirements.txt, 446–447
        routes.yaml, 450
        services.yaml, 450
        settings.py, 380–381, 415–418, 420, 424, 434, 451, 457–460
        topic.html, 398–399, 443–444
        topics.html, 395–396, 442–443
    ongoing development, 460
    pages, 391
        edit entry, 412–414
        home page, 388–394
        login page, 416–419
        new entry, 408–411
        new topic, 404–408
        registration, 420–423
        topic, 397–400
        topics, 394–397
    writing a specification (spec), 374
len() function, 44–45
library, 184
Linux
    Python
        checking installed version, 8
        setting up, 8–12, 465–466
    terminals
        running programs from, 12
        starting Python session, 9
    troubleshooting installation issues, 10
    VS Code, installing, 9
lists, 33
    as arguments, 142–145
    comprehensions, 59–60
    copying, 63–64
    elements
        accessing, 34
        accessing last, 35
        adding with append(), 37–38
        adding with insert(), 38
        identifying unique, 104
        modifying, 36–37
        removing with del, 38–39
        removing with pop(), 39–40
        removing with remove(), 40–41
    empty, 37–38
    enumerate() function, 331

lists *(continued)*
errors
indentation, 53–56
index, 46
`for` loops, 49–56
nested, 108–109, 261–262
indexes, 34–35
negative index, 35
zero index, 34–35
`len()` function, 44–45
naming, 33–34
nesting
dictionaries in lists, 105–108
lists in dictionaries, 108–109
numerical lists, 56–60
`max()` function, 59
`min()` function, 59
`range()` function, 58–59
`sum()` function, 59
removing all occurrences of a value, 125
slices, 61–62
sorting
`reverse()` method, 44
`sorted()` function, 43–44
`sort()` method, 43
square brackets, 34
logical errors, 54
`lstrip()` method, 22–23

M

macOS
.DS_Store files, ignoring, 453
Homebrew package manager, 499
Python
checking installed version, 7
setting up, 7–12, 464–465
terminals
running programs from, 12
starting Python session, 7
troubleshooting installation issues, 10
VS Code, installing, 8
magicians.py, 49–56
magic_number.py, 74
making_pizzas.py, 150–152
mapping earthquakes, 342–352. *See also* Plotly
downloading data, 343, 352
GeoJSON files, 342–347, 350–351
latitude-longitude ordering, 345
location data, 346–347
magnitudes, 346
world map, 347–348
Matplotlib
axes
`set_aspect()` method, 313–314
removing, 317
`ax` objects, 303
colormaps, 310–311
`fig` objects, 303
`figsize` argument, 318
formatting plots
`alpha` argument, 337–338
built-in styles, 306
custom colors, 310
labels, 303–304
line thickness, 303–304
plot size, 318
shading, 337–338
tick labels, 309–310
gallery, 302
installing, 302
`plot()` method, 303–306
`pyplot` module, 302–303
`savefig()` method, 311
saving plots, 311
`scatter()` method, 306–311
simple line graph, 302–306
`subplots()` function, 303
methods, 20
helper methods, 237
modules, 149–152, 173–179. *See also* classes: importing; functions: importing
modulo operator (%), 116–117
motorcycles.py, 36–41
mountain_poll.py, 125–126
mpl_squares.py, 302–306
my_car.py, 174–175
my_cars.py, 176–179
my_electric_car.py, 176

N

name errors, 17–18
name_function.py, 211–217
name.py, 20
names.py, 211–212
nesting. *See* dictionaries: nesting; lists: `for` loops
newline (`\n`), 21–22
`next()` function, 330–331
`None`, 98, 140
number_reader.py, 202
numbers, 26–28
  arithmetic, 26
  constants, 28
  exponents, 26
  floats, 26–27
  formatting, 291–292
  integers, 26
  mixing integers and floats, 27–28
  order of operations, 26
  `round()` function, 291–292
  underscores in, 28
number_writer.py, 201

O

object-oriented programming (OOP), 157. *See also* classes
`or` keyword, 76. *See also* `if` statements

P

pandas, 320
parameters, 131
parrot.py, 114, 118–121
`pass` statement, 198–199
paths. *See* files: file paths
PEP 8, 68–69
person.py, 139–140
pets.py, 125, 132–136
pip, 210–211
  installing Django, 374–376
  installing Matplotlib, 302
  installing Plotly, 320
  installing Pygame, 228
  installing pytest, 211
  installing Requests, 357
  Linux, installing pip, 465–466
  updating, 210
pi_string.py, 187–189
pizza.py, 146–148
Platform.sh. *See* Django: deployment
players.py, 61–62
Plotly, 302, 319. *See also* mapping earthquakes; rolling dice
  chart types, 322
  customizing plots, 323, 325–326, 364
  documentation, 368
  `fig.show()` method, 322
  `fig.write_html()` method, 327
  formatting plots
    axis labels, 323
    color scales, 349–350
    hover text, 350–351, 365–366
    links in charts, 366–367
    marker colors, 349–350, 367
    tick marks, 325–326
    titles, 323
    tooltips, 365–366
    `update_layout()` method, 325–326, 364
    `update_traces()` method, 367
  gallery, 320
  histograms, 322
  installing, 320
  `plotly.express` module, 322, 347, 368
  `px` alias, 322
  `px.bar()` function, 322–323, 363–367
  saving figures, 327
  `scatter_geo()` function, 347–352
`pop()` method, 39–40
positional arguments, 131–133. *See also* functions: arguments
POST requests, 406. *See also* Django: forms
printing_models.py, 143–145
Project Gutenberg, 196–197
prompts, 114–115
.py file extension, 15–16
Pygame. *See also Alien Invasion*
  background colors, 231–232
  `clock.tick()` method, 230–231
  collisions, 266–267, 270–271, 289–290
  creating an empty window, 229–230
  cursor, hiding, 282–283

Pygame *(continued)*
  displaying text, 278–280
  ending games, 274–275
  event loops, 229–230
  frame rates, 230–231
  fullscreen mode, 245
  groups
    adding elements, 249–250
    defining, 248–249
    drawing all elements in, 249–250, 257–258
    emptying, 268–269
    looping through, 249–251
    removing elements from, 250–251
    updating all elements in, 248–249
  images, 234–236
  installing, 228
  levels, 283–285
  Play button, 278–283
  `print()` calls in, 251
  quitting, 244–245
  `rect` objects, 234–235
    creating from scratch, 247–248
    `get_rect()` method, 234–235
    positioning, 234–235, 238–243, 247–248, 256–262, 278, 286–298
    `size` attribute, 261
  responding to input, 230
    events, 230
    keypresses, 238–242
    mouse clicks, 281–283
  screen coordinates, 235
  surfaces, 230
  testing games, 268
pytest. *See* testing code
Python
  >>> prompt, 4
  built-in functions, 467
  checking installed version, 466
  installing
    on Linux, 465–466
    on macOS, 7–11, 464–465
    on Windows, 5–6, 463–464
  interpreter, 15–16
  keywords, 466
  Python Enhancement Proposal (PEP), 68
  standard library, 179–180
  terminal sessions, 4
    on Linux, 9
    on macOS, 7–8
    on Windows, 6
  versions, 4
  why use Python, xxxvi
python_repos.py, 357–362
python_repos_visual.py, 362–367

Q

quit values, 118

R

random_walk.py, 312–313
random walks, 312–318
  `choice()` function, 313
  coloring points, 315–316
  `fill_walk()` method, 312–313
  generating multiple walks, 314–315
  plotting, 313–314
  `RandomWalk` class, 312–313
  starting and ending points, 316–317
`range()` function, 58–59
`read_text()` method, 185, 195–196
refactoring, 204–206, 237–238, 260, 269–270
remember_me.py, 202–206
`removeprefix()` method, 24
`removesuffix()` method, 25
Requests package, installing, 357
return values, 137–141
rollercoaster.py, 116
rolling dice, 319–327. *See also* Plotly
  analyzing results, 321–322
  `Die` class, 320
  different-size dice, 326–327
  `randint()` function, 320
  rolling two dice, 324–326
rubber duck debugging, 478
`rstrip()` method, 22–23
rw_visual.py, 313–318

S

scatter_squares.py, 306–311
sets, 103–104
sitka_highs_lows.py, 336–338
sitka_highs.py, 330–336
`sleep()` function, 272
slices, 61–64
`sorted()` function, 43–44, 102–103
`sort()` method, 43
`splitlines()` method, 186–187
`split()` method, 196–197
SQLite database, 376–377
square_numbers.py, 58–59
squares.py, 59–60
Stack Overflow, 479
storing data, 201–204. *See also* JSON files
    saving and reading data, 202–204
strings, 19–25
    changing case, 20
    f-strings, 20–21, 291–292
    methods
        `lower()`, 20
        `lstrip()`, 22–23
        `removeprefix()`, 23–24
        `removesuffix()`, 25
        `rstrip()`, 22–23
        `split()`, 196–197
        `splitlines()`, 186–187
        `strip()`, 22–23
        `title()`, 20
        `upper()`, 20
    multiline, 115
    newlines in, 21–22
    single and double quotes, 19, 24–25
    tabs in, 21–22
    variables in, 20–21
    whitespace in, 21–23
`strip()` method, 22–23
`strptime()` method, 333–335
style guidelines, 68–69
    blank lines, 69
    CamelCase, 181
    classes, 181
    dictionaries, 96–97
    functions, 153
    `if` statements, 89
    indentation, 68
    line length, 69
    PEP 8, 68
survey.py, 218
syntax errors, 24
    avoiding with strings, 24–25
syntax highlighting, 16

T

tab (`\t`), 21–22
templates. *See under* Django
testing code, 209–223
    assertions, 213, 217–218
    failing tests, 214–216
    full coverage, 212
    naming tests, 213
    passing tests, 212–214
    pytest, 209–223
        fixtures, 221–223
        installing, 210–211
        running tests, 213–214
    test cases, 212
    testing classes, 217–223
    testing functions, 211–217
    unit tests, 212
test_name_function.py, 212–217
test_survey.py, 220–223
text editors and IDEs. *See also* VS Code
    Emacs and Vim, 475
    Geany, 474
    IDLE, 474
    Jupyter Notebooks, 475
    PyCharm, 475
    Sublime Text, 474
third-party package, 210
toppings.py, 74, 82–83
tracebacks, 10, 17–18, 192, 195–196
`try-except` blocks. *See* exceptions
tuples, 65–67
    defining, 65
    `for` loop, 66–67
    writing over, 67
type errors, 66

U

underscore (_)
    in file and folder names, 10
    in numbers, 28
    in variable names, 17

unit tests, 212
user_profile.py, 148–149

V

values() method, 103–104
variables, 16–19, 28
    constants, 28
    as labels, 18–19
    multiple assignment, 28
    name errors, 17–18
    naming conventions, 17
    values, 16
venv module, 374–375
version control. *See* Git
virtual environments, 374–375
voting.py, 78–80
VS Code, 4–5
    configuring, 470–473
    features, 469–470
    installing
        on Linux, 9
        on macOS, 8
        on Windows, 6
        Python extension, 9–10
    opening files with Python, 185
    Python extension, 9
    running files, 10
    shortcuts, 473–474
    tabs and spaces, 471

W

weather data, 330–341. *See also* CSV files; Matplotlib
while loops, 117–126
    active flag, 120–121
    break statement, 121
    continue statement, 122
    infinite loops, 122–123
    moving items between lists, 124
    quit values, 118
    removing all items from list, 125
whitespace, 21–23. *See also* strings
Windows
    file paths, 186
    Python
        setting up, 5–6, 9–12, 463–464
        troubleshooting installation, 10
    terminals
        running programs from, 12
        starting Python session, 6
    VS Code, installing, 6
word_count.py, 197–199
write_message.py, 190–191
write_text() method, 190–191

Z

Zen of Python, 30–31
ZeroDivisionError, 192–195

RESOURCES

Visit *https://nostarch.com/python-crash-course-3rd-edition* for errata and more information.